CESAR - Cost-efficient Methods and Processes for Safety-relevant Embedded Systems

Ajitha Rajan • Thomas Wahl
Editors

CESAR - Cost-efficient Methods and Processes for Safety-relevant Embedded Systems

Springer

Editors
Ajitha Rajan
Thomas Wahl
Dept. of Computer Science
University of Oxford
Oxford
United Kingdom

ISBN 978-3-7091-1386-8 ISBN 978-3-7091-1387-5 (eBook)
DOI 10.1007/978-3-7091-1387-5
Springer Wien Heidelberg New York Dordrecht London

Library of Congress Control Number: 2013931232

Printed on acid-free paper

Springer is part of Springer Science+Business Media (www.springer.com)

Foreword

The *ARTEMIS Joint Undertaking* is a European public-private partnership in the field of embedded systems. It aims to achieve effective coordination and synergy of resources to strengthen Europe's future growth, competitiveness, and sustainable development in this critical ICT sector.

In the implementation of the ARTEMIS Strategic Research Agenda, the Joint Undertaking has observed and encouraged a cluster of projects in the domain of safety-critical applications that is creating remarkable industry impact. Among those projects, CESAR has been the main catalyst of new technology and innovation: It has laid down the industry innovations and the technology foundations for the enhancement of ES design in the safety-critical applications.

Next to the impressive contribution to the various technology domains, CESAR has strongly contributed to overcoming the fragmentation of the underlying industry sectors: The CESAR reference technology platform is an industry-ready platform that represents a so far unachieved level of integration in ES design for the aerospace, rail, automotive, and automation sectors. Easy integration, reusability, and fast certification process are today within reach in a real industry context. This not only helps the ARTEMIS JU reach one of its major policy objectives, it also provides real solutions to some barriers to industrial innovation.

In this respect, CESAR—a flagship ARTEMIS project—is a very successful example of the ARTEMIS philosophy, where public and private resources join to create the critical mass necessary to drive European industry's innovation leadership, making best use of collaborative innovation: the ARTEMIS engine for the creation of industry and market impact.

Brussels, Belgium — Antonio Vecchio

Foreword

Embedded systems are indispensable enablers of most products and services in our modern world. A special category of embedded systems can be found in safety-critical applications, such as in control functions of cars, airplanes, trains, or industrial machinery. Failure or malfunction of such systems may cause loss of life or significant material damage. Today's development, evolution, and operation of safety-critical embedded systems are confronted with four main challenges:

- Meet user's requirements and expectations, both functional and nonfunctional
- Manage the (ever-growing) complexity
- Provide provable dependability (including safety, security, availability, and integrity)
- Find efficient methods for the development and evolution (resulting in adequate cost and short time to market)

The CESAR project has contributed impressively to the advancement of the state of the art in all four areas: requirements engineering methods, component-based development, reference technology platform including an industry-ready platform, formal modeling of systems, interoperability standards, and a first step in the realization of building systems out of pre-certified components. Last but not least, the viability of all methods and tools were demonstrated in a large number of pilot applications from the automotive, aerospace, railway, and industrial domains. Many results and deliverables of the CESAR project are comprehensively described in this book.

This book is a valuable gift from the CESAR project to the embedded systems community, both to academia and to industry.

Stein am Rhein, Switzerland

Dr. Frank J. Furrer

Preface

This book describes the technical outcomes of CESAR, a project funded in partnership between the ARTEMIS Joint Undertaking and national authorities. From March 2009 through June 2012, CESAR combined the forces of 55 industrial and academic partners from 10 countries to develop "Cost-efficient methods and processes for safety relevant embedded systems." The ubiquitous presence of embedded devices in society has led to a plethora of methods and tools for their design and development. Given the common scientific basis of these approaches, there is a pressing need for standardization to enable the reuse of tools across domains. The main goal of the CESAR project was to address this need with the creation of a highly customizable systems engineering instrument, named the "Reference Technology Platform" (RTP). The RTP provides companies with an efficient, streamlined, component-based system development process and as a by-product, creates a de facto European standard for embedded systems design.

Another important goal of CESAR was to make significant technological advances in two of the engineering disciplines underlying many of the embedded system design process:

Requirements engineering, with an emphasis on standardizing both functional and nonfunctional requirements, and requirements across development stages, from conceptual design to implementation

Component-based development, with an emphasis on building highly customizable, reusable, and verifiable components, along with a composition strategy that allows the inference of reliability guarantees from component specifications in a scalable manner

Both the development of the RTP, and the desired technological advances in requirements engineering and component-based development are intended to assist system engineers with solving concrete design challenges in their respective domains. To demonstrate the general applicability of the CESAR approach to system design, the work plan contained significant work on pilot applications from various classical embedded system domains, namely, the aerospace, automotive, automation, and railway sectors. The presence of these sectors made the CESAR consortium a very large and multifaceted community, composed of academic and industrial partners from a variety of domains and countries. All partners contributed to CESAR through individual and sometimes comparatively independent work packages and reported their results in numerous deliverables. The purpose of this book is to harmonize the overall results of CESAR, and to interpret them in a reusable way that abstracts from the "cultural specialities" of individual domains or partners. It is the hope of the editors that this book will ensure the sustainability of the project outcomes, by providing critical assistance, far beyond what the deliverables can achieve, in making the technical results of CESAR available *and usable* both to CESAR partners, as well as to companies outside the CESAR consortium, within Europe and worldwide.

Acknowledgments

The research leading to these results has received funding from the ARTEMIS Joint Undertaking under Grant Agreement no. 100016 as well as by the national authorities of the countries involved respectively in the CESAR project: Austria, France, Germany, Greece, Italy, Norway, Portugal, Spain, Sweden, and the UK. The responsibility for the content of this publication lies with the authors.

We would like to use this opportunity to thank all project participants, on average 450 people over 40 months for contributing to this extraordinary project and also leading to the creation of this book. Unfortunately, the available space does not allow us to mention them all individually.

The editors, Ajitha Rajan and Thomas Wahl, would like to express their gratitude to the team at the University of Oxford for their invaluable help organizing the work on this book:

- Pamela Farries
- Matt Cherill

Contents

Chapter 1
Introduction

Eric Armengaud, Stephane Bonnet, Laurent Duffau, Gianfranco Fenu, Gerhard Griessnig, Christian Hein, Andreas Keis, Ingrid Kundner, Kung-Kiu Lau, Marc Malot, Markus Örtel, Fabien Paganelli, Marc Pantel, Magnus Persson, Nikolaos Priggouris, Ajitha Rajan, Tom Ritter, Alain Rossignol, Adeline Silva, Martin Törngren, Parham Vasaiely, and Thomas Wahl

E. Armengaud (✉) · G. Griessnig · I. Kundner
AVL List GmbH, Graz, Austria
e-mail: eric.armengaud@avl.com

S. Bonnet · M. Pantel
Centre National de la Recherche Scientifique, Paris, France

L. Duffau
Airbus Operations SAS, Toulous, France

G. Fenu
Universita degli Studi di Trieste, Trieste, Italy

C. Hein · T. Ritter · A. Silva
Fraunhofer Gesellschaft zur Förderung der Angewandten Forschung e.V., Munich, Germany

A. Keis · P. Vasaiely
EADS Deutschland GmbH, Munich, Germany

K.-K. Lau
The University of Manchester, Manchester, UK

M. Malot · F. Paganelli
Sagem Défense Sécurité, Paris, France

M. Örtel
OFFIS e. V., Oldenburg, Germany

M. Persson · M. Törngren
Kungliga Tekniska Högskolan, Stockholm, Sweden

N. Priggouris
Hellenic Aerospace Industry S.A., Tanagra, Greece

A. Rajan · T. Wahl
Oxford University, Oxford, UK

A. Rossignol
ASTRIUM SAS, Toulouse, France

A. Rajan and T. Wahl (eds.), *CESAR - Cost-efficient Methods and Processes for Safety-relevant Embedded Systems*, DOI 10.1007/978-3-7091-1387-5_1,

The embedded safety-critical system design and development industries are facing ever-increasing demands regarding the variety and flexibility of systems and devices from society. At a technical level, these demands result in more and more complex solutions that, at the same time, need to abide by stringent regulatory requirements and economic challenges, such as cost, performance and time to market.

These constraints have led to a profusion of enablers (new processes, methods and tools), mostly developed independently, and often with a focus on a particular concern, such as safety. The resulting solutions are neither integrated with each other into comprehensive development processes, nor are they even designed to be widely interoperable, thus preventing reuse of existing technology and methodology. This complicates sharing of information between different tools, and developers are forced to manually transfer information between many different tools. An immediate consequence of this is a huge amount of repeated work leading to unnecessary costs. The risk of inconsistencies between models in different tools, which also may compromise safety, is evident [73].

A limiting factor in designing reusable, customizable solutions is the absence of internationally recognized open standards for safety-critical embedded system design, implicitly facilitating the proliferation of individualized approaches. Thus, any attempt at remedying the status quo in the industry must (i) provide technological advances that *enable* such customizable solutions, as well as (ii) develop and advocate standardisation mechanisms that *catalyse* the use of such solutions across the embedded systems community.

For the development of integrated tools, there is need for a platform that controls and protects the metamodel and orchestrates the connected tools according to a predefined development process. Such an approach and platform would allow significant savings of development effort and costs by reducing repeated work.

Meeting the need for standardization in embedded system design requires a multi-domain project involving all major safety-critical embedded systems domains. For this reason, the CESAR project was started to provide improved methods, tools and processes to meet the demands in embedded systems development in the domains of Aerospace, Automotive, Automation and Railways [72]. A total of 54 partners[1] from 10 countries have collaborated for over 3 years within this European project, which started in March 2009 and lasted until June 2012. The major objective was to decrease the costs of safety-critical embedded systems development by up to 50 %.

CESAR is an EU research project, co-funded by the ARTEMIS Joint Undertaking and National Funding Authorities. ARTEMIS projects are designed to follow a multi-domain strategy with strong involvement of industrial end-users to ensure that their needs are addressed, academic partners to provide state-of-the-art development and tool providers, often small and medium-sized enterprises (SMEs). To address

[1]Some partners have left or joined the project with the project lifetime. A full list of organisation is available in the annex.

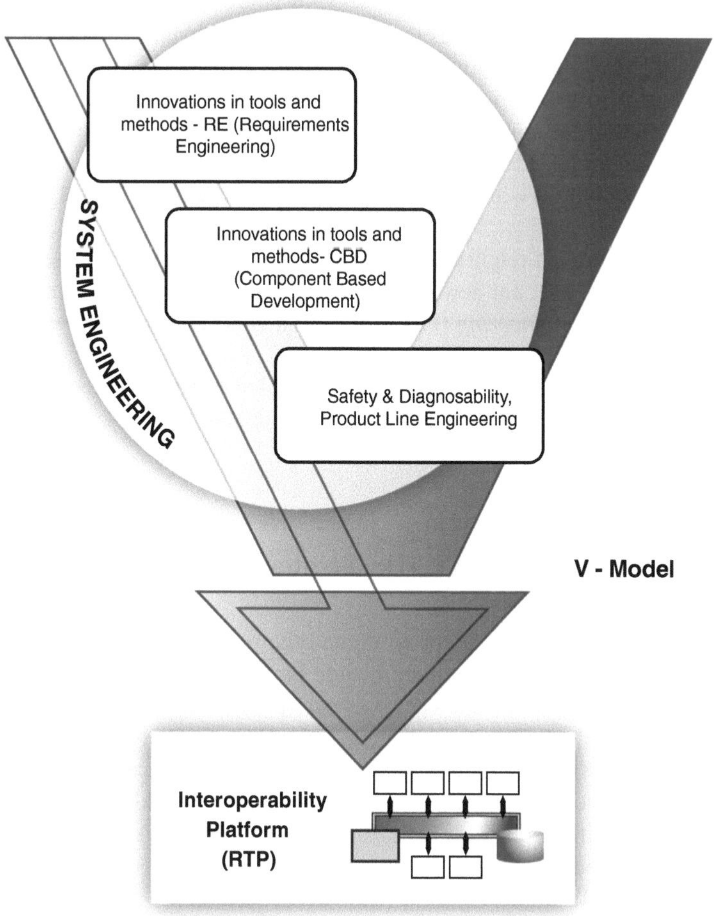

Fig. 1.1 The CESAR systems engineering idea

challenges on a European scale, ARTEMIS favours consortia "thinking big" with an involvement of many partners and countries.

In the spirit of the aims of ARTEMIS, the CESAR consortium brought together partners from four different domains. One of the major reasons for this is to find commonalities between the different domains and to bring together state-of-art and state-of-practice.

To meet the challenging project objectives, CESAR addresses the entire system engineering life cycle (Fig. 1.1) by improving its disciplines and implementing fundamentals for interoperability in a *Reference Technology Platform (RTP)* as an integrated tool platform. The RTP is a set of interoperable tools, methods and processes designed to together improve development work.

The expected improvements include:

1. Innovation in tools and methods for *requirements engineering* (RE), in particular through formalization of multi-viewpoint, multi-criteria and multi-level requirements. The focus here is on the full traceability and analysability of a requirement throughout the development chain and even the entire supply chain.
2. Innovation in tools and methods for *component-based development* (CBD) and extending capabilities with multi-view, multi-criteria and multi-level architecture trade-offs.
3. Combining improved requirements engineering and component-based development, taking into account new disciplines such as *safety & diagnosability* (SD) and *product line engineering* (PLE) given that close cooperation between these is necessary.
4. Only an integration of the disciplines accompanied by adequate tool support into a seamless tool chain (the CESAR RTP) can unleash the full potential of the CESAR approach.

1.1 CESAR and the ARTEMIS Strategic Research Agenda

ARTEMIS is an industry-driven program and the partners engaged in this framework are all European entities which include large enterprises, academic partners and tool providers (which can be large enterprises and SMEs). According to the ARTEMIS Strategic Research Agenda [10], European industries are urged to strengthen their position in the secondary IT sectors where leadership and market dominance are achieved by innovative and high quality embedded system functionalities in safety-critical domains. The share of embedded system components in products is expected to increase by the order of at least 30 % over the 10 years in which the ARTEMIS Joint Technology Initiative runs.

At the same time, forecasts for 2015 show a strong global increase (around 150 %) in systems and software development costs for all domains addressed within CESAR, which is well above the market growth and almost twice the growth of general R&D costs. There is thus an obvious evolution towards increased systems and software development in these sectors, making it a strategic factor for industry competitiveness.

Technically, CESAR is positioned within the ARTEMIS Industrial Sector and directly addresses key ARTEMIS challenges within the Sub-Programme No. 1, *Methods and Processes for Safety-Relevant Embedded Systems*:

- Strengthening Europe's SMEs, by creating new market opportunities, improving interoperability, facilitating access to the embedded systems market, and spinning off new technologies,
- Supporting an integrated chain of European-sourced tools,
- Reducing non-recurring costs and time to market in embedded systems development,

- Curbing the exponential complexity increase in embedded systems while maintaining or reducing development costs,
- Reducing the time and effort required for certification and qualification upon changes and during system evolution,
- Improving product maturity at entry into service, and
- Pushing interoperability of components within and across the domains covered, thus contributing to cross-sectoral re-usability of embedded components.

1.2 What to Expect in the Book

This book is meant to give a technical overview of the CESAR project, its contributions, and the lessons learned. The book documents the outcomes of the project in the different system engineering disciplines, the technical choices we made, and their rationale. The book is not meant to serve as a tutorial for using the tools developed as part of the project or how to manage a project. The target audience for the book are potential users of the CESAR project outcomes, embedded systems developers and vendors across different domains.

1.3 Organization of the Book

The content of the CESAR book is organized as follows (see Fig. 1.2): Chaps. 2–5 describe improvements with respect to systems engineering for the development of safety-relevant embedded systems. In particular, Chap. 2 presents the system design life-cycle and serves as backbone for the methods described in Chap. 3 – Requirement Engineering, Chap. 4 – Architecture exploration, and Chap. 5 – Component-based development. The proposed methods and processes are implemented in tools that require further integration into toolchains in order to cover all or part of the design life cycle. The Chaps. 6–8 present the CESAR outcomes relating to tool integration and interoperability. Chapter 6 – The Reference Technology Platform – presents the main concepts and architecture of the CESAR RTP. In Chap. 7 – Implementation of the RTP – the RTP implemented within the project during three innovation cycles is presented. Drawing on this experience, Chap. 8 – Interoperability – describes the CESAR interoperability specification, which has been formulated as a proposed first step towards the harmonisation of interoperability and tool integration approaches at European level. Closing the innovation loop, Chap. 9 describes the activities of the end users in order to (a) present the pilot applications as a description of the industrial challenges, thus illustrating the basis for requirements on processes, methods and tools, and (b) summarize the evaluation performed by the end-users against the CESAR desired technical outcomes. Finally, Chap. 10 – CESAR results – performs a mapping from the detailed technical evaluation back to the main CESAR goals.

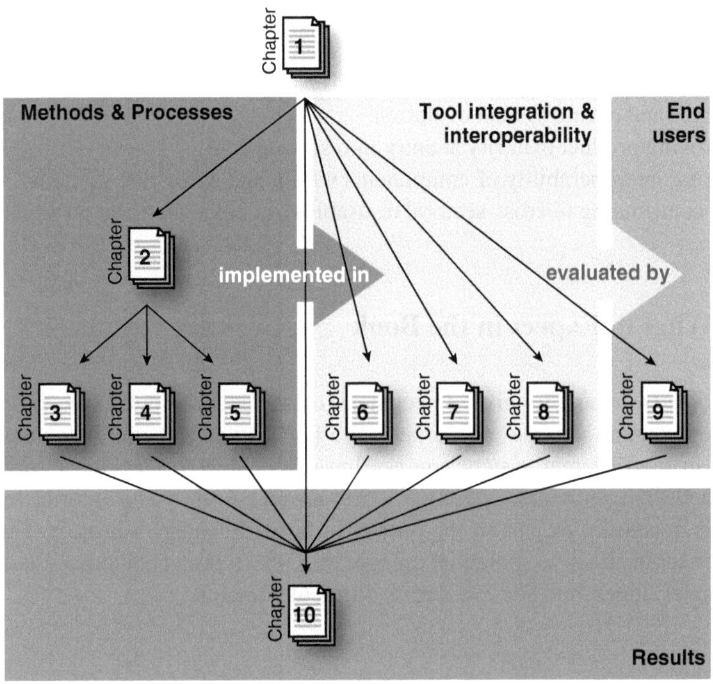

Fig. 1.2 CESAR book organisation

1.3.1 Systems Engineering

Chapter 2 summarises CESAR progress towards enhancing activities within the overall design and development life-cycle of safety-critical systems, starting from the well-known V-cycle. The approach recommended is completely model-based with different views used to represent the various aspects (e.g., safety, real-time) and perspectives (e.g., functionality, physical architecture, operations) of the system. The use of (reusable) components is suggested as throughout the development life cycle.

The suggested approach and corresponding innovations attempt to over come, as far as possible, the inherent differences characterizing the CESAR industrial domains of Aerospace, Automotive, Rail and Automation by providing a common approach of addressing architecture design at various levels of abstractions. The chapter highlights the need for a system engineering data model – which can enable sharing of information between development phases and reduce inconsistencies and ambiguities between engineering teams of different disciplines when working in parallel – and defines the main concepts that it must support at various levels of system analysis. Important modelling aspects adopted by CESAR such as the use of views defined by standard viewpoints and the design-by-contract paradigm are briefly outlined since they are regarded as cornerstone elements to better comprehension of the life-cycle enhancements that focus on the use of models and components during all design phases.

The chapter explicitly addresses safety process activities and explains the importance of enhancing system models with safety-related information as a means to automate the generation of safety analysis data. Moreover, Product Line Engineering (PLE) principles, their importance, and the way variability is addressed in the RTP platform through appropriate tool-chains, is covered in a dedicated section.

The chapter concludes by describing how traceability is achieved between the artefacts of a system developed during the various engineering phases, whether these concern requirements, models, test cases, specifications, components, and so on. This was deemed necessary since in safety-critical systems, much of the overall certification credit is based on the ability to demonstrate efficient traceability management.

Important concepts in this context are the CESAR Meta-Model and the Common Meta-Model (CMM). The CESAR Meta-Model is a semantically founded conceptual meta-model that is independent from a concrete implementation. It defines a common terminology with conceptual elements and relationships. As reference ontology it allows a common understanding of information that is exchanged among system engineers from different domains using various tools, data formats and standards. It contains modelling concepts that are of common interest for the different industrial domains Aerospace, Automotive, Automation and Railway. These core concepts reflect contract-based requirements engineering, component-based design, traceability and verification/validation for multiple design levels and viewpoints. Together, they provide the basis for the following engineering disciplines: systems engineering, safety engineering, product line engineering, and performance engineering. A meta-model provides a concrete solution for achieving the core concepts. This is the foundation for the implementation of the common Meta-Model in the RTP to enable tool-interoperability and integration with CESAR methods. This RTP solution will be demonstrated in Chap. 7.

The objective of Chap. 3 is to describe the activities related to the improvement of requirements management and engineering processes, in order to enhance interchange in the development and supply chain and to ease the definition and identification of safety critical requirements in line with safety standards. Supported by traceability mechanisms, verification and validation of requirements will be performed through improved techniques and methods. From a technical point of view the aim is:

- To provide formalized multi-criteria requirement specification languages (RSLs).
- To ensure complete traceability of requirements from concepts to products across supply chain boundaries.
- To support validation and verification of formal multi-criteria requirements with respect to consistency, completeness and correctness.

- To manage requirements effectively by providing mechanisms to measure the quality and progress of requirements capturing, as well as to analyse the impact of requirement changes.

A precondition of automated verification and validation of requirements is that these requirements are given a formal notation. Pure formal languages as first or second order logic are well-defined from a mathematical/theoretical point of view, with a clear and unambiguous semantics, which enables the application of formal analysis. Such pure formalisms are however rarely accepted in industry (if at all). The main reason is that the syntax is typically not user friendly and the semantics is often hardly accessible.

In order to deal with the management and traceability of requirements, a Requirement Engineering framework was previously built on some existing concepts including those established in SysML [121], EAST-ADL2 [42], metamodels from the MeMVaTEx project [2] or in the traceability reference model by Jarke and Ramesh [132].

This framework is integrated into the CESAR meta-model and allows tool-independent analysis of requirements. Based on this framework, complex management activities like the measurement of requirements based on the progress of verification tasks have been implemented. Furthermore it is possible to reliably compute the affected system artefacts if a requirement changes. This impact analysis can be performed across tool boundaries to reflect realistic results. This integrated CESAR Requirements Engineering approach with its sub-disciplines *Requirements Elicitation*, *Requirements Definition*, *Requirements Analysis* and *Requirements Management* is described in detail in Chap. 3.

Chapter 4 introduces the concept of architecture exploration. Architecture exploration is an approach to the task of architecture design, involving the development of several different models of architectural design alternatives in order to perform full or partial design optimization according to some selected design criteria. This chapter relates on the concept of variability/product line engineering introduced in Chap. 2.

The chapter presents architecture exploration by introducing a few basic concepts in the area: metrics, design criteria and design goals, design and search space, and finally trade-offs. The chapter then continues by describing a new methodology for architecture exploration of safety-critical distributed embedded systems which was developed during the CESAR project, and explains and exemplifies the methodology through a case study based on one of the CESAR public pilot applications: the Simplified Doors Management System – SDMS (described in Chap. 9). The chapter concludes with a section on current tool support, followed by a discussion about the challenges in performing architecture exploration with the current tool support. Finally a few possible directions for future work are explored.

Chapter 5 focuses on Component-Based Development (CBD). CBD targets the development of reusable system parts, called components, that can then be assembled to build new systems or components. CBD is the standard practice in many engineering domains such as electronics, mechanics, hydraulics and chemistry, which are strongly constrained by physical laws. Standardized component catalogues ("repositories") are available, which precisely describe the qualitative and quantitative properties of both the ***service provided*** by the component and the ***services required*** by the component from the environment in order to run efficiently. By nature, system and software are more virtual, less constrained by physics, and thus allow a lot more freedom in the design and implementation space. This has not eased the natural emergence of components as in the other engineering disciplines. CBD is an important focal point in the CESAR project, as it requires new processes, methods and tools for its application to software and systems. In particular, appropriate description of ***provided and required services*** for safety-critical systems are mandatory to ease the development of pre-verified components. Regarding the new processes, methods and tools, two issues must be managed: on the one side the design and implementation of new components, and on the other the design and implementation of systems from existing components. This chapter introduces CBD principles and their impact on processes and methods, and especially the W life-cycle for system development, an adaptation of the common V cycle for CBD. Then it summarizes the two main contribution in CESAR related to CBD:

- The integration of the X-MAN component model and infrastructure in the RTP, exemplified through its use for the design of an aircraft refuelling system;
- The component aspects at the various levels of the CMM and RTP and its application to the development of a simplified doors management system.

1.3.2 Tool Integration and Interoperability

Chapter 6 presents the main concepts and architecture of the CESAR RTP. The CESAR Reference Technology Platform is a generic model-based tool integration platform composed of a set of interoperable tools, methods and processes designed to improve the development of safety critical embedded systems. This generic integration platform will host tools from different vendors and must be able to host additional tools from end users. Interoperability between these tools is a key feature. It should be based on stable and non-proprietary standards.

This chapter also introduces how the different notions of meta-models were defined and used in the CESAR RTP process, from basic concepts in the Core Meta-model through to the specification, implementation and customization of current RTP versions. Finally this chapter describes the different processes defined for RTP development, quality, assessment and instantiation.

Chapter 7 presents the challenges of building a running Reference Technology Platform (RTP). Here, the general principles of RTP tool chain implementation are introduced, including distributed development, information management and model-driven development. This section is followed by an overview of the typical RTP implementation phases and the different prominent roles of developers within this engineering task. The three RTP implementation phases: Generic RTP, RTP tailoring and RTP instantiation, are described in detail. The chapter mainly focuses on the RTP tailoring aspect, as this aspect is probably the most ambitious engineering task undertaken with respect to the RTP implementation. Therefore, numerous best practice recommendations are given in order to guide potential RTP developers in making qualified design decisions for RTP implementation work. This chapter refers to different concepts which are described in the preceding Chap. 6 – Reference Technology Platform.

Chapter 8 aims to summarise the experience gained in CESAR regarding tool integration and describes the outcomes contributing to the CESAR Interoperability Specification (IOS). This specification is intended to meet the needs for interoperability and collaboration between tools across the entire engineering lifestyle. However, in order to provide a common approach for exchanging information as well as basic technologies which connect that information in order to achieve full traceability, the CESAR IOS has adopted interoperability technologies proposed by the Open Services for Life-cycle Collaboration (OSLC). In addition the content of this section will provide a reference architecture, which allows a consistent implementation of RTPs. Furthermore the IOS allows different RTPs based on the same IOS to be connected and integrated to produce a multi-domain engineering environment.

1.3.3 Evaluation by End-Users

Chapter 9 – Evaluation of CESAR: Pilot Applications – presents the industrial needs, described in terms of "pilot applications" as well as the domain specific tailoring of the technical outcomes and their evaluation. This chapter illustrates (using examples) the work performed by the end-users (1) to communicate their needs to the technical domains by refining requirements for methods, processes and tools according to real-world scenarios, and (2) to evaluate and take on the technical outcomes in order to adapt relevant solutions to their specific needs and development processes.

This chapter describes one typical application for each industrial domain (Automotive, Aerospace, Railway and Automation), illustrating the harmonization work performed during the elicitation of the requirements, introducing the most important requirements (with respect to the RTP development) and finally illustrating the outcomes achieved during the evaluation performed by end-users with such pilot applications.

Some of the applications presented are used as motivating examples in other chapters of the book: the so called "Simplified Doors Management System" (SDMS) is the most cited and used. The SDMS is cited as an example in Chap. 3, in particular when the requirement definition – applying patterns, boilerplates etc. – is described, and as an example for the application of SysML. It is also a case-study for Chap. 4, explaining the concept of "architecture exploration" as an approach

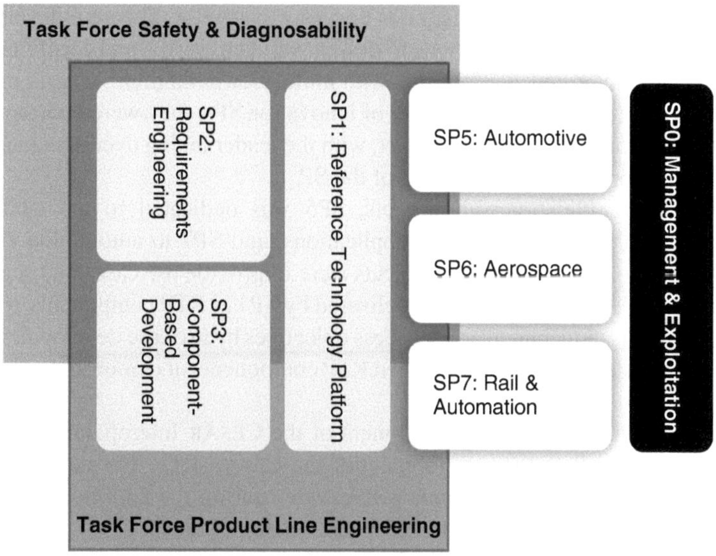

Fig. 1.3 CESAR project organisation

to the design of distributed embedded safety-critical systems. In Chap. 5, it is the application used to introduce the X-MAN tool and for instantiating the CMM. Features and details on requirement elicitation and outcomes, related to the other pilot applications, are also described in this chapter.

1.4 Project Organization

The organisation of this large projectis a key feature. At the top level, the CESAR project was partitioned into *sub-projects* (SPs), the arrangement of which is shown in Fig. 1.3. The sub-projects form two main groups:

Innovation/technical sub-projects – SP1, SP2, SP3 (TSPs) jointly achieving the harmonized technical requirements, and

Application/domain sub-projects – SP5, SP6, SP7 (DSPs) providing requirements to the TSPs and performing evaluation of and feedback on technical solutions provided by these.

Among the innovation sub-projects, SP2 **Requirement Engineering** was dedicated to a formalized multi criteria requirement capturing language, to traceability and consistency of requirements, and to the validation of formal multi-criteria requirements with respect to consistency and completeness. Sub-project SP3 **Component-Based Development**, on the other hand, focused on guiding systems/multi-systems architecture choices against industrial and operational criteria, on providing analysis methods covering system-architecture development and detailed design, and on methods to design, validate, verify, and certify products made of available or new components with multi-viewpoint multi-criteria interface specifications. Reflecting the character of innovation SPs, they were coordinated by both an industrial and a research partner, with the leader taking overall management responsibility for R&D developments of the SP.

Among the application sub-projects, SP5 was dedicated to the automotive domain, SP6 to avionics and space applications, and SP7 to automation and rail. The objectives of application sub-projects were to provide use cases and scenarios, to incorporate CESAR innovations as delivered by SP1 as RTP components, to assess the achievements of technical and business objectives by use case developments, and to prepare industrial take-up of validated RTP components of demonstrated business relevance.

SP1 was dedicated to the establishment of the CESAR Interoperability Specification (IOS) as well as the development of the CESAR RTP. The development of the RTP also gave rise to four major *milestones* during the CESAR project, each milestone signifying the publication of a new version of the RTP.

SP0 **Management and Exploitation** comprised the overall project management and all dissemination and exploitation activities, such as the definition of a dissemination strategy, including interaction with relevant standardization bodies, relevant industrial user groups, etc. Each sub-project consisted of work packages, many of which had associated deliverables. Each work package was assigned to a partner as a work package leader, responsible for the coordination of tasks and activities making progress towards the work package objectives, including progress monitoring, and reporting of delays or changes in procedures.

Task Forces CESAR Task Forces were designed to provide cross-SP work throughout the life of project. The Task Forces Safety and Diagnosability as well as the Task Force Product Line Engineering, staffed with experts from the relevant fields, were organized as Working Packages of SP1. One Task Force was created during project execution – Task Force Business Model, organizationally linked to the SP0 Work package dealing with Dissemination and Exploitation of results.

Working Groups Despite the organizational efforts to improve project collaboration and the innovation cycle, during project execution it became evident that a further cross-SP layer was necessary to work on central topics for the project. To this end, Working Groups were created on demand. Their intention was to work on a specific topic for a specified period of time. The working groups created for this purpose are listed in the following Table 1.1.

CESAR Innovation Cycle The work in CESAR was following a clear project strategy which is illustrated in Fig. 1.4.

The rightmost arrow of Fig. 1.4 illustrates the CESAR multi-domain approach with Aerospace, Rail, Automotive and Automation working in the domain Sub-Projects. These Domain SPs reflect the current state of industrial practice development and deliver information on their needs to the Innovation SPs (Technical SPs)

Table 1.1 CESAR working groups

Name of working group	Mission of working group
Requirements Management Team (RMT)	The Requirements Management Team (RMT) was set up to define a process to author the domain and technical specification (DTS) document and subsequently to operationalise and manage the authoring process
Convergence	RTP roadmap management: 1. End users' needs coverage status 2. SE process convergence 3. RTP roadmap adjustment
Requirements (RQ)	Define a process and tool landscape to efficiently collect the requirements from the domains and the technical subprojects
Meta-Model Harmonization	Various different meta-models have been defined in the different CESAR work packages and sub-projects, such as meta-models for PLE, for SP3, for SP2 (RMM) and SP5 (EAST-ADL). The objective of this working group is to align all meta-models and to define a common meta-model
Objectives	Define a systematic and structured way to measure the CESAR objectives
Principles	1. Improvement of traceability of technical items to deliverables, artefacts 2. Completion of list of CESAR principles 3. Description sheet for TI description relating them to principles
Business Model	Provide orientations to support the RTP, after the completion of CESAR contract
System Engineering Process	Define a general System Engineering Process
Integration and Testing Team (ITT)	1. Test operational capacity of RTP 2. Favour close interaction between RTP technical contributors
Process Description Landscape	Agree on the common use of a process description language for CESAR
Interoperability specification (IOS)	1. Definition of Interoperability Specification 2. Exchange with OSLC and research initiatives

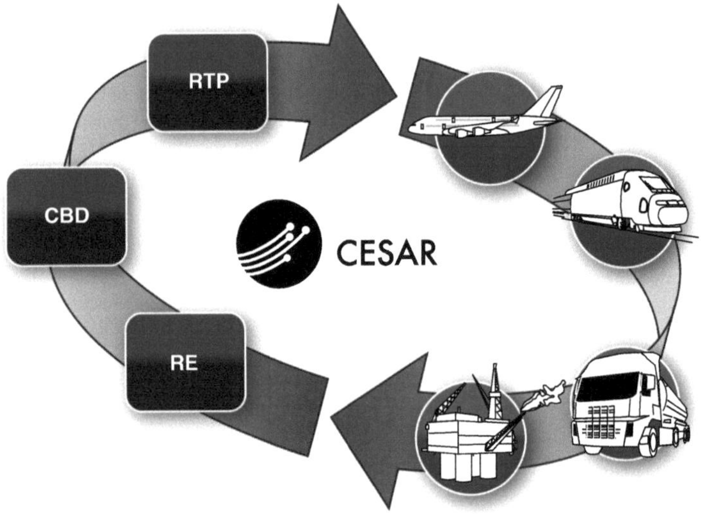

Fig. 1.4 Project strategy: Iterative innovation cycles between domain subprojects and technical subprojects

which are represented on the left hand arrow of the figure. The Innovation SPs provide solutions, supported by the Reference Technology Platform, to the Domain SPs to address their needs as far as possible. The RTP was then evaluated by the Domain SPs performing practical industrial applications to refine the needs and improve subsequent versions of the RTP. This process, called the innovation cycle, was repeated twice.

Chapter 2
The System Design Life Cycle

Nikolaos Priggouris, Adeline Silva, Markus Shawky, Magnus Persson, Vincent Ibanez, Joseph Machrouh, Nicola Meledo, Philippe Baufreton, and Jason Mansell Rementeria

This chapter focuses on a generic process for developing (safety-critical) systems. After a reminder concerning the current development process including the safety aspects, the "CESAR-proposed" development process, based on multi-views and a component-based approach, is highlighted.

Despite the fact that the system design life-cycle in each CESAR domain (aerospace, automotive, rail and automation) is characterised by many commonalities, there are also inherent differences, prescribed by domain standards, which are usually reflected in the overall engineering activities. Convergence on a generic development process is crucial for CESAR because this process is the cornerstone of the RTP. More specifically, it will condition the RTP instantiation capabilities and the data which must be manipulated at all stages of system development. These data are the main drivers for sharing information between the development phases on which the interoperability principles rely.

N. Priggouris (✉)
Hellenic Aerospace Industry S.A., Tanagra, Greece
e-mail: priggouris.nikolaos@haicorp.com

A. Silva
Fraunhofer Institute for Experimental Software Engineering IESE, Kaiserslautern, Germany

M. Shawky
Centre National de la Recherche Scientifique, Paris, France

V. Ibanez · J. Machrouh · N. Meledo
Thales Group, Neuilly-sur-Seine Cedex, France

M. Persson
KTH Royal Institute of Technology, Stockholm, Sweden

P. Baufreton
SAGEM Defense Securite, Paris, France

J.M. Rementeria
Fundación European Software Insitute, Donostia – San Sebastián, Spain

A. Rajan and T. Wahl (eds.), *CESAR - Cost-efficient Methods and Processes for Safety-relevant Embedded Systems*, DOI 10.1007/978-3-7091-1387-5_2,

The cross-domain generic process depicted in this chapter is the result of close cooperation between the domain applications. This proposed process is subject to evolution throughout the CESAR innovations cycles. Indeed, only the experiments led by the application domains will be able to demonstrate its adequacy and its adaptability to meet the industrial needs.

The main breakthrough of the proposed process is that it tries to focus on, and impose the use of, models and components during all design phases so that the system description moves from a document-based approach to a more intuitive one where formal methods can be used and early validation techniques can be applied in a more rigorous way.

The chapter also includes a short presentation of important modelling aspects that are regarded as a prerequisite for addressing the "CESAR-proposed" development process.

2.1 The "Existing" Development Process

All industrial partners involved in CESAR deal with the development of embedded safety-critical real-time systems. This includes in the majority of cases the consideration, during the development and design life-cycle, of mechanical, electrical and software parts. At a high level of abstraction, the complete product life-cycle can be described by the activities depicted in Fig. 2.1.

The focus of the CESAR project within this generic process is on systems development, starting with Requirements Engineering and ending with the System Test. Software related steps, which are the main focus at detailed design level, are explained in more detail below.

The V-Model depicted in Fig. 2.2 offers a finer grained view of the various steps and interactions pertaining to the development process and can be regarded as the most commonly used work-flow that applies at system or software level. It is also the basis for safety standard compliant development processes (as in for example the generic IEC 61508 [85] or in the domain specific standards IEC 50128 [49] for railway, ISO 26262 [152] for the automotive and ARP 4754 [141] for aeronautics).

In the context of software intensive system development, the mechanical, hydraulic and electrical parts of the system become environmental constraints for the electronic part of the system.

The left branch of the *V-Model* addresses mainly design and analysis phases. Although variations of the model do exist, in general the main activities performed can be summarized in the following figure (each activity is referenced by the number drawn on Fig. 2.2):

1. Requirements Analysis & Functional Safety Concept The customer requirements are analysed, elicited and refined. The missions and the functions of the system are highlighted. In parallel to this functional approach, a first level of safety analysis is conducted. This safety related activity is elaborated in the safety process related subsection under the term FHA (Functional Hazard Analysis).

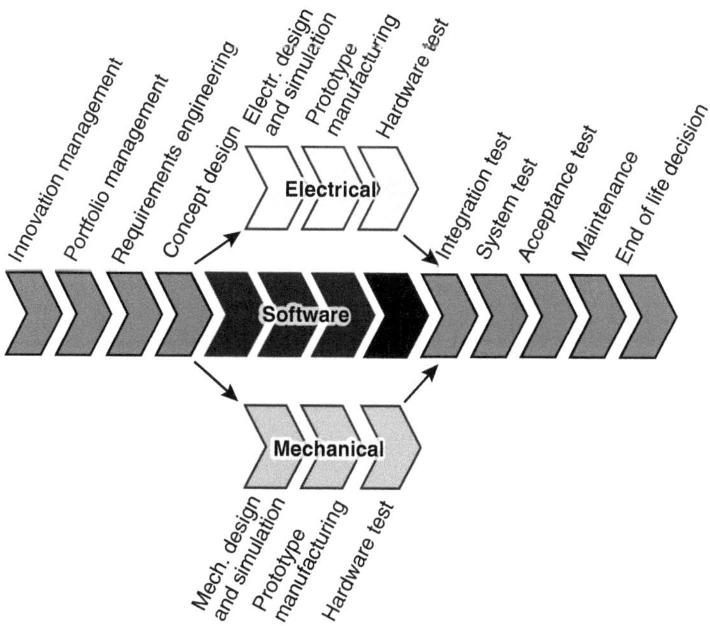

Fig. 2.1 Generic life-cycle of system development

The requirements elicitation and refinement are generally supported by tools like DOORS [81] while the functional analysis and FHA mainly rely on engineering judgement performed manually. Even if some models can be developed for functional aspects integrating sometimes dysfunctional behaviours this is the exception and not the rule.

2. System Architecture Design & Technical Safety Concept The system architecture design is one of the most crucial phases of the system design. It consists of several steps such as refining the functions of the system and allocating them to the different components of the system that can be either physical or software. In parallel to this design phase, the safety constraints are analysed through the PSSA (Preliminary Safety System Assessment). The system architecture design and the PSSA rely, in most cases, on manual activities and experience from previous developments. At this stage, in addition to safety constraints, the architecture solution is assessed against individual criteria such as weight, performance, interface complexity, HMI (Human Machine Interface) usability. However, these evaluations are generally supported by legacy methodologies and are not defined in a systematic manner.
3. SW & HW Requirements Safety Evaluation In this phase of the development, the hardware and software requirements of the system are detailed. The requirements are either textual requirements, or models. At this stage, the suppliers appear in the development process taking as inputs these software and hardware

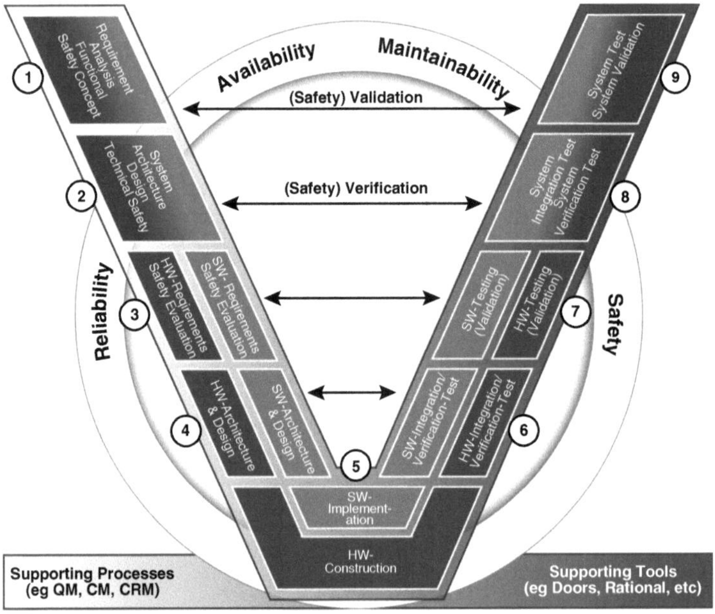

Fig. 2.2 The different phases of the current V life-cycle

requirements for designing the parts of the system they are responsible for. In parallel to this refinement specification phase, the safety activities are deepened through the SSA (Safety System Assessment). Even if the detailed requirements of the system in some domains are specified in formal languages, current practices show that the great majority of them are still informal and natural language based. The phase may include limited validation activities that rely, in the majority of cases, on simulation. Static analysis techniques are applicable, especially for verifying safety properties. These early validations, even if limited, allow detection of potential errors before the final product is produced thus reducing the cost and time of development.

4. SW & HW Architecture & Design In this phase of the system development, the hardware and software of the various equipment items constituting the components of the global system are designed. The activities that apply are similar to those conducted at the system architecture phase, but focusing on a specific component of the system: (1) refinement of the functional and non-functional hardware and software requirements including the derived requirements and (2) allocation of the software functions to the hardware components. FMEA (Failure Modes and Effects Analysis) for addressing safety concerns is also performed,
5. SW Implementation & HW Construction Once all the components of equipments are designed, the hardware items are physically developed and assembled and the software modules are implemented in parallel and finally integrated with

the hardware. At the end of this step the software and hardware elements should be available for verification activities. Some unit testing may be performed in parallel with implementation, especially for software. In some cases, when software is generated by a qualified automatic code generator from models, the unit tests can be suppressed (but this is still quite a limited practice).

The right branch of the *V-Model* mainly includes activities related to integration, verification and validation. Integration is performed at various levels (SW, HW, System). V & V activities rely mainly on testing, when the product or a part of the product to be verified is available. Sophisticated test benches may be used, depending on the tests to be run and on the stage of that verification is carried out at. The main activities are:

6. SW & HW Integration/Verification Test In this step hardware and software components are assembled. Verification tests are executed to prove compliance with the design objectives. These tasks mostly rely on reviews and analysis that are conducted by the components' suppliers.
7. SW & HW Testing In this step, functional tests are conducted against the hardware and software components that have already been verified against the design and passed a first level of integration. The tests aim to ensure that the software and hardware components fulfil software and hardware requirements. They usually rely on sophisticated test benches that provide a good approximation to the system functional environment.
8. System Integration and System Verification Test In this step, the various elements of the system (SW, HW) are combined and verification against functional and non functional requirements at system level takes place. Integration testing activities are performed, by the system integrator with the suppliers' support, on test benches that are representative of the real system environment.
9. System Test and System Validation This final verification phase consists of verifying that the system meets its functional and non functional requirements in its operational environment. In general, this phase is called product validation because it ensures the right product has been built.

In addition to the design and V&V activities, transverse tasks such as configuration management, traceability management and quality assurance are also carried out in order to support the exchange and coordination of work-product artefacts between the various phases.

2.2 The "CESAR-proposed" Component Based Development Process

As it becomes evident from the "existing" development process, a significant number of design activities are not fully supported by models. In addition, re-usability is generally not very well managed in the development process as the components, whatever their nature, have not been designed with reuse in mind. This situation impedes the deployment of real product line capabilities and consequently acts as a deterrent in reducing development time and cost. An exception to this rule is the automotive domain. Here the state of industrial practices have demonstrated significant re-usability. Therefore, advancements in the automotive domain can bring valuable experience to other application domains.

The aim of the "CESAR-proposed" component-based development process is to provide a methodology that allows leveraging the productivity gains offered by uniting model-based development with component-based development. While the latter can contribute significantly to re-usability and minimize the need for re-verification or re-qualification activities, the former, if supported by appropriate tools, can support automation of certain tasks and identification of errors at an early design stage.

Based on the above some of the expected characteristics of this enhanced development process can be summarized as follows:

- Use of models as a basis for the development process,
- Definition of components as primary and mandatory artefacts throughout the development life-cycle,
- Traceability between development steps, requirements and various types of artefacts in general,
- Possibilities for early verification and validation (at design stage),
- An enhanced safety process based on models that are fully linked and/or synchronized with system design models.
- Adoption of product line principles during component design that can promote re-usability

Since the objective of this CESAR development process is to rely on existence of the appropriate models comprised of components, it is necessary to identify each activity of the V life-cycle process depicted on Fig. 2.2 that is to be questioned. In this section and corresponding subsections, we will propose adaptations to the design activities and the V&V activities of the previous paragraph so that they can rely on a fully component model based approach. Early validation will be promoted, as design models will be available prior to implementation of the product's components. The new approach for safety related activities is addressed in the safety process subsection of this chapter (Sect. 2.5).

2.2.1 Design Activities Relying on a Component Model

The three primary activities that constitute the design process are:

- Functional analysis
- Requirements refinement
- Architecture design

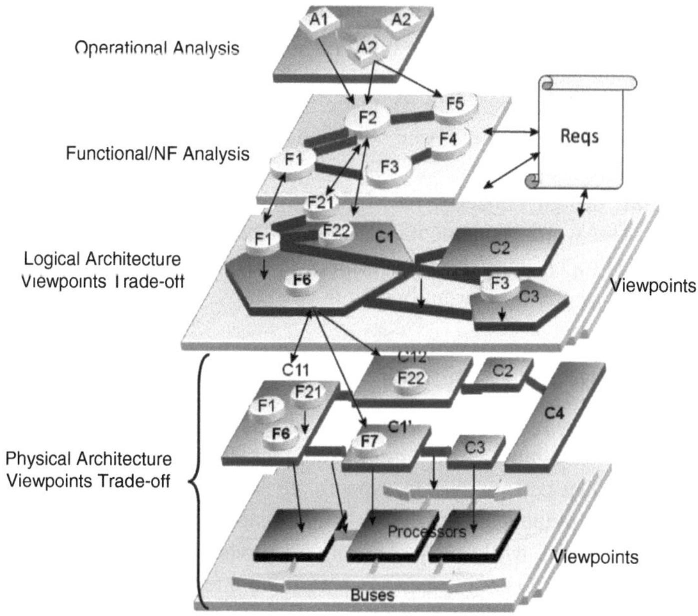

Fig. 2.3 The functional analysis and the architecture design phases

These activities (addressed by steps 1, 2 and 3 respectively in Fig. 2.2) are crucial for the system development and, until today, are not supported by clear and efficient engineering practices. Therefore in CESAR we paid particular attention to these steps and tried to converge on a common cross-domain vision in mutual agreement with all industrial partners.

In CESAR during system design and in order to address both functional analysis and architecture design we consider a multi-view approach where different representations of a common underlying model can be used for depicting certain features of the system under design, Fig. 2.3 gives an overview of this component-based multi-view approach. More details on how to define such a system model as well as on the proposed views and their implementation (viewpoints) can be obtained in subsequent sections (see Sect. 2.3)

In most application domains, functional analysis is formally preceded by an operational analysis where the aim is to highlight the missions of the system and to decompose these missions into operational scenarios. These operational scenarios reflect the way in which the system will be operated by the customer or generally interact with its environment. This step is considered optional. Models developed at this stage could be based on activity or use case diagrams or similar formalisms.

The functional analysis step allows definition of the functions through functional models that are built from existing elementary functions. Taking operational scenarios as inputs, it is possible to verify whether the functions defined are sufficient to accomplish the system missions. These functional models allow refinement and formalisation of the functional and non functional requirements of the system. These requirements will preferably be formalized with the use of a RSL(Requirement Specification Language). The CESAR project defines the basic concepts and possible syntax of such an RSL (see Chap. 3). Once the functional model's associations to non functional constraints have been specified, the system architecture design can start.

Architecture design concerns the definition of the fundamental organization of a system. The word fundamental derives from the fact that this organization determines key properties such as performance and cost of the system. The system architecture design phase is formally split in two steps:

- The logical architecture definition
- The physical architecture definition

These two steps are supported by multiple views. Indeed, during system architecture design, in complex systems, many engineering teams are involved and each team is usually interested in certain system aspects. To give an example, in the avionics domain, a safety engineer is usually only interested in the functional and safety aspects of the system while a system engineer who is in charge of Integrated Modular Avionics (IMA) concepts focuses more on interfaces aspects, performance constraints, etc. Each discipline examines the system under its own prism, the so called *viewpoint*. These different viewpoints, are defined following a fully component-based approach for both logical and physical level analysis.

During the logical architecture definition, the functions are refined in sub-functions and are allocated to logical components. This process may include multiple steps of hierarchical refinement. At each step, the logical models are validated against the functional and non functional requirements, outputs of the functional analysis phase. Some different architecture solutions can be envisaged and trade-off analysis can be conducted in order to choose the best compromise. This trade-off analysis, usually referred to as *architecture exploration*, is based on multi-criteria assessment and could be considered either as an extension or as an integrated part of the overall architecture design (architecture exploration concepts are briefly presented in Sect. 2.3.4 while for more details please refer to Chap. 4).

The second step of the architecture design is the physical architecture analysis phase that consists of developing models that represent the physical parts of the system and then allocate the logical components on the physical ones. At this stage, a complete executable model of the system architecture is available. New trade-offs based on multi-criteria can be performed to compare different physical allocations and to exhibit the preferred allocation solution. Some of these trade-off criteria are proposed in Table 2.1.

Moving from system-level design to item (detailed) design we should consider:

- The hardware and
- Software architecture design and implementation.

These steps constitute significant parts of the detailed design that correspond to activities 4 and 5 in Fig. 2.2.

Table 2.1 Example of criteria that may apply based on a particular viewpoint

Viewpoint	Homogeneous criteria definition
Safety	Minimal cut sets Fault probabilities, Absence of unexpected state
Performance	Response time, Latency, Worst Case Execution Time (WCET), Remaining bandwidth of the network, Remaining memory
Functional/logical	Number of functions, Depth of functional decomposition, Invariant functional property, Number of functions activated to accomplish missions, Number of outputs/inputs per function
Physical	System weight, Average power consumption of the physical components, Adequacy of physical link related to functional exchange between components, Number of connectors, Maximum number of parameters transmitted per physical media, Geometrical dimensions

At the end of the architecture design phase, the software and hardware requirements are completely specified. In particular, the software requirements are specified in formal models and can serve as inputs for the software parts of the equipment. In this case, and provided that development tools support it, the embedded code of the equipment(s) can be automatically generated from the software detailed models.

The software and hardware architecture design can be decomposed into logical and physical exactly as described above for the system architecture design. Indeed, equipment might be composed of different hardware components and of a set of software functions. The allocation activity from logical components to physical components follows the same principles as the ones described above. Then, the hardware and software implementation can be derived from the hardware and software models. The detailed design may also be supported by different views and corresponding viewpoints, similar to those described at system level.

The process described relies on fully formalized component models, thus avoiding ambiguities at all stages the development process. In addition, this fully model-based and component-based approach opens new perspectives in terms of V&V. This topic is discussed in the next section.

2.2.2 The "CESAR-proposed" Process for V&V Activities

One of the main interests of model based design is the ability to conduct validation based on models, offering the capability to detect errors early in the design process and thus avoid costly rework at later stages of the product life-cycle. A model-based approach can support an integrated V&V strategy encompassing simulation activities and static analysis on models in addition to direct testing and static verification on the final product. Figure 2.4 depicts the V life-cycle with particular focus on the V&V activities.

The main characteristics of these V&V activities are:

- All artefacts are traceable
- Lower level models are verified against upper level models
- The design is verified against requirements relying on simulation and analysis activities
- Lower level requirements are validated against upper level requirements and stakeholder requirements, to check if the requirements correctly and completely describe the needs
- Testing activities are conducted through out the development life-cycle to verify the product fulfils the design requirements and the functional product requirements.

We should also distinguish between V&V activities at detailed design and V&V at system (architecture) level. At system level, analysis of system architecture correctness is performed against the designed model by checking a set of predefined criteria that depend on the particular view of interest. This *model checking* technique needs to be supported by appropriate tools. In CESAR and in order to address the V&V process the multi-view design approach is augmented with a set of criteria that are specific to one or more views and can be regarded as part of a corresponding viewpoint. System architecture assessment against a set of criteria is a major requirement to guarantee/prove that a system fulfils its non-functional constraints while meeting its functional objectives. This type of architecture assessment is called *multi-criteria assessment.*

Architecture assessment multi-criteria can be classified in two different types:

- Homogeneous multi-criteria, criteria that can be applied on the same view,
- Heterogeneous multi-criteria, criteria that have to be applied on different views.

An example of such criteria, categorized per viewpoint and identified during an initial gap analysis is given in Table 2.1.

Within CESAR, given the fact that we address safety critical embedded systems, it seems reasonable to expect multi-criteria assessment based mainly only on safety and performance criteria. Such a restriction does not mean that the other criteria will not be assessed, but it will necessitate that different multi-criteria assessments will be done one after the other. *Simulation* remains the main dynamic V&V technique but is strongly dependent on the presence of tools that can support behavioural definition of a given system and executable models.

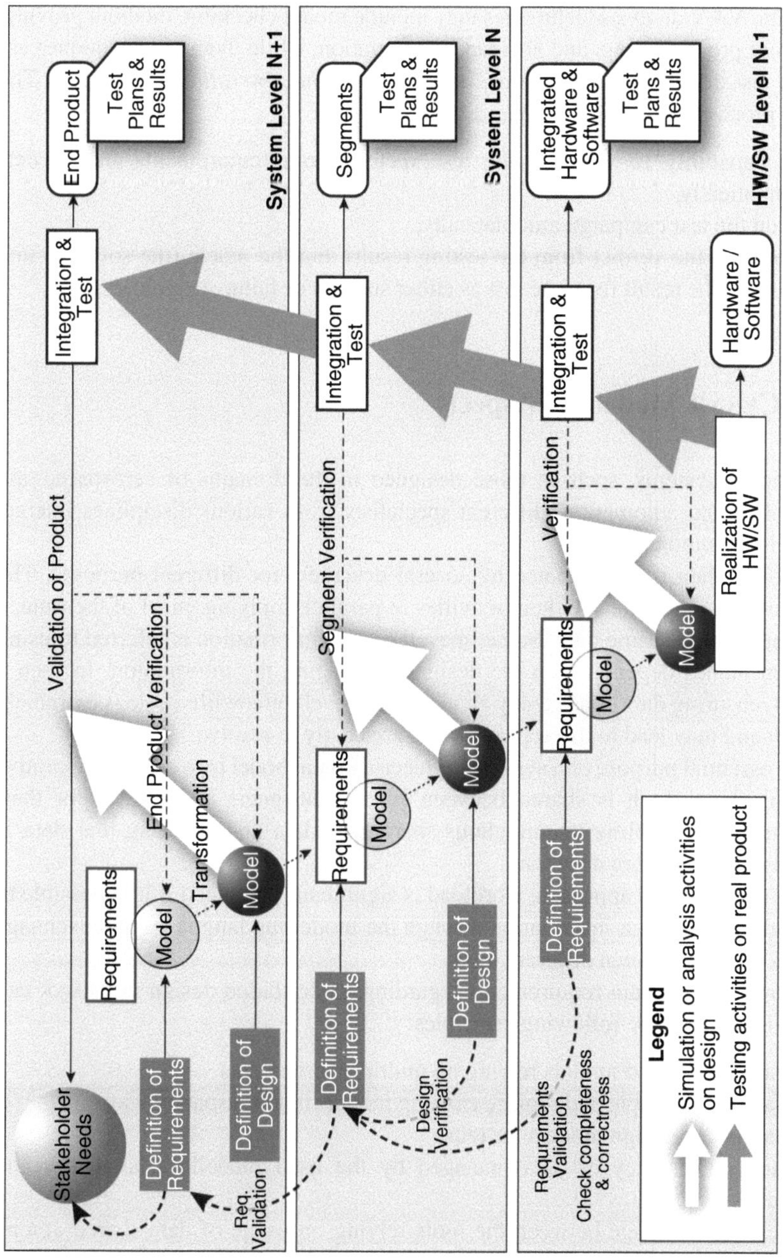

Fig. 2.4 A generic model-based development process with particular focus on testing

At detailed design stages, formal V&V can apply through various types of static analysis. The V&V methodologies that apply at component detailed design level are analysed in another chapter of this project handbook. Here we briefly state that static V&V analysis techniques may include model checking, theorem proving, constraint programming, and abstract interpretation, while dynamic techniques are usually test-driven and focus mainly in *automating the generation of test cases*. The testing process automation encompasses

- The capability to generate the test vectors from requirements and models automatically,
- To run the test campaign automatically,
- To deduce the verdict from the testing results and the oracle (the software that evaluates the result from the test as either success or failure) automatically

2.3 CESAR Modelling Aspects

In complex systems, such as those designed in the domains of aerospace, rail, automotive and automation, different specialists from various disciplines interact to build a common product.

A lot of data is manipulated by several designers for different purposes. The different designers manage their activities in parallel implying, most of the time, a duplication of the same data. Sometimes, the same information is referred to using different names depending on the designer managing the information. In such a context, ensuring data consistency all along the development life cycle is extremely difficult and may lead to discrepancies that are costly to resolve.

The essential purpose of a system engineering data model is to ease collaborative work on data which is shared between several designers and crucial for their disciplines, by enabling unambiguous sharing of data and ensuring that data is properly updated when necessary.

In a model-based approach, workload is significantly reduced if it is possible to manipulate the design information through the modelling languages and exchange various models between different tools.

Therefore, the main requirements regarding model-based design in CESAR can be summarized in the following principles:

- Capability to build models relying on multiple formalisms
- Capability to couple and replace existing models in a transparent way
- Capability to design models for reuse
- Ensure consistency of data managed by the used modelling languages and formalisms
- Ease data exchange between the tools relying on usage of data shared among each other (i.e. the same global model or common set of artefacts)

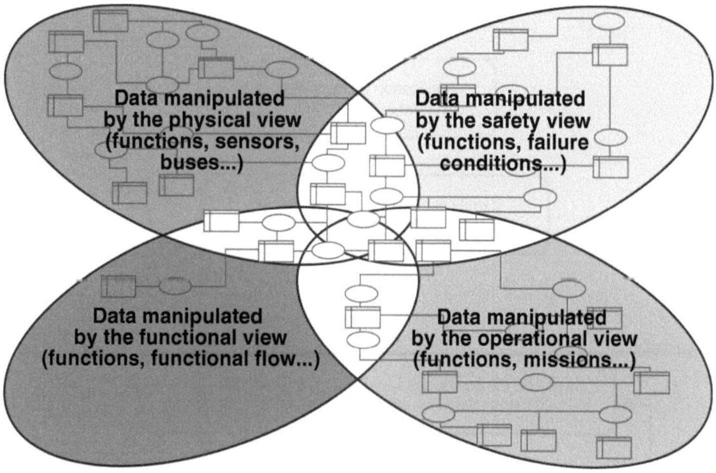

Fig. 2.5 An example of a system engineering data model

All of the above imply that a generic system engineering data model should exist. This data model should be capable of managing the system design information, support multiple views for architecture modelling but also provide the ability to assess various architecture solutions based on multiple criteria. This model in its core will include all the common part of development data managed by the different disciplines throughout the development life cycle activities but should also be extensible in order to easily adapt to future needs. The core part should reflect the most critical data aspects, that is the ones that if modified may have a strong impact on the whole product. In Fig. 2.5 an example of a system engineering data model for aeronautics is shown. The data in white sectors correspond to data shared between at least two views. The core data model part corresponds to the elements residing in the common intersection between all views.

Such a system engineering data model though, capable of manipulating development data managed by different engineering teams and process activities, is also dependent on the application domains and on the industrial processes used. Generally, it seems very difficult to build up a system engineering data model that covers all the domains involved in CESAR: Aerospace, Automotive, Rail and Automation. Even inside each domain, the system engineering data model may vary depending on company specific industrial processes. Therefore, a generic approach must be set up to specify a system engineering data model. *This approach should be customizable by each company according to domain, internal procedures and specific project needs.*

The different steps needed to define a system engineering data model are shown on Fig. 2.6.

The first step consists of identifying the different skills that have to work in tight cooperation. From this identification of disciplines, the views and the associated

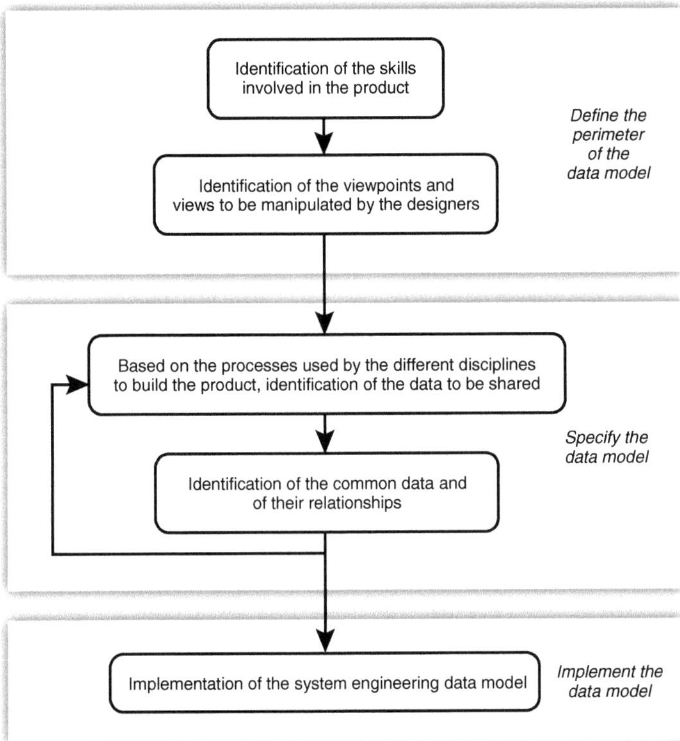

Fig. 2.6 Workflow for defining a system engineering data model

kinds of models to manage in a model based engineering approach are deduced. This phase allows pointing out all the actors that will have to work together.

The second step is crucial as it deals with the identification of the potential data to be shared. This potentially shared data is identified by analysing the processes used by each discipline to develop the product. In particular, the outputs of activities that are produced by one discipline and consumed by a set of activities led by other disciplines are potential candidates for sharing.

After having identified the potential data to be shared, the core data used to build the foundations of the system design activities are selected as common shared data and will be part of the system engineering data model. In order to succeed in this approach, all the stakeholders (or at least the majority of them) have to be involved.

Once the data to be shared have been specified, the system engineering data can be implemented. The implementation details depend on the language used to describe the data model but may also be affected by the environment in which this data model will be used.

As it becomes evident that the data model is based on a viewpoint driven approach, different views are defined that can be used to represent a system from the perspective of a related set of concerns.

The following subsections will try to further expand on many of the concepts pertaining to the CESAR engineering data model.

2.3.1 An Engineering Data Model to Support Architecture Modelling

In CESAR a high level system engineering data model has been proposed. This conceptual data model has the ambition to cover the activities led in the system architecture design as already presented in Fig. 2.3. Therefore, this data model describes the different concepts and details the expected relations between these concepts, to support the design activities along the steps of a system development process. It addresses the following four types/levels of analysis:

- Operational analysis
- Functional and non functional need analysis
- Logical Architecture analysis
- Physical Architecture analysis

The expected model elements for each type of analysis together with their relationships are summarized in the next subsections. A "UML class diagram like" formalism is used. Any implementation meta-model (as supported by RTP tools) shall allow the user to develop, transform, analyse, exploit a model defined with these concepts and their relationships. Any used model transformation tool shall be able to extract these concepts and links from models built using the RTP. Support for most of these concepts is provided in the CESAR Common Meta Model (CMM). The CESAR Common Meta-Model (CMM) (see Sect. 6.4) includes all these system modelling concepts as part of its core meta-model. CMM also provides additional meta-model concepts to address other engineering aspects such as management of requirements (see Sect. 3.5).

2.3.1.1 Operational Analysis

This step focuses on analysing the customers' needs, the goals that shall be reached by the system users, and the activities being performed by the users.

Outputs of this step leads to an "operational architecture" describing and structuring the needs, in terms of actors/users and their operational capabilities and activities, operational use scenarios giving dimensioning parameters, operational constraints including safety, security, system life cycle, and others.

The elements that are introduced in this step are:

Operational actor Person or entity, including system end users, involved in activities and missions related to the use of the system

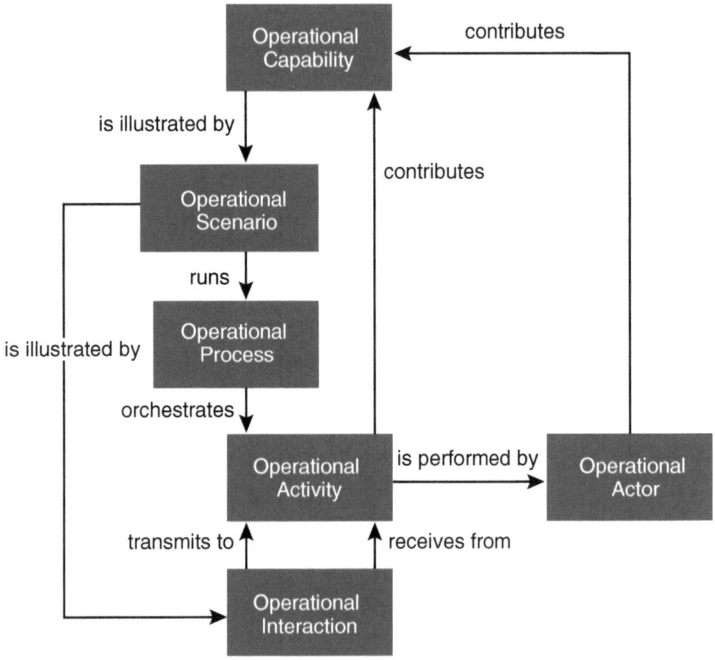

Fig. 2.7 Relation between elements in operational analysis level

Operational capability Quantified overall operational goals, objectives, services that are expected from actors

Operational activity Any process step or function, both mental and physical, performed by actors toward achieving some objective

Operational interaction Generic exchange or information flow between operational activities

Operational scenario description of the sequence of interactions between actors in a given context, usually to achieve a given capability

Operational process Sequence of activities contributing to an operational capability

The relation between all of these elements is described in Fig. 2.7.

2.3.1.2 Functional and Non-functional Need Analysis

The role of this step is to focus on the system itself in order to define how it can satisfy operational needs (prescribed in operational analysis) along with its expected behaviour and qualities. This includes the definition of functional as well as non-functional requirements of the system, e.g., safety, security, and performance. At this step, they are focused on system boundary level and not on individual system components. Furthermore, role sharing and potential interactions between system and operators are of concern here.

Outputs of this step mainly consist of the system functional need description, descriptions of interoperability and interaction with the users and probable external systems (functions, non-functional constraints, and interoperation), and system/SW requirements. Note that these two steps, which are a prerequisite for architecture definition, have high impact on the future design being developed in the forthcoming steps, and therefore should be approved with and validated by the customer.

The elements that are introduced in this step are:

System An organized set of elements functioning as a unit, which is the target of the engineering.
System actor External actor interacting with the system via its interfaces
System capability Contribution of the system to an operational capability
System function Actions or services defined on a system or an actor
System port "entry point" or means to interact with a function or an actor
Data Data related to functional and non-functional need, especially functional exchange contents
Functional exchange Data exchange between functions, linking their ports
System functional chain Sequence of Functions, activated through a path and carrying non functional properties such as latency, criticality level
System scenario Description of the sequence of interactions between system and actors in a given context
State or mode The state characterizes the system behaviour in given environmental conditions. A mode or state is activated by external conditions and by operators

The relation between all these elements is described in Fig. 2.8.

The relation between the operational analysis and the functional analysis is described in Fig. 2.9. As a general rule, each element in the functional analysis has to support the equivalent element in the operational analysis.

2.3.1.3 Logical Architecture Analysis

The role of this step is to identify the system's parts (hereafter called components), their roles, relationships and properties, while excluding implementation or technical issues. This constitutes the logical architecture of the system.

This is achieved by grouping or segregating functions according to functional and non-functional constraints, and afterwards allocating them to logical components. During this step major functional and non-functional constraints (safety, security, performance, etc.) are taken into account. Usually, not all of the non-functional requirements may be supported in an equally good manner, so different variants have to be analysed and the best compromise has to be taken.

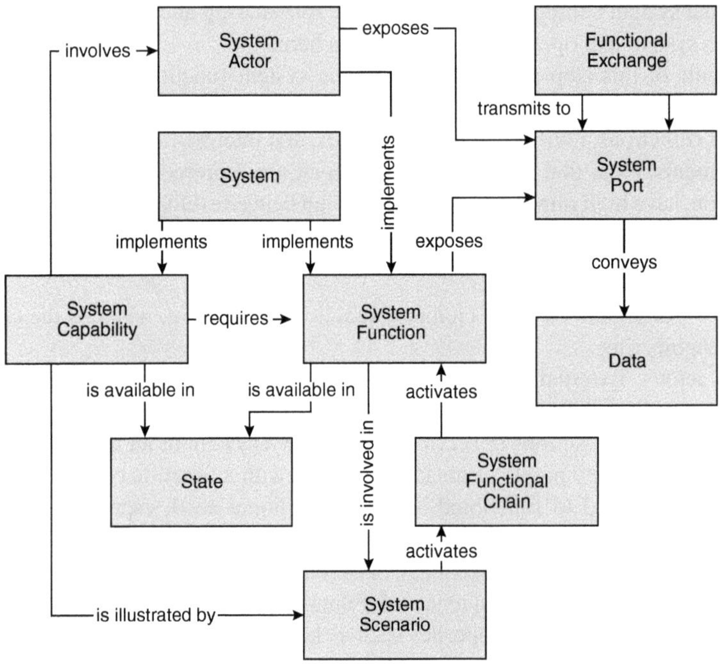

Fig. 2.8 Relation between elements in the functional analysis phase

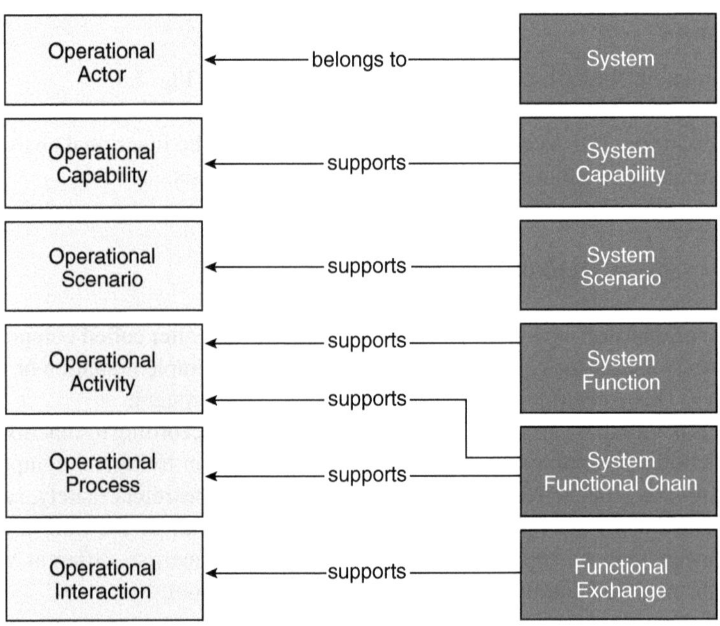

Fig. 2.9 Relation between operational analysis and functional analysis

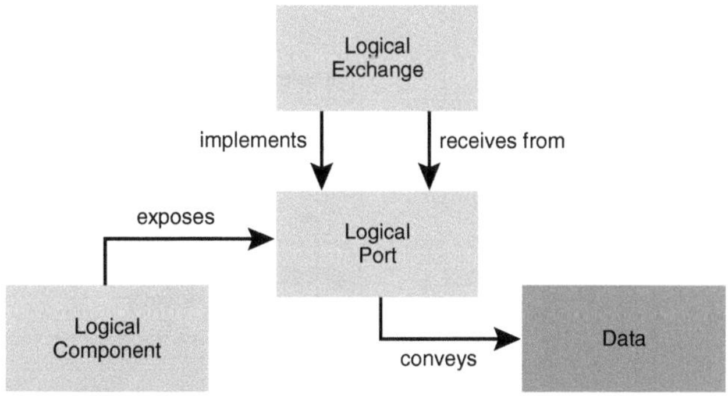

Fig. 2.10 Relation between elements in the logical analysis phase

The output of this step is a logical architecture consisting of components, their interface definitions, and functions allocated to them (along with their functional exchanges).

Since the architecture has to be validated against the needs identified during operational and functional analysis, traceability links to the requirements and operational scenarios are established. The elements that are introduced in this step are:

Logical component component present in an abstract, notional breakdown of the system. This kind of component is independent of technology choices (i.e selection of hardware architecture).
Logical functions functions that are allocated on logical components
Logical exchange exchanges between logical components
Logical scenario scenario between actors and logical components
Logical functional chain functional chain allocated on logical components

The relation between all these elements is described in Fig. 2.10.

2.3.1.4 Physical Architecture Analysis

This step defines the "final" architecture of the system at physical level, ready to be developed (by lower engineering levels). Therefore, it introduces rationalization, architectural patterns, new technical services and components, and makes the logical architecture evolve according to implementation, technical and technological constraints or choices (at this level of engineering). Again, non functional aspects are taken into account as in logical architecture, and multiple different variants are being analysed and checked against each other.

Output of this step is the chosen physical architecture consisting of components to be produced and the way they are taken into account in the subsystem design. As is the case with logical analysis, traceability links towards the requirements and operational scenarios are established.

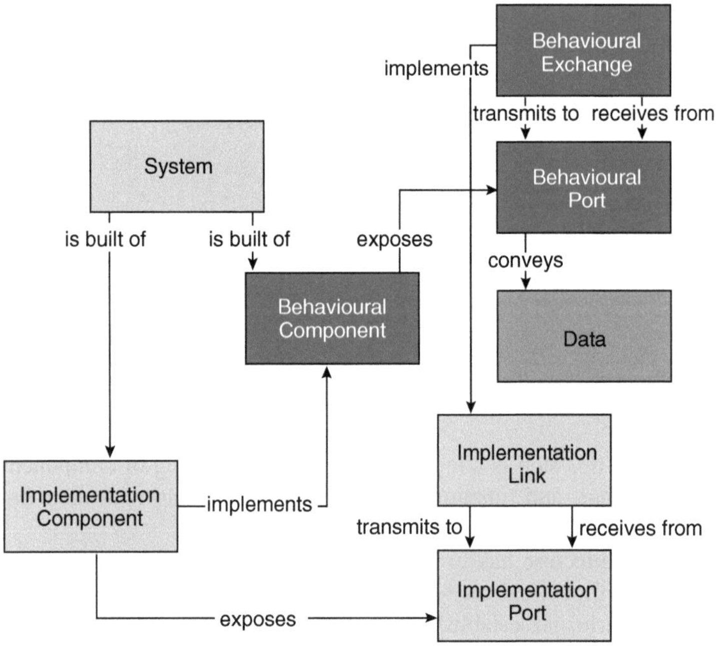

Fig. 2.11 Relation between elements in the architectural analysis phase

The elements that are introduced in this step are:

Behavioural components part of the system that realizes a set of functions. Examples include software component, FPGA VHDL program etc.

Behavioural exchange exchange between behavioural components, linking some of their ports

Implementation components resource hosting behavioural components. Example: Processing unit, display unit

Implementation link Link between implementation components

Physical scenario description of the sequence of interactions between actors and physical components in a given context

Physical functional chain Functional chain allocated on physical components

The relation between all these elements is described in Fig. 2.11.

2.3.2 Views and Viewpoints

The envisioned system modelling approach is viewpoint-driven. According to IEEE 1471 [86], the definition of views and viewpoints are as follows:

Views A representation of a whole system from the perspective of a related set of concerns.

Viewpoints A specification of the conventions for constructing and using a view. A pattern or template from which to develop individual views by establishing the purposes and audience for a view and the techniques for its creation and analysis.

Thus, a view normally conforms to a viewpoint and may consist of one or more architectural models. Moreover, an architectural model may participate in several views. These different models describe the view in different perspectives. In CESAR the four main views of interest are defined according to the four types of analysis described in Sect. 2.3.1.

The IEEE1471 standard however says little about how to choose viewpoints from which to construct the system views. The reason is simple: it is hard to decide on a standardized set of views that will cover all stakeholders' concerns, and still not leave many systems horribly over-specified (and hence also making architecture development more expensive due to work overhead). Many different viewpoint frameworks have been proposed as such standard sets, e.g., the $4+1$ view model [101], MODAF [162] or DODAF [168], but none of them have been established as a de-facto standard across several domains.

Further, many issues such as safety concerns, may be approached in different manners depending on the application domain. Hence, there are many possible selections of sets of viewpoints, and the choice of views ultimately depends on the type of system under development and the stakeholders that the system developers consider most relevant. The viewpoints that will be used to describe a system are typically chosen early during the design, and maybe according to some standard, be company-specific or even produced specifically for the exact system under design. For the types of system in relevance to CESAR, clear candidates for choice of viewpoints, besides the ones defined to directly support the architecture analysis steps (operational, functional, logical, physical), are for example safety, cost, performance, product line, and so on.

It becomes evident that in order to identify a first set of viewpoints it is important to collect and categorise the issues addressed along a typical engineering cycle. Table 2.2 provides a summary of such issues based on brainstorming performed among CESAR industrial partners. Each column represents an example of viewpoint (such as safety or performance) and each row describes the engineering phase and the content that must be included in this viewpoint at each engineering phase. The proposed set may not be exhaustive and will be insufficient for particular designs or domains, however the purpose is merely to illustrate the concepts of a possible framework that can be extended and customized rather than be exhaustive.

2.3.2.1 Viewpoints Standardized in the CMM

Before concluding this section, a brief description will be given on how system modelling is achieved in CESAR through the use of views and viewpoints. The CMM defines *abstraction levels*, *perspectives* and *aspects*. Abstraction levels cover the certain refinements of the system beginning with the first very abstract represen-

Table 2.2 Main concepts handled by example viewpoints along the engineering cycle

		Viewpoints			
		System description			
		Decomposition	Behavior	Safety	Performance
Engineering phases	Operational Analysis	• Actors • Activities • Exchanges	• Scenarios between actors • Phases, missions • Modes and states	• Feared events, failure conditions	• Operational latency constraints on activities
	Functional and Non-functional Need Analysis	• Expected functions • Data • Data exchanges between functions	• Scenarios between system and actors • Functional chains • Modes and states	• Criticality of functional chains, data, and data exchanges • Failure and recovery scenarios	• Latency on functional chains • Processing complexity of functions (or full behavior description)
	Logical Architecture Analysis	Same plus: • Added functions • Components implementing functions	• Scenarios between system components and actors • Functional chains allocated to components • Modes and states of components	• Redundancy paths • Vote, monitoring components behavior • Dysfunctional behavioral model for functions and components	• Estimation of function resource consumption • Estimation of data and exchanges complexity, rate

Physical Architecture Analysis	Same plus: • Implemented functions (incl. technical) • Behavioral components (e.g. software) implementing functions • Implementation components (e.g. computers) delivering resources to behavioral components • Interfaces between behavioral components • Physical links between implementation components	• Scenarios between system components and actors, using interfaces between components • Functional chains allocated to components • Modes and states of components	Same plus: • Component failure scenarios	• Estimation of function resource consumption • Estimation of data and exchanges complexity, rate • Resource available for each implementation component • Bandwidth available for each physical link • Resulting resource consumption of functions, behavioral components . . .

tation and ending with the final concrete representation. Perspectives and aspects define different views for the system under design. Each abstraction level generally contains a (non-empty) set of perspectives. Perspectives represent the system in certain development phases. The CMM differentiates five different perspectives that map to the type of analysis defined in Sect. 2.3.1:

1. *Operational perspective* addresses the operational analysis view
2. *Functional perspective* addresses the functional analysis view
3. *Logical perspective* addresses the logical analysis view
4. *Technical perspective* addresses a subset of elements covered by the physical analysis view
5. *Geometric perspective* (experimental) maps to another subset of elements covered by the physical view

For concerns which are cross-cutting to abstraction levels and perspectives the term aspect is used. Aspects contain representations which might belong to different abstraction levels and/or perspectives. Through aspects dynamic behaviour of a component can be classified and described.

Therefore, in a model-focussed world we can regard the various models that describe a real world system to evolve along three axes covering the analysis levels of the process steps and the viewpoints. This is illustrated in Fig. 2.12). The first axis corresponds to an ordered set of abstraction levels. These abstraction levels generally start with the definition of very coarse models and end with models that address low level details of the system. Models defined at a lower level of abstraction refine models of a higher abstraction level. Several kinds of composed models can be regarded on one abstraction level, which cover different perspectives of the system as a second axis of the model evolution. The models of the different perspectives may be related to each other via allocation links (check Fig. 2.28). Furthermore within the models several non-functional aspects are regarded which are covered by different artefacts of the models. Such non-functional aspects may include safety, performance, costs or interfaces depending on the kind of model and the level of abstraction. The coverage of aspects in the models therefore provides a third axis of model evolution.

Figure 2.13 provides a conceptual example from the avionics domain that illustrates how abstraction levels, perspectives and aspects are related to the development items. Only perspectives which are relevant in a certain abstraction level are contained in the specific abstraction level. The same for aspects, only the relevant items from several perspectives are associated to the specific aspect.

2.3.3 Components and Design-by-Contract

CESAR design methodologies and suggested modelling approach also strongly promotes component-based engineering (CBE) principles during design and construction of safety-critical systems. This will lead to reusable elements that adhere to

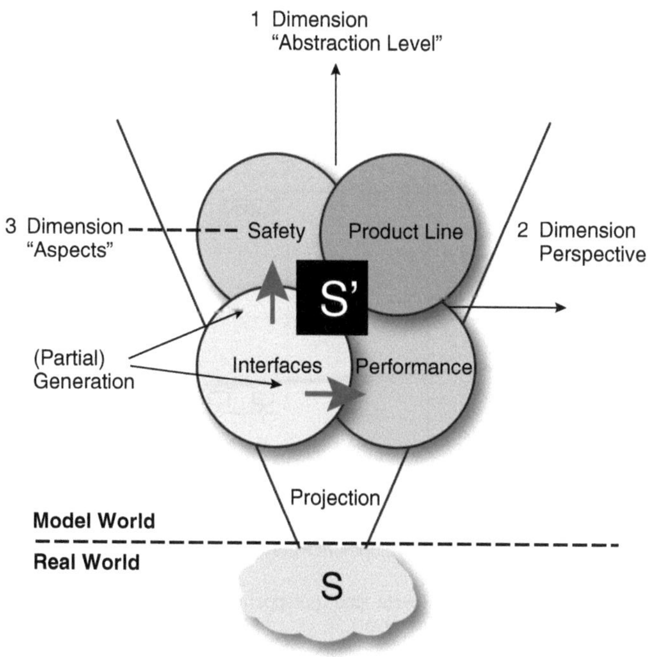

Fig. 2.12 System model evolution along three axis (perspectives, abstraction levels, aspects)

clearly defined specifications and will eventually provide for significant benefits in terms of cost, time and reliability. The components can be maintained in a repository where they are constantly improved and validated in order to attain a level of pre-qualification.

According to [128] the CBE paradigm should encourage the use of predictable architectural patterns and standard infrastructures that improve overall product quality. It encompasses two parallel engineering activities, domain engineering and systems' component-based development (CBD). Domain engineering explores the application domain with the specific intent of finding functional, behavioural, and data components that are candidates for re-use and place them in re-use libraries. CBD elicits requirements from the customer and selects an appropriate architectural style to meet the objectives of the system to be built. The next steps are to select potential components for re-use, qualify the components to be sure they fit the system architecture properly, adapt the components if they must be modified to integrate them, and then integrate the components into subsystems within the application. Therefore the expected process contains two parallel life cycles, one that relates to the building of a component and its deposition in a repository and another one that address system development through assembling and reuse. Chapter 5 related to component-based development provides additional information on CBD concepts and proposes an adaptation of the classical V cycle (called W model) to efficiently address the two life-cycles mentioned above.

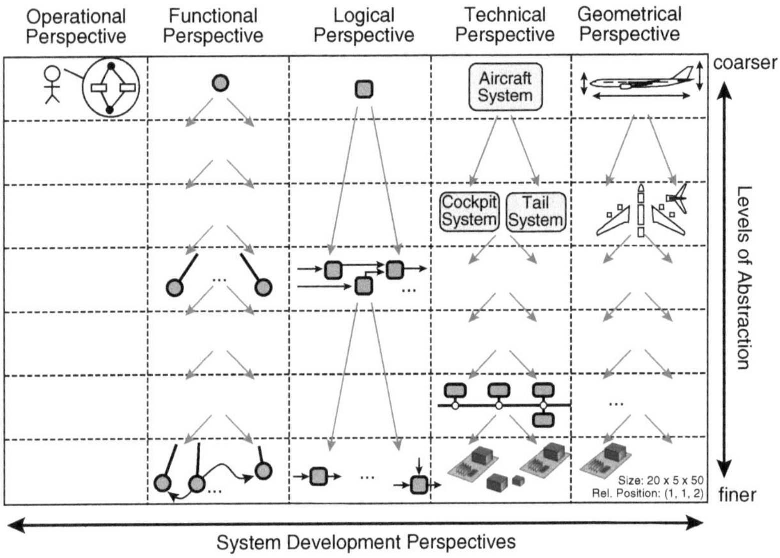

Fig. 2.13 Example of abstraction levels and perspectives structuring (Graphic taken from [SPES] [39])

Component-based approaches "gain from" the definition of unambiguous component interfaces, and the separation of component interfaces, and their realization. Furthermore, components need to be self-contained, realizing exactly their offered services. These realizations need to strictly conform to the defined component interfaces – based on this principle, component realizations may be replaced without affecting other components, as long as the new realization does not affect a component's interface. Interfaces therefore have a very important role in component-based approaches. They do not only cover operation names, parameters, return values, and exported variables, but also need to provide a description of the offered services. For safety-critical systems the sole existence of an interface is not enough. There are convincing arguments for adding further semantic annotations to component interfaces, to make their usage safer. Essentially, an interface signature alone does not tell us what the interface methods are supposed to do and that this information is necessary to use the interface safely. It becomes evident that in a component-oriented development approach, there is a need for a way of determining beforehand whether we can use a given component within a certain context. Ideally, this information would take the form of a specification that tells us what the component does without entering into the details of how. Furthermore, the specification should provide parameters against which the component can be verified and validated, thus providing a kind of *contract* between the component and its environment (which can include interactions with other components).

The contract theory and the notion of contract were originally introduced for object oriented designs (mainly class methods but in general they can be applied

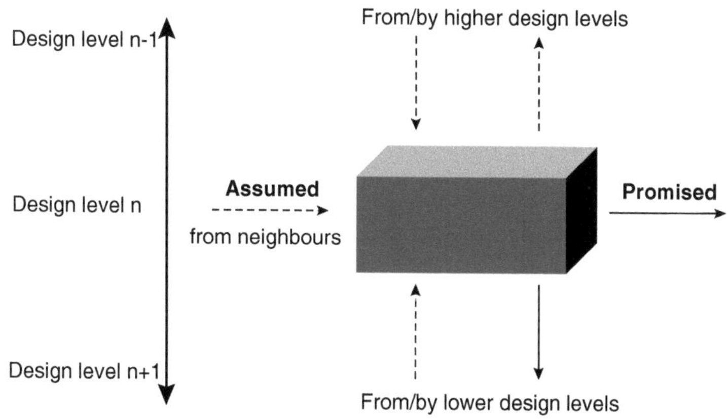

Fig. 2.14 Assumptions and promises for a component's contract

to functions) in the design-by-contract principle [116, 117]. One could summarize design by contract by "three questions" that the designer must repeatedly ask:

- What does it expect?
- What does it guarantee?
- What does it maintain?

Contracts therefore do not only define detailed component specifications, but also define necessary assumptions (pre-conditions) that need to hold in order to ensure proper behaviour for a given component (post-conditions) . These assumptions and promises are usually specified on a per component basis. An assumption specifies how the context of the component (i.e. the environment from the point of view of the component) should behave. Only if the assumption is fulfilled, the component will behave as promised. Figure 2.14 shows the relations between assumptions and promises.

In the CESAR project, the component models proposed (i.e. HRC model or X-MAN model) support the notion of contracts through the definition of pre- and post-conditions that can also be used for component formal verification and validation. More details are provided in Chap. 5.

2.3.4 Architecture Exploration

A useful feature when designing system architecture is the ability to explore different architecture alternatives in order to choose the one that corresponds to a reasonable compromise between different system properties. Considering system development with given end-user requirements and use cases, a designer may have many options at hand, including the choice of suitable functional artifacts and behaviors, choice of hardware elements and structure, and many solutions for mapping behavior to the structure.

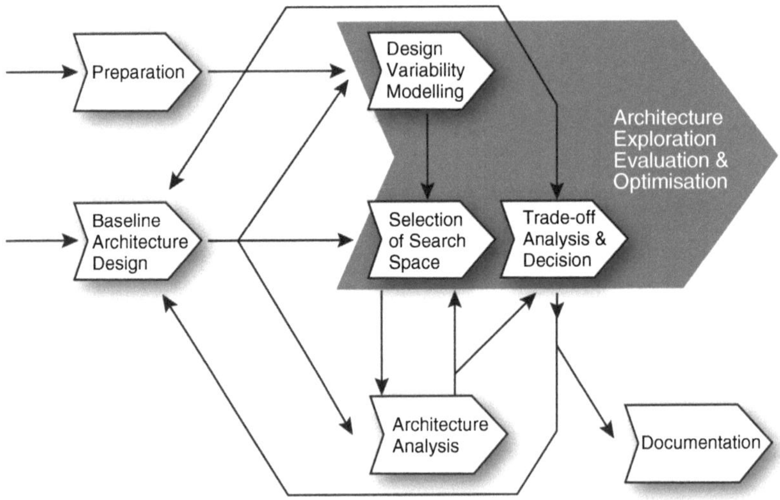

Fig. 2.15 A simplified view of the architecture exploration methodology developed in CESAR (the full version is described in Sect. 4.3).

In general, each requirement can be met through several different alternative solutions in system design and implementation. In practice however, the choices can be more restricted due to earlier design decisions. For example, a fixed network infrastructure will set limitations on the feasible allocation of software components onto hardware nodes due to bandwidth and delay constraints. Further, some functions may already be associated with suppliers that deliver integrated components (hardware and software), implying no freedom of allocation.

Whenever alternative behaviours, structures or mapping solutions are possible, it is also the task of architecture designers to elicit and describe the design options, and to provide the selection of which option(s) is(are) preferred. While many design alternatives can often be identified and resolved based on stakeholder consensus, heuristics, or design patterns, explicit architecture exploration and evaluation allows more systematic investigation of the merits of different alternative architecture solutions.

Architecture exploration is a design approach aiming to support the deduction of architectural design alternatives for full or partial system optimization. It is covered in further detail in Chap. 4. It allows a designer to systematically examine viable and alternative design options for a particular system, and thereby complements the application of heuristics and domain experiences for justifying design choices and performing design optimization. Figure 2.15 offers an overview of the related work products and activities, which are further elaborated in the next paragraphs.

In architecture exploration the system architect is responsible for selecting choices from a number of alternatives based upon criteria based on metrics such as performance, cost, memory resource usage, mass, failure rates and so on. When the

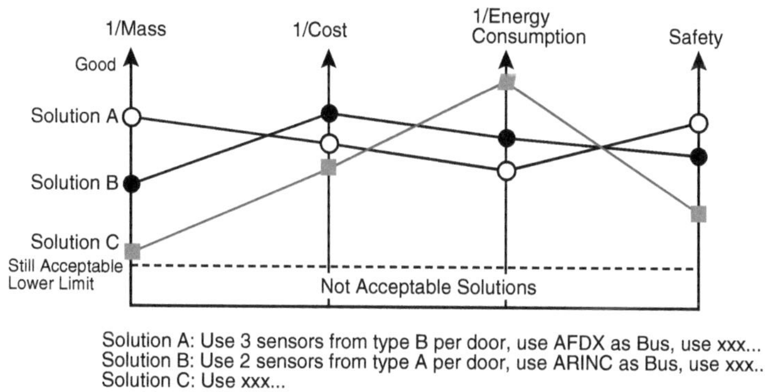

Fig. 2.16 An example trade-off analysis illustrating how different solutions affect metrics and hence indirectly criteria

selection has taken place, the architect should be able to understand the implications of certain design choices and do some trade-off analysis. Figure 2.16 attempts to visualize how different solutions (on the vertical axis) can impact the metrics associated with certain trade-off criteria (on the horizontal axis). The final choice will depend on the aggregation of these metrics and the possible weights assigned to them.

To put architecture exploration into its context in the CESAR project, architecture exploration can easily be coupled with a multi-view approach to modelling, since the architecture exploration typically has to consider several different properties of the proposed architecture that are typically not (all) strongly coupled with each other. Example properties that need to be addressed in safety critical systems may relate to reliability/safety, performance, complexity/extensibility, cost, weight, power consumption, or any other well-known or custom metric that can be defined for a system. All of them should be considered in the context of the various CESAR viewpoints as already presented in the previous subsection.

The design space available to a developer depends on the particular scenario, and technical and strategic constraints. Several options may be evaluated in parallel or sequence, especially if it is unclear which the most suitable product design might be. For example, consider the common case of adding a new function to an existing product range. The question then arises whether new computer units need to be added, or whether the new function can be accommodated by existing computer units. The overall design space in this case consists of alternative hardware structures (including options with new hardware elements and their physical placement and cabling) and alternative allocations of the functions to existing and/or new hardware. On a more detailed level, the new functions need to be structured, partitioned and scheduled on existing and new resources; there alternative realizations might also exist.

By describing design variability and thereby capturing the design variants and their impacts on system attributes, the design space for the exploration work will be described. In any real-world setting, a selection and prioritization between these design variants will need to be done, since it is not feasible to evaluate them all. A smaller search space (a subset of the design space) will need to be selected, either at the start of the process, or iteratively using heuristics. Formalization of the design-space allows the automation of exhaustive exploratory search of some or all possible alternative architectures resulting from different design variants in scope. This kind of exploratory search of alternative architectures can for example be supported by genetic algorithms.

To support architecture exploration, efficient tools for architecture analysis, producing values for selected metrics for evaluation, will be needed. These tools will assess the viability of each architectural design from the chosen viewpoints and provide metrics for comparing them in regards to the chosen functional and quality goals. Analysis can be performed at different levels of abstraction or different development life-cycle stages, for system level features or operations, functional and logical design, software and hardware solutions. The analysis tools may require dedicated analysis models (e.g. component behaviours, end-to-end timing, system safety and reliability). For example, while a timing analysis requires models that capture the triggering, periodicities, execution time of behaviours in functional chains, a safety analysis requires models that capture the subsystems' failure modes, errors and error propagation. Each architecture analysis normally assesses only one architecture instance at a time.

Trade-off is a natural part of any multi-criteria architecture exploration, just as any design task with conflicting goals.

Whereas in current practice, many architectural trade-off decisions are taken on the basis of best practices and application domain experience, resolving design variations in safety- or mission-critical systems often would be improved by substantiated decisions with explicit evidence supporting the rationale. In order to compare and choose among substitutable architectures, architecture design needs to define the optimization goal(s), evaluation criteria and metrics.

For alternative architectures, trade-off analysis aids designers in establishing a choice. There are both quantitative (mathematical optimization) and qualitative approaches to trade-off analysis.

2.4 CESAR Reference System Engineering Process

Considering the cross-domains dimension of CESAR, it is necessary to offer flexibility in the definition of the development process. One of the foundations of the RTP is the ability of its adaptation to industrial needs while promoting sound engineering practices. Thus, within CESAR a development process definition shall be able to be built either from scratch using process elements or relied on pre-existing process model possibly enhanced and tailored depending on the industrial context.

The method chosen to compose a reference system engineering process is by assembling "method fragments" or "chunks" defined and stored in an appropriate method library repository. The language chosen to support this compositional process approach was SPEM 2.0 [122]. The Eclipse Process Framework (EPF) [47] Practice Library (EPL) [54], a UMF-compliant [165] method library that is available from the Eclipse project, was used as the base for the expression and collection of these "methods chunks". These process fragments collection, along with other material, constituted the CESAR Practices Libraries (CPL).

In order to represent the CESAR practices in a modular, extensible and reusable way, the concept of "Library" has been investigated. A "Library" of practices represents an organised and easily accessible collection of the techniques, methodologies and processes that are being developed in the project. The Unified Method Architecture (UMA) [164], used in EPF, is a process engineering meta-model that defines schema and terminology for representing methods consisting of method content and processes. UMA provides a clear separation of Method Content definitions from its application in Processes. This is accomplished by separately defining:

- Reusable core *Method Content*, in the form of general content descriptions such as *Roles*, *Task*, *Work Products* and *Guidance*;
- Project-type specific applications of Method Content in the form of *Process* descriptions that reference Method Content.

Method Content provides step-by-step explanations of how specific development goals are achieved independent of the placement of these steps within a development life-cycle. Processes take these Method Content elements and organize them into a sequence that can be customized to specific types of projects. For example, a software development project that develops an application from scratch performs development steps similar to a project that extends an existing software system. However, the two projects will perform similar steps at different points in time with a different emphasis and perhaps individual variations.

UMA allows each Process to reference common Method Content from a common Method Content pool. Because of these references, changes in the Method Contents will automatically be reflected in all Processes using it. However, UMA still allows overwriting certain method-related content within a Process as well as defining individual process-specific relationships for each Process Element (such as adding an additional input Work Product to a Task, renaming a Role, or removing Steps that should not be performed from a Task). Further technical details and descriptions, useful to deeply understand the CPL structure and to gain knowledge on how to author it, can be obtained from [47, 164, 165].

Using the tools and frameworks described above, several "method chunks" were authored in EPF and were used in order to define an example development process instance. The purpose was not to define a de-facto reference process applicable to all CESAR domains but rather to propose an example process covering the whole system engineering life-cycle and comprising of activities and tasks. This process would act as a starting point for application domains to create their own custom

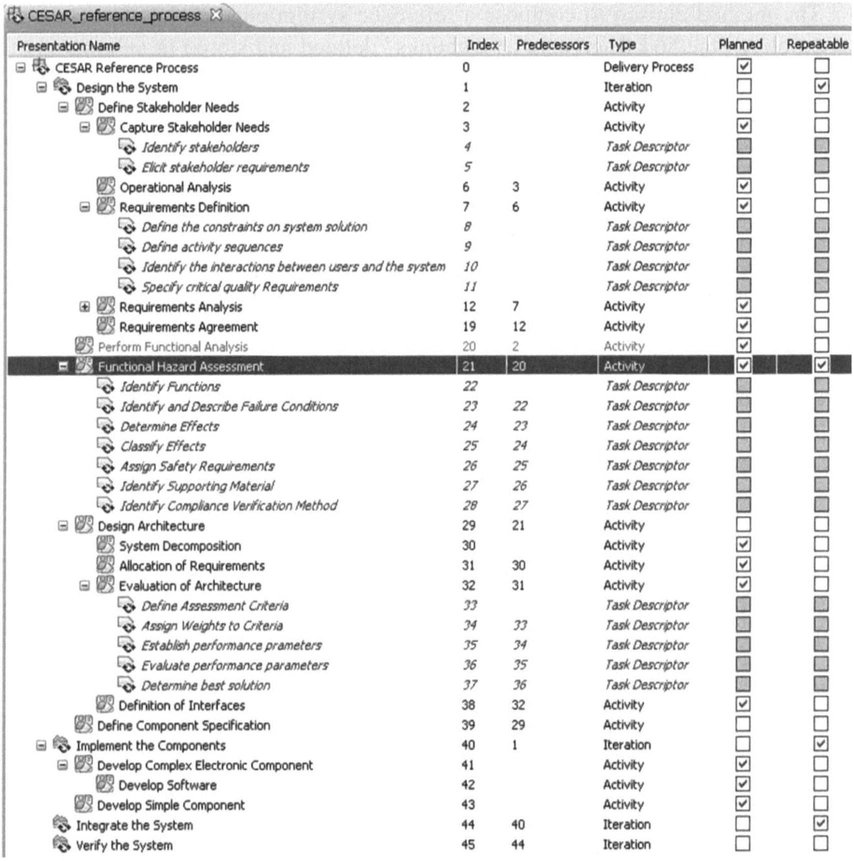

CESAR_reference_process

Presentation Name	Index	Predecessors	Type	Planned	Repeatable
CESAR Reference Process	0		Delivery Process	☑	☐
Design the System	1		Iteration	☐	☑
Define Stakeholder Needs	2		Activity	☐	☐
Capture Stakeholder Needs	3		Activity	☑	☐
Identify stakeholders	*4*		*Task Descriptor*	■	■
Elicit stakeholder requirements	*5*		*Task Descriptor*	■	■
Operational Analysis	6	3	Activity	☑	☐
Requirements Definition	7	6	Activity	☑	☐
Define the constraints on system solution	*8*		*Task Descriptor*	■	■
Define activity sequences	*9*		*Task Descriptor*	■	■
Identify the interactions between users and the system	*10*		*Task Descriptor*	■	■
Specify critical quality Requirements	*11*		*Task Descriptor*	■	■
Requirements Analysis	12	7	Activity	☑	☐
Requirements Agreement	19	12	Activity	☑	☐
Perform Functional Analysis	20	2	Activity	☑	☐
Functional Hazard Assessment	21	20	Activity	☑	☑
Identify Functions	*22*		*Task Descriptor*	■	■
Identify and Describe Failure Conditions	*23*	*22*	*Task Descriptor*	■	■
Determine Effects	*24*	*23*	*Task Descriptor*	■	■
Classify Effects	*25*	*24*	*Task Descriptor*	■	■
Assign Safety Requirements	*26*	*25*	*Task Descriptor*	■	■
Identify Supporting Material	*27*	*26*	*Task Descriptor*	■	■
Identify Compliance Verification Method	*28*	*27*	*Task Descriptor*	■	■
Design Architecture	29	21	Activity	☐	☐
System Decomposition	30		Activity	☑	☐
Allocation of Requirements	31	30	Activity	☑	☐
Evaluation of Architecture	32	31	Activity	☑	☐
Define Assessment Criteria	*33*		*Task Descriptor*	■	■
Assign Weights to Criteria	*34*	*33*	*Task Descriptor*	■	■
Establish performance prameters	*35*	*34*	*Task Descriptor*	■	■
Evaluate performance parameters	*36*	*35*	*Task Descriptor*	■	■
Determine best solution	*37*	*36*	*Task Descriptor*	■	■
Definition of Interfaces	38	32	Activity	☑	☐
Define Component Specification	39	29	Activity	☐	☐
Implement the Components	40	1	Iteration	☐	☑
Develop Complex Electronic Component	41		Activity	☑	☐
Develop Software	42		Activity	☑	☐
Develop Simple Component	43		Activity	☑	☐
Integrate the System	44	40	Iteration	☐	☑
Verify the System	45	44	Iteration	☐	☐

Fig. 2.17 Work breakdown structure of Reference Development Process defined in EPF composer

instances by extending or modifying existing practices or even defining their own based on specialized domain needs. In the Fig. 2.17 the work breakdown structure for the CESAR System Engineering reference process. In this example a practice is highlighted and it is shown how this practice is used (or better re-used) in a wider process. For more details on the elements comprising the Reference Development Process please refer to [31].

2.5 Safety and Diagnosability Process

Safety activities are an essential part of safety critical system's design which seeks to identify certain system constraints that must be addressed in order to avoid situations that may cause faults in system operation and consequently endanger life of human beings.

A first step in the safety process is to identify hazards and risks associated with new concepts or product changes. If the risks are unacceptably high, then the first steps would be to (1) try to eliminate the hazard altogether; (2) to reduce the risk; and finally (3) to provide alternative measures that minimize the impact and consequences of an accident. Given steps (1) and (2) it is clear that safety considerations strongly influence the system development including the architecture design, see e.g. Storey [149].

System requirements, including high level safety requirements, constitute inputs to the architecture design process. Initial hazard and risk analysis will yield a categorization of functions and systems into so called safety integrity levels (SILs) that characterize the severity (worst case consequences) of potential failures. The various domains' safety standards have somewhat different approaches on how to do this. The SILs imposes a number of requirements and recommendations for the consequent development that relate to:

- *Quantitative* (probabilistic) targets corresponding to the desired risk reduction (e.g. in terms of 10^{-9} failures/h for highly critical failure modes) and for error detection coverage (to prevent latent faults). This kind of quantification and reliability prediction is accepted for random faults having only hardware as source.
- *Qualitative* techniques that have to be applied and judgement that have to be made in order to demonstrate and ensure a proper development and verification process with prescribed design styles such as the use of redundancy and diversity. By these means systematic safety (hardware design and manufacturing/software/installation errors) is addressed.

The high-level safety requirements (including the SIL level) will be used to define more detailed architectural requirements, and it is then the objective of architecture design to develop solutions that meet those requirements. As part of the architecture design, the safety standards will also require that independence between functions/systems of different criticality can be argued and ensured.

On the other hand, *diagnosability* is also an essential property that determines how accurate any diagnostic reasoning can be on a system given any sequence of observations. The diagnosability process can be seen as an extension of the notion of *observability* when considering the presence of potential faults but also as a generalization of *testability* since it makes use of tests that provide data for all measurable variables, e.g., for all sensors of a given system. However, limiting diagnosability to a particular observability property – defined as the capability to distinguish any two different cases of faults in a specified fault model (or at least, any two ones for which the subsequent recovery or repair processing is different) is likely to be insufficient to be usable in practice because the needed observable information depends on the diagnosis procedure. Within CESAR diagnosability issues have been slightly tackled. In the corresponding diagnosability process subsection we present some principles of a proposed analysis that can apply.

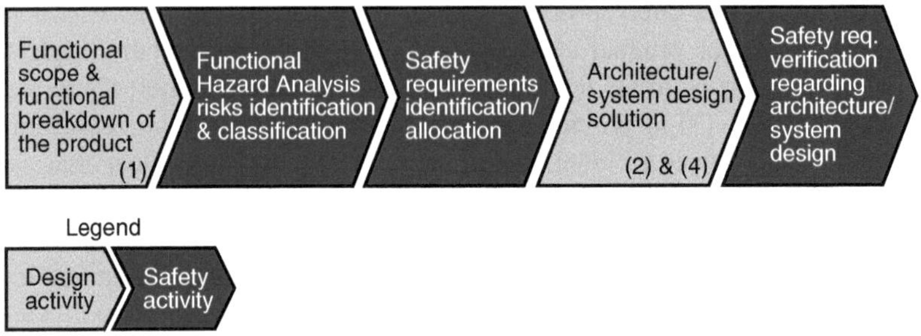

Fig. 2.18 The safety process and design process

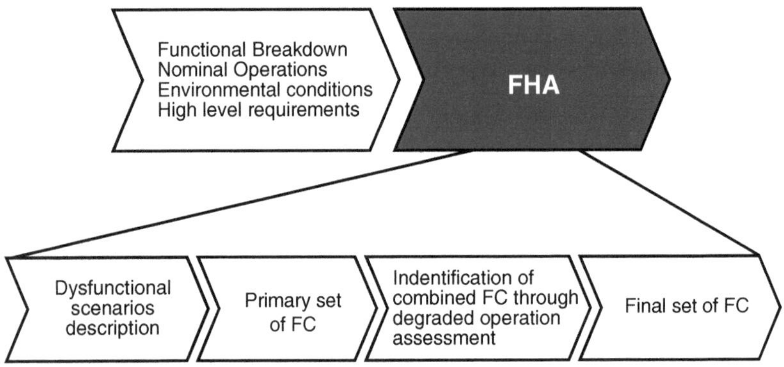

Fig. 2.19 FHA steps

2.5.1 An Overview of the Safety Process

Safety tasks are an area where significant divergence exist between the various CESAR domains especially regarding the rigour of the applied process (e.g. in avionics and rail safety regulations and standards impose many design restrictions while in the automotive and automation the safety process is considerably less restrictive).

Figure 2.18 describes the main steps of a generalized safety process interlaced in the design process (the numbers drawn on Fig. 2.18 refers to the V life cycle process phases depicted in Fig. 2.2). Each safety task is described in more details in the next paragraphs.

2.5.1.1 Functional Hazard Analysis

The main steps of Functional Hazard Analysis (FHA) are described in Fig. 2.19.

The first step consists in identifying for each function, the scenarios that lead to a failure mode considering each product mission and under each relevant environmental condition. For each functional failure mode, the repercussion on the function of the failure mode is assessed and classified according to a number of criteria including:

- Total Loss of the function,
- Partial Loss of the function,
- Inadvertent functioning of the function,
- Erratic functioning of the function.

Then, an exhaustive description of the functional scenarios that induce abnormal behaviours is elaborated. The description of the abnormal behaviour encompasses the description of the functional failure, the effects on the system, the procedure to deploy in order to minimize the failure consequences and the safety classification that depends on the safety standard. Finally scenarios are classified according to their repercussions and factorized per effects in order to get the failures conditions (FC).

2.5.1.2 Safety Requirements Allocation

During this phase, the failure conditions identified in the previous phase are allocated to systems (that constitute the global product) which support the function. These failure conditions(FCs) define the safety requirements for each system. The system shall be designed in a way to comply with the qualitative (e.g. no single failures) and quantitative (probability values) safety requirements defined by the FC classification.

2.5.1.3 Safety Requirements Verification (PSSA/SSA)

Once the system is designed, a demonstration that the system fulfils the qualitative and quantitative safety requirements is required. This demonstration should highlight that failure combinations for system components, leading to an allocated failure condition, are acceptable. Fault tree or reliability block diagrams are usually used to get cut sets and verify qualitative criteria and quantitative objectives.

During the system design life-cycle, there are two main assessment activities that try to assess system safety based on various defined qualitative or quantitative criteria. These are Preliminary System Safety Assessment (PSSA) that usually applies directly after the initial architecture has been defined and System Safety Assessment (SSA) that is the final activity that takes place after a system is integrated by its sub-elements to ensure that all safety constraints have been correctly addressed.

Note that for systems with high reliability requirements it is generally impossible to demonstrate by testing that the requirements have been achieved. Acceptance and certification are based on reliability models for the hardware and use of appropriate development techniques for the software. Process measures and special design techniques are employed to provide confidence in the software.

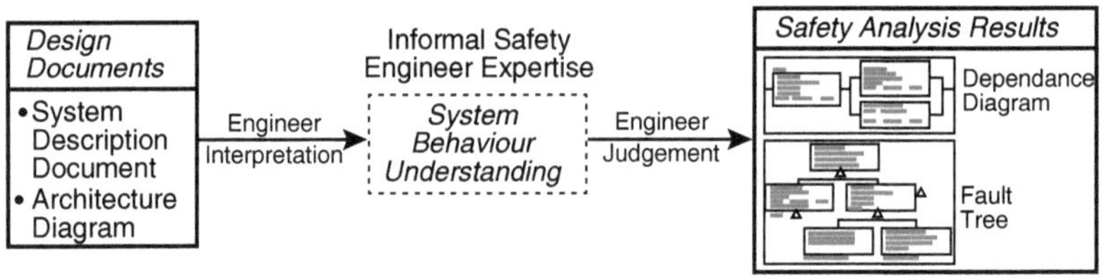

Fig. 2.20 The current practice for safety engineering process

2.5.2 *"Existing" Safety Process*

Today the majority of safety activities, risks identification and safety requirements verification are performed in a completely manual manner with minimal support of tool automation. The safety engineer has to handle non formalized data and also write dysfunctional scenarios or fault trees relying mainly on engineering judgement as depicted in Fig. 2.20:

This manual safety engineering process has important repercussions on the system development in terms of cost and time. Indeed, due to increase in system complexity, the number of failures combinations becomes very huge and difficult to manage without adequate tools. As a consequence, the PSSA and SSA activities are very complex to lead and are more and more time-consuming and costly. Due to these complexity factors, it is risky to modify an existing design when developing new system features and conservative assumptions are often preferred to innovative solutions. After design changes, the update of the safety assessment induces a huge work that has a negative impact on development time and cost.

2.5.3 *"*CESAR*-proposed" Safety Process*

Due to the strictness and constraint imposed by the various safety standards it is difficult to develop a completely new safety methodology within CESAR. Therefore focus was mainly given in:

- Enhance traceability and consistency of safety requirements to design elements
- Enhance system modelling with safety related information and dysfunctional models (models that describe erroneous behaviour)
- Provide means to automate the generation of safety analysis data

2.5.3.1 Enhance Traceability and Consistency of Safety Requirements to Design Elements

The RTP provides appropriate infrastructure services such as the *link repository* (see Sect. 6.3), implementing a lightweight integration mechanism that enables full traceability between requirements, design elements, test procedures and results. Although this feature is not limited to safety requirements only, the safety process greatly benefits. In addition to that a couple of CESAR tools and tool-chains have demonstrated features like architecture and functional consistency checking of formalized safety requirements against a given architecture design.

2.5.3.2 Enhance System Modelling with Safety Related Information and Dysfunctional Models

In order to support the safety process objectives as they have already been analysed, the CESAR modelling approach has prescribed a safety view (see Sect. 2.3.2). The safety view is supported as an extension package to the core CMM specification and it should be regarded as an *aspect* since its concerns span different perspectives. The elements of the safety view extension provide means for the analysis of potential component failures including among other concepts that allow specifying *failure conditions*, *safety cases*, *cut sets*, etc. The CMM Validation and Verification package provides also elements like *VerificationCase*, *VerificationProcedure*, *VVStatus* and others that are an important part of a safety viewpoint (safety standards always prescribe verification and validation as parts of a safety analysis). The *Verification-Case* allows specifying which component is verified and how this is done via the *VerificationProcedure*. Figure 2.21 provides an overview off all the safety concepts specified in the CMM.

The safety view is applicable at different stage of the development process. In particular in the concepts and requirements phase, in the design phase and in the validation and verification phase. Three different categories of models are involved which address three main categories of objectives of safety analysis:

- *Structural safety models*: Structural models, used to represent the organisation of the various elements composing a system, can be extended to represent how the faults and failures affect the various elements of the structure, and how these faults and failures propagate along the structure. Such augmented models are good candidates to support, possibly through coupling with classical existing safety and dependability analysis tools, the automatic analysis of fault propagation and demonstration of the related requirements. Of course, not all the characteristics of fault propagation can be easily represented on structural engineering models (e.g., thermal, EMC, etc.). Therefore these models are not expected to address all possible safety issues.
- *Behavioural safety models*: behavioural models are often used for formal verification of behavioural properties i.e., the correctness of the behaviour even in case of complex behaviour with many interactions between a large number of elements and a large number of possible combinations of events and states. It is worth investigating the applicability, efficiency and maturity of behavioural modelling approaches as a support to demonstrate e.g., fault

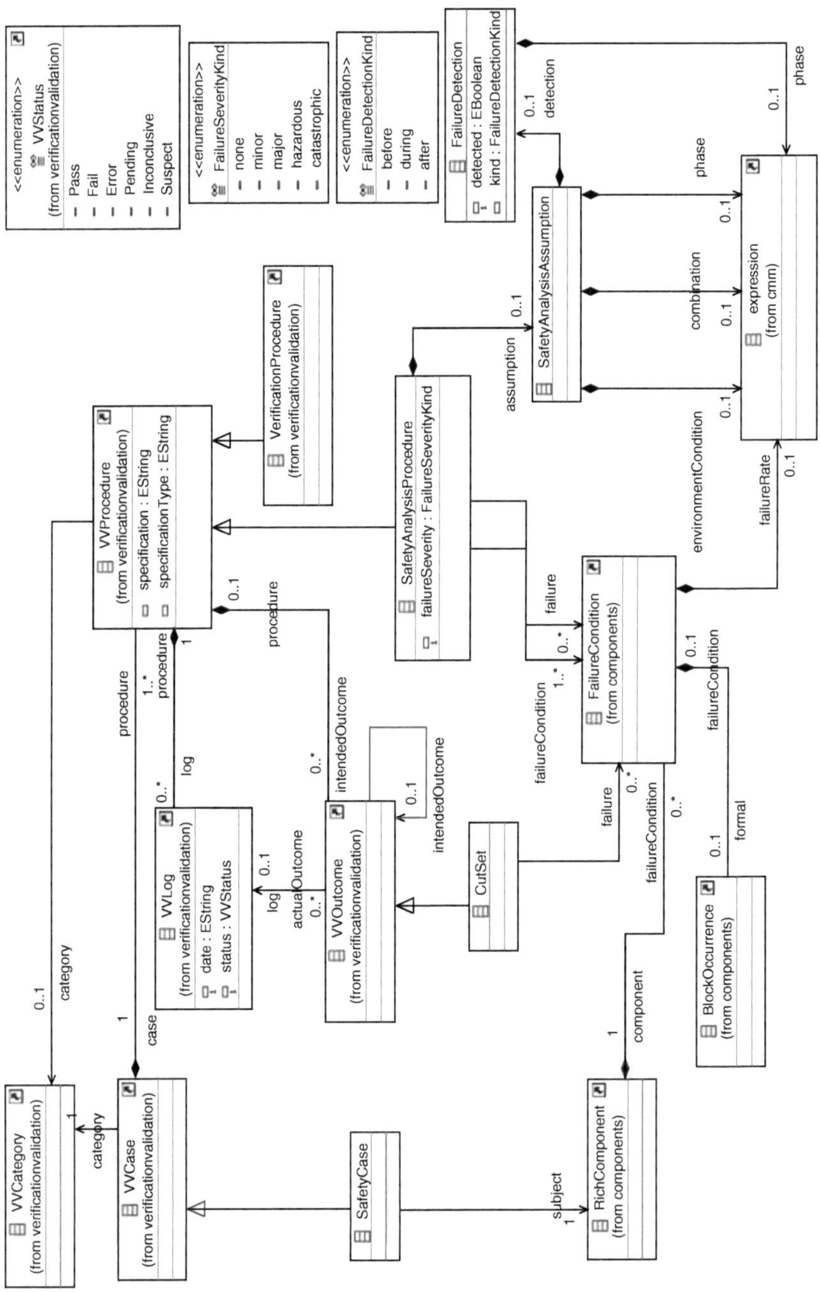

Fig. 2.21 Safety related concepts in the CESAR meta-model

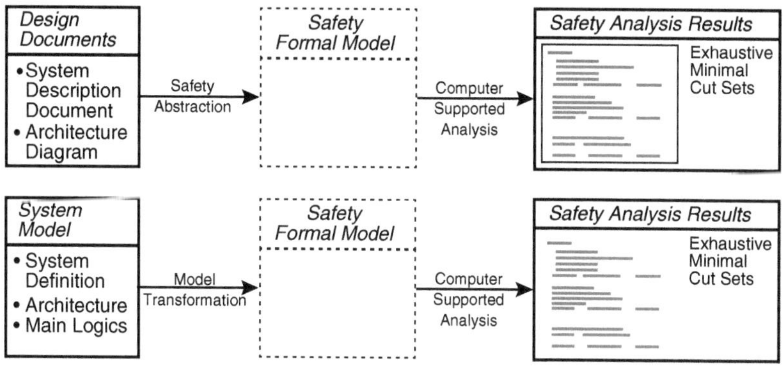

Fig. 2.22 Example concept of automating generation of safety analysis data

tolerance, correctness, liveness and safety properties; despite the fact that fault tolerance is generally regarded as a particularly difficult case in this respect, and moreover the severity of consequences of potential failures of the fault tolerance mechanisms may be very high.
- *Logical safety models*: within the scope of the CESAR RTP, logical safety models can be used to describe the logical structure of the arguments (safety objectives, claims, assumptions etc.) involved in a safety analysis.

A complete safety view should encompass structural, behavioural and logical models and also provides the means to capture the safety abstraction concerning system dysfunctional behaviour that is the basis for analysing failure and faults propagation.

2.5.3.3 Provide Means to Automate the Generation of Safety Analysis Data

Formalising system requirements and extending existing system models with safety concepts and dysfunctional behaviour, are the main prerequisites in order to enable automation of certain safety tasks including among others:

- Partially or completely generation of failure propagation models from system description
- Creation of initial PHA and FHA table
- Minimal cut sets/fault tree generation

These could be achieved by first translating system models (augmented with safety data) to formal safety models and then applying computer aided analysis tools to generate the safety analysis results (see Fig. 2.22). Some of the expected benefits of such an automatic translation are:

- Reduction of iteration time improving the interaction between safety and design activities,
- Earlier safety assessments improving architecture design trades-offs,
- Possibility to assess a greater number of alternative architecture solutions,
- Capability to generate automatically minimal cut sets,
- Reduction of errors risk due to the automation of fault tree generation and the initialization of the dysfunctional models,
- Ability to focus on system dysfunctional behaviour that can be shared easily with other domains
- Facility to update the cut sets after modifications to the model design.

Materialization of such automations are still quite limited in the CESAR project. The RTP platform includes tools chains that can be used for model-based safety analysis. This includes, e.g., a model checker which identifies the minimal cut sets from Simulink models exported in CMM format. Creation of initial PHA and FHA tables is possible from simple Rhapsody Models as well as partly generation of fault trees. However the overall provided functionality is still quite immature and should be regarded more as a proof of concept rather than being applicable in real system development.

2.5.4 An Overview of the Diagnosability Process

Usually in safety critical systems we refer to *fault diagnosability*, which includes:

- The ability of a system to support the identification of information related to its potential faults/errors or
- Capability of a system and its monitors to exhibit different observables for different anticipated faulty situations

The system designer must ensure that the system is diagnosable, i.e. must ensure that the errors that will appear are identifiable by appropriate procedures. A diagnosability procedure can be considered as a form of test that checks/measures the values of multiple variables and applies algorithms to determine if the system (whether it is SW, HW, etc.) functions correctly. Diagnosability should be addressed during design in order to provide a more reliable system, with predictable maintenance costs.

In order to perform a functional diagnosability analysis, a behavioural model of the system is needed. Multiple models can be envisioned such as continuous-state-based, event-based, or hybrid.

When dealing with the diagnosability of discrete event systems, the most widely accepted analysis method is the "diagnoser" approach Sampath [134]. The input model is a finite deterministic state machine modelling the system behaviours (a representative example is given in Fig. 2.23). The analysis is performed by augmenting that model with non-normal or fault states and checking if the augmented and transformed model is diagnosable.

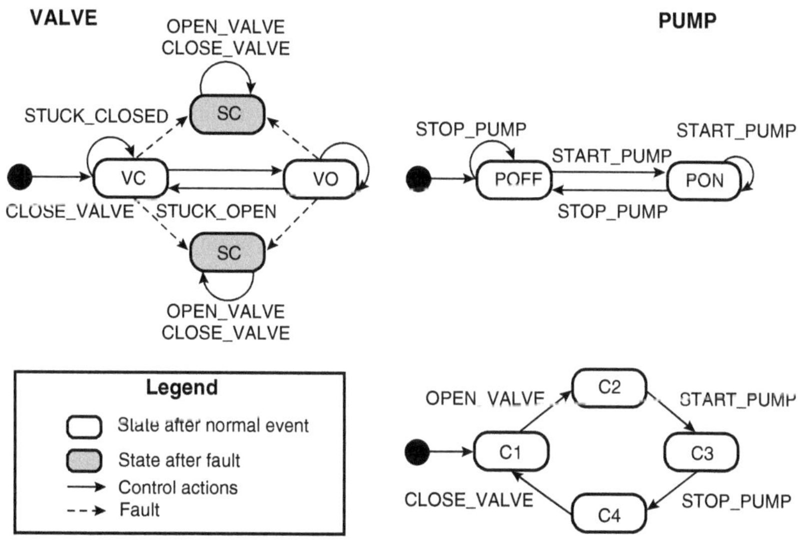

Fig. 2.23 Example behavioural components models of a PVC (Pump, Valve, Controller) subsystem of HVAC (Heating, Ventilating, Air Conditioning) system

While the system model being functionally diagnosable and the existence of a diagnoser are necessary conditions for the final embedded system to be diagnosable by itself, they are not sufficient. The interaction between the functions and the diagnoser as well as their mapping to the underlying architecture has to be considered. It is thus necessary to define checkable properties that the architecture has to demonstrate, with respect to the functions and the diagnoser. To give an example, in order to check properties of the functional architectural diagnosability, the system architecture should be described in an architecture description language such as AADL (Architecture Analysis and Description Language) or a hardware description language such as VHDL (Very high speed integrated circuit Hardware Description Language) or SystemC. The main idea is to enable parsing of the architecture description model to extract and identify all architectural features necessary for the analysis, such as embedded computing units, sensor devices, communication links, data flows as well as inputs/outputs connections representing state and event variables.

In fact, each state and event variable found in the functional models (automata) can be matched to the corresponding data flow from the AADL model, including origin, destinations, physical links supporting the data or message transmission, etc. The architecture properties are evaluated using that matching. For instance, to check the accessibility property for a given event variable, one has to check that a data flow exists for that variable, (i.e. it is actually output by some component), that the data flow actually reaches the observer (i.e. there exists a connection between its source and the observer) or that its source and the observer are mapped to the same computing unit. The global accessibility property is evaluated by repeating that process for each event variable, one at a time. Checking for other properties follows a similar process.

This approach can possibly be automated, making its integration into a tool-chain possible. The CESAR projects provides only suggestions on how to address diagnosability issues. In this sense tools, such as the COSITA prototype Khlif [99], (a tool that enables analysis of iterated co-simulations traces in order to evaluate the interaction of functions with the architecture according to selected critical scenarios), although still a work in progress, are good candidates if diagnosability is to be considered as a metric during the architecture exploration process.

2.6 Product Line Engineering

Product line engineering (hereafter referred to as PLE) aims at building products for a well-defined problem space in shorter time, with less costs and with higher quality through a proper utilization of platform assets, including product line architectures (blue-prints for creating families of related applications), reusable components and design artefacts as well as process related artefacts and certification arguments.

In CESAR PLE activities, the book *Software Product Line Engineering* by Pohl et al. [126] is used as a foundation, as it represents a well-established reference in the area of software PLE and is also aligned with the EU-supported PLE projects ESAPS [53], CAFE [65] and FAMILIES [60] of the ITEA programme [87]. It should be noted that, despite that the focus of [126] is on software, the presented approaches have been applied to hardware products as well.

The rationale behind applying PLE is to save cost and improve efficiency by two main effects:

- *Improving reuse*: The reuse of development assets was also the goal of previous reuse approaches. Compared to previous approaches, however, PLE goes further: it involves the *business case of the products* and plans explicitly for reuse. Previous approaches have covered reuse on a technical level only; organizational hurdles of the company or particular demands and restrictions resulting from the business case have not been taken into account. PLE now means engineering for reuse that covers not only technical aspects but also integrates organizational, management and business case views.
- *Managing variability*: A product line provides the basis for products that differ for each customer and/or application. These products are built from a given asset base – that is, what forms the product line. However, since each customer will require specific modifications making the product a (more or less) unique solution, the producing enterprise must efficiently build the product according to the given requirements. The more efficient a product can be derived from the product line, the more competitive can the product be sold. The main advantage

for applying PLE is the structured and systematic handling of variability that improves the efficiency of the derivation (configuration) process.

To cover these two issues, PLE aims at the systematic development of a set of similar systems by understanding and controlling their common and distinguishing characteristics.

PLE is based on a development process including both, development *for* reuse (domain engineering) as well as development *with* reuse(application engineering). These two processes are detailed in the next sections.

2.6.1 Product Line Engineering Life-Cycle

Pohl et al. [126] have established a representative framework addressing the PLE life-cycle which was used as a reference for analysing the current practice within CESAR. According to this framework, the product line engineering activities are divided into:

Domain engineering is the part of the product line engineering process in which the commonality and the variability of the product line are defined and realized.

Application engineering is the part of the product line engineering process in which the applications of the product line are built by reusing domain artefacts and exploiting the product line variability.

Figure 2.24 shows the general set-up of this framework which is explained in details in the next paragraphs.

The PLE life-cycle starts with a product management activity that defines the concrete boundaries of the product line in terms of the envisaged product or the product categories that are to be supported. This activity is in close relation with *domain engineering* and often considered as part of it.

2.6.1.1 Domain Engineering

Domain engineering is the part of the product line engineering process, in which the basis for the reuse is developed. This means, that the commonality and the variability of the products in the product line are identified or defined and that the common parts of the product line products are realized. Domain engineering comprises therefore the following activities:

- Product line scoping
- The development of the product line variability model
- The development of the building blocks for the product line's products – the core assets
- The definition of the production plan

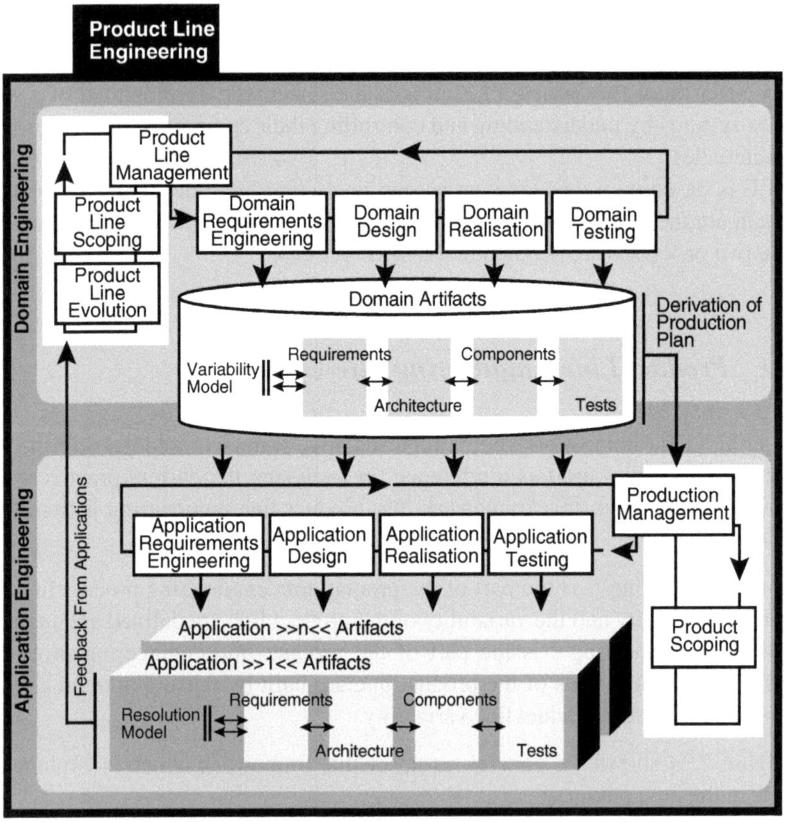

Fig. 2.24 Software Product Line Framework (Pohl et al. [126])

The scoping of the product line will define the products planned to be built by the product line and decide on the approach to be followed. There are two primary strategies, proactive and reactive. In the case of a pure proactive approach, all of the products needed on the foreseeable horizon are supported by the product line. In the case of a pure reactive approach, only the products needed in the immediate term are supported by the product line and new products are incrementally added as the needs change. There is, of course, a continuum between the two. The scope is not static; it is continuously changing in order to allow the product line evolution.

The variability model is created based on the scope of the product line and continuously evolves throughout the assets' base creation, which is supported by usual development steps, including requirements engineering, design, realization and testing, with the entire product line in mind. Concurrently with the development of the domain assets the production plan is derived that contains a description on how the different assets can be used to yield a product.

All the entities developed in the domain engineering process are called domain artefacts and form the product line's base of reusable assets (also known as *core asset base*).

2.6.1.2 Application Engineering

Application Engineering is the part of the product line engineering process, where the products are actually built, by reusing the set of assets defined during the domain engineering process. The way the applications are built will depend on the development process. Two extra activities, however, need to be performed:

- Variability model resolution, according to the product scope
- Core assets adaptation, according to the production plan

As Fig. 2.24 shows, assets developed in particular domain engineering tasks influences the corresponding application engineering tasks and its resulting artefacts. Reusable artefacts produced during domain engineering (that is, the domain artefacts), are made available for the different application engineering instances that yield the different products as members of the product line.

Resulting from different areas that variability covers, resolving the variability is considered possible at different stages during a system's life-cycle:

- *At Modelling Time*: Variability resolution is done when generating a model that either includes a determined feature or not. The generated model can serve as a building plan for later creation stages as well as for simulations.
- *At Assembly Time*: Variability is resolved at the time when either (a) the source code is written or generated, or (b) the physical construction of a device takes place.
- *At Deployment Time*: At deployment time, variability can be handled by deployment settings that configure the system itself. Or, binding at deployment time refers to settings that configure the deployment environment.
- *At Run Time*: At run time, a system can adapt dynamically to an environmental change or a new setting it has detected. For example, for software systems a plug-in mechanism is possible that allows binding functionality during run time.

There is also a *feedback path* from application to domain engineering where specific application engineering findings are propagated to the domain models and used to update them. This may happen when problems emerged during the reuse of certain domain artefacts, that is, if defects were observed or change requests occurred. Another possible reason may be that it becomes necessary to incorporate newly identified generic aspects of the domain models or new variability characteristics. Thus, the domain and application engineering tasks are *iterative* – meaning that they can be performed more than once in order to achieve the desired results. This completes the development cycle of product line engineering and makes it evolutionary.

2.6.2 Variability

Variability plays a major role in product lines while many of its aspects apply to other engineering activities addressed by CESAR, such as architecture exploration. Therefore, it is important to explain some major aspects of variability that will help to better understand CESAR introduced improvements through the RTP platform PLE related tool-chains.

2.6.2.1 Variability Characteristics

Individual products of a product line share common parts. These common parts are the commonalities. But naturally, the products show also differences, that is, they show variability, which is expressed by so-called variation points. Actually, variation points identify the points where products of a product line family show differences. From a system development point of view, *variation points* thus identify situations where decisions about the configuration of the individual products have to be made. In this context, a *variant* denotes then a particular setting for a variation point – after a certain decision was made. Variation points can occur in many ways in products since variability may span system characteristics related to structural architecture, behaviour, functional and non-functional aspects.

Pohl et al. characterise variation points by three main dimensions [126]:

- *What does vary?* This question should lead to the identification of the item or property of a system that represents a variation point.
- *Why does it vary?* This question should deliver the benefit or reason of the variation. Examples of reasons are regulatory demands in different market regions.
- *How does it vary?* The answer to this question leads to the variability subjects and denotes the actual occurrence of the variability.

Variability can also be characterised as negative or positive [35]. *Positive variabilities* leave the underlying commonality untouched – they always add something to the commonalities. *Negative variabilities* break commonalities – they remove properties that are usually common. Negative variabilities embody the "same as-except" concept [14], which means that there exists a quasi-common part which is overridden by a negative variability in exceptional cases. This results in an optimization, because less variation points are needed. The choice between negative or positive variability should be driven by the effort that is required to build the product line members.

Pohl et al. also distinguish between *internal* and *external* variability, denoting that internal variability is not visible to the customer while external is.

For embedded safety relevant systems, hardware and software variability should not be considered separately for software and hardware. Thus, it may prove worthwhile to reduce hardware variability and provide the needed variation in functionality through software variants (or vice versa).

2.6.2.2 Variability Management

One of the main activities in PLE is variability management. Variability management can in general be divided into two areas: *variability realization* and *variability specification*.

Variability realization deals with the implementation of the different types of variation points in the development artefacts (e.g., specifications, architecture, code, V&V artefacts). For example, variability realization may entail the implementation of generic test cases that contain variation points in terms of parameters and decision nodes. It may also entail the creation of generic specifications in terms of documents that contain place-holders for the points that may vary.

On the other hand, variability specification concentrates on describing, structuring and managing the variation points that are spread out across the different development artefacts. Variability specification can be seen as a configuration layer over all variation points in the development process. In other words, the variability specification manages all information that allows the artefacts' variabilities to be controlled. This configuration layer can be seen as a front end for navigation through the different variations in the development process and for resolution of the variation points. Hence the variability specification shows only configuration information, that includes:

(a) The type of variability implemented by a variation point,
(b) The position of the variability in the global hierarchy of variations and, most important,
(c) The dependencies between variations

In order variability specification to be beneficial, it is necessary to establish clear traceability links between the configuration possibilities shown in the variability specification and the different variation points arising in the development process.

There are several definitions and approaches to deal with Variability Management, most of them applied or envisioned to be applied in Software Product Line (SPL) engineering processes:

- *Feature modelling (FM)*: this approach is used in order to model both the commonalities and variabilities of product variants. The main idea behind this formulation is the representation of the variation through the classification of features within different categories (e.g. mandatory, optional, alternative) using a graph-based model [18]. This approach is useful in high abstraction engineering levels, since it provides a global perspective, serving as a basis for both, end-customer and engineering configuration. However, the main drawback of this approach is the increasingly complexity that it is inherent to features definition.

As soon as the feature model covers several aspects of the domain, the model itself becomes unmanageable.

- *Decision modelling (DM)*: Decision models focus only in the variable part of the product variants. This means that they guide the derivation of new products, and control their development through the definition of inter-dependencies and relationships among the decisions. Indeed, decision modelling provides great flexibility and precision regarding the capture of decisions and their relationships at different levels of abstraction in a more precise way compared to feature modelling. In this sense, complex dependencies can be easily addressed and at the same time, the complexity grows less than in the previous approach since only the variable is managed. The main idea behind decision modelling is to define the domain based on decisions (business or technical ones).
- *Common Variability Language (CVL) [140]* is a generic and separate language for modelling variability in models in any Domain Specific Language (DSL) that is defined based on Meta Object Facility (MOF) [115]. In the CVL approach three kind of models are used:
  - The base model (a model described in a DSL).
  - The variability model (the model that defines variability on the base model).
  - The resolution model (the model that defines how to resolve the variability model to create a new model in the base DSL).

It must be noted that there are some overlaps among the identified approaches, which means that in certain cases is feasible to use multiple of them, although not recommended. Feature Models are generally preferred for those phases where there is an need to have a complete view of the whole system in order to handle the variability while Decision Models can be used when the whole system view is not needed and the SPL manager needs just to concentrate in the variants, because the commonalities are implicit. CVL is nearest to the physical models due to its direct link to DSLs and therefore it is more suitable to deal with actual physical components.

In order to address variability among engineering phases Feature Models or Decision Models can be used, since they can exploit their generic nature to easily manage the dependencies across different phases and with different involved engineers.

2.6.3 CESAR *Approach and Enhancements in PLE*

Within the CESAR project, a combination of several techniques in each of the analysis layer presented in Sect. 2.3.1 is proposed in order to solve specific PLE issues. This is depicted in Fig. 2.25 where it is shown the techniques that can be used at the same analysis level and those that can be used among different analysis levels.

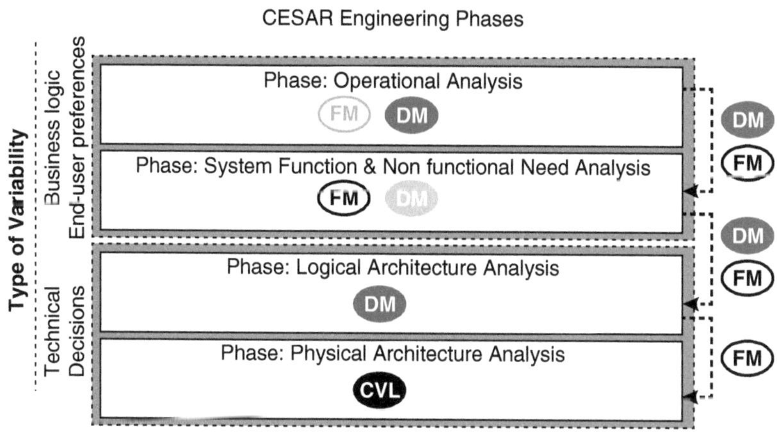

Fig. 2.25 Mapping of the variability approaches to engineering analysis phases

Operation analysis level: DM, thanks to its generic nature, is suitable to deal with business and customer needs and it is easier to see this variability needs as decisions. Moreover, the full description of the system is not really needed in order to take any decision on the variants, making DM the ideal solution for addressing variability. FM can also apply and a possible mixture of both could be interesting due to their different representational capabilities over a domain.

Functional and non-functional need analysis level: In this phase the engineer usually needs the whole view of the system in order to solve the variability, making FM the best choice. Some parts of the variability management can also be addressed by DM in order to speed up the global performance of model management (concentrating only in the variable parts). In this particular phase, the selection of the approach to be applied will largely depend on the exact type of system and the type of variability.

Logical Architecture analysis level: At this level, the variability can be tackled with all three options. However, the use of decision modelling will simplify and hide complexity of the whole system, by not showing what is actually not needed in order to solve the variability. The power of DM at this level lies in the ability to express in a more human intuitive way variability through questions and choices.

Physical Architecture analysis Level: This level is usually tight to a specific domain and, probably, to a specific DSL. Although the three options can be used, CVL seems the most suitable providing the mechanisms to solve the variability closer to the specific DSL.

In order to support variability management along the design process the CESAR RTP platform includes tools that can be used for Feature Modelling, Decision Modelling and the Common Variability Language approach. These tools are CVM (Compositional Variability Management) [34], PLUM (Product Line Unifier

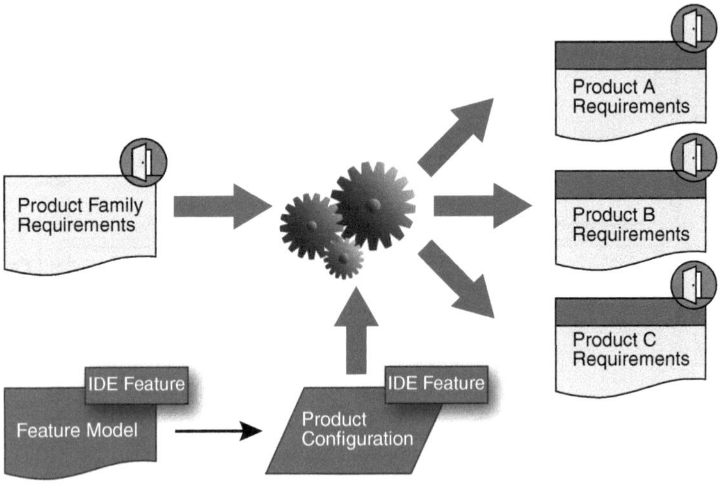

Fig. 2.26 Feature modelling approach for product family management

Modeller) [153] and CVL tool respectively. Appropriate tool-chains have been defined to support enhancements during PLE that aim to:

- Address complexity of handling variations points in complex system through the use of features models.
- Automate the transition from domain models (containing unsolved variation points) to resolution models

In the first case, each variability point of a product family is modelled in a feature model. Then product family requirements are developed and allocated to features and a subsequent configuration of the feature model identifies requirements for a specific product. The main idea of the corresponding tool-chain is to separate product configuration from requirements management as depicted in Fig. 2.26.

In the second case, system design models (defined, for example, using the Papyrus tool [125] in EAST-ADL [42] or other UML-like language) after being extended with variability data, they are processed by variability management tools like PLUM or CVL so that variation points are solved and specific resolutions models are generated. Both high level and detailed design can be addressed. The whole concepts is depicted in Fig. 2.27.

2.7 Traceability During Design Phases

Before closing the chapter related to the CESAR system life cycle, we should also say a few words related to traceability during design since it is considered a key element for any structured process development and a major prerequisite for critical system design. Much of the overall system certification credit is based on the

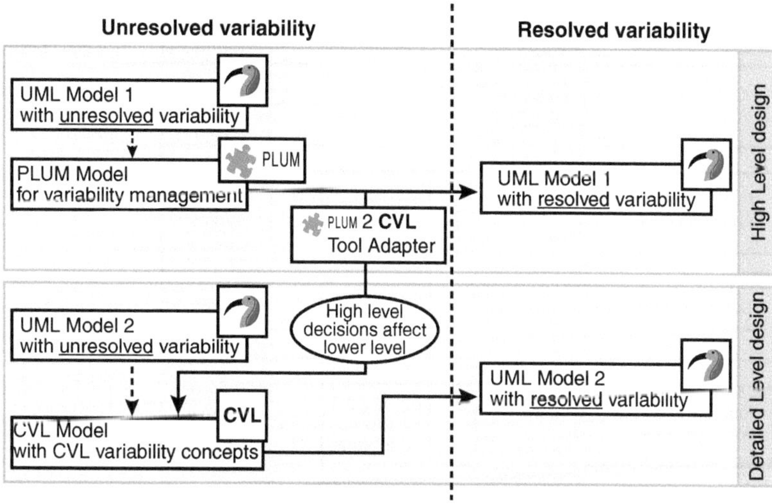

Fig. 2.27 Example tool chain for automating resolution model generation

ability to demonstrate efficient traceability management between the artefacts of a system developed during the various design and development phases whether these concerns requirements, models, test cases, specifications, components etc.

The notion of traceability can be considered primary related to requirements as they are refined, satisfied or verified within the system life cycle, while we move from high level architecture analysis to detailed system specification. This is supported in CESAR since the envisaged CESAR common meta-model (CMM) foresees concepts like the "refine" link, which indicates that a new set of requirements can safely be substituted for the original set, or "satisfy" – "verify" links that express conformance of a design to a (set of) requirement(s), either because the designer asserted so or because V&V actions have substantiated it. Other requirement related link types are also possible. This part of traceability, related mainly to requirements, is described in more details in Chap. 3 (see Sect. 3.5.1).

Another form of traceability can be considered related to PLE. There it is usually necessary to establish clear traceability links between the configuration possibilities shown in the variability specification and the different variation points arising in the development process. This is considered part of product line management and is usually achieved through special PLE tools like PLUM.

A third form of traceability, of great interest during system design and especially in what concerns architecture analysis and iterative system refinement, will be briefly outlined in this section. It addresses the need to:

- Link models of the same system representing different aspects,
- Link elements referring to different types of analysis (i.e. logical architecture analysis elements with physical architecture elements) at the same abstraction level,

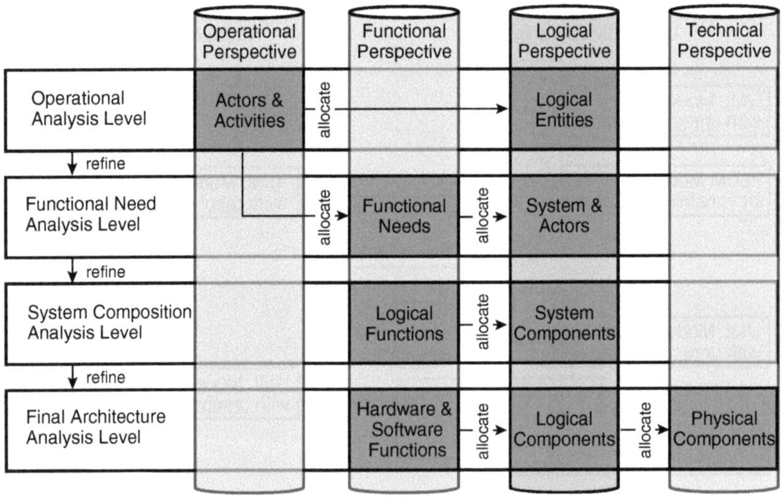

Fig. 2.28 Elements and their relationships during system analysis

- Link elements referring to same or different types of analysis at different abstraction levels (usually in a top down analysis where hierarchical refinement applies)

In CESAR's design life cycle, it has been identified that these links may need to be established (a) between elements residing in the same model description file as well as (b) between elements in different models even when defined and developed using different tools.

The intra-model design traceability of (a) is supported through the CMM special "mapping" meta-class. This mapping can be used to describe the mapping solution between two elements which may belong to analysis models that are defined in different perspectives (of the same abstraction level) or in different abstraction levels. An "allocation" link is used in the former case while the link is described as "realization" in the latter case. Figure 2.28 illustrates the "allocate" relationships that are presumed to be defined between the most commonly used perspectives during the various system analysis levels as presented in Sect. 2.3.1. The examples presented in Chap. 5 (DOORS Management system) may provide more concrete insight on how traceability is enabled between the perspectives.

To address inter-model design traceability between elements defined in different model files, a necessity when designing large systems where many different teams may be involved, CESAR has identified the need for a lightweight data integration layer that will enable to manage the creation/edition of links between any models and/or internal model elements stored in a model repository. This has become possible in the RTP by the development of a special linking service, which provides uniform access to models and their internals elements in a manner that is independent on how these models are defined and implemented, and the existence of an appropriate repository that can be used to store the defined links. Through this link repository service, different kinds of link, such as "satisfy", "refine", "implement" and so on can be defined and managed (actually the exact types of links supported is left open and may vary depending on each domain needs). More details on the link repository mechanism can be obtained from Chap. 6.

The two approaches described above are completely independent one from the other. The inter-model one is more generic since it may address elements of any kind developed in completely different tools. However, there is no prohibiting factor in using both of them in a complementary function.

Chapter 3
Requirements Engineering

Markus Ortel, Marc Malot, Andreas Baumgart, Jan Steffen Becker, Ralf Bogusch, Stefan Farfeleder, Nora Gerber, Oystein Haugen, Stefan Häusler, Bernhard Josko, Jason Mansell, Andreas Mitschke, Roopak Sinha, Tor Stalhane, Niina Uusitalo, and Philip Rehkop

3.1 Introduction

Requirements engineering is a fundamental part in the development of safety related embedded systems. Requirements engineering comprises eliciting, developing, analysing, validating, communicating and managing requirements. As the development of systems is based on the identified and developed requirements it is important that requirements accurately capture the stakeholder needs, are

M. Ortel (✉) · P. Rehkop
OFFIS, Oldenburg, Germany
e-mail: markus.oertel@offis.de

A. Baumgart · J.S. Becker · S. Häusler · B. Josko
OFFIS e. V., Oldenburg, Germany

R. Bogusch
CASSIDIAN

S. Farfeleder, Language, Boilerplate Analyzes · N. Gerber · N. Uusitalo
Infineon

O. Haugen
Stiftelsen SINTEF, Trondheim, Norway

M. Malot
Sagem Défense Sécurité, Paris, France

J. Mansell
Fundación European Software Insitute, Donostia-San Sebastiàn, Spain

A. Mitschke
EADS Deutschland GmbH, Munich, Germany

R. Sinha
Institut National de Recherche en Informatique et en Automatique, Paris, France

T. Stalhane
Norwegian University of Science and Technology, Trondheim, Norway

A. Rajan and T. Wahl (eds.), *CESAR - Cost-efficient Methods and Processes for Safety-relevant Embedded Systems*, DOI 10.1007/978-3-7091-1387-5_3,

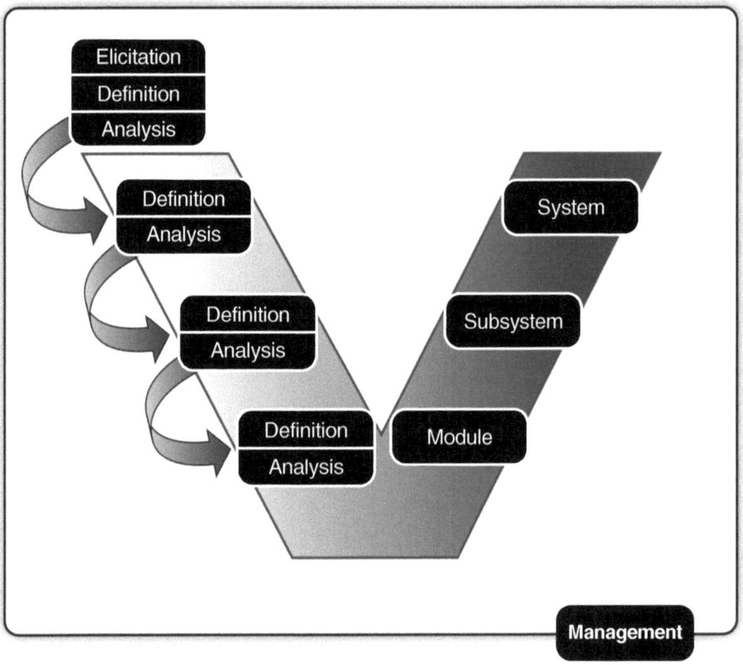

Fig. 3.1 Basic understanding of the RE process

well-understood and verifiable. Due to that, requirements engineering guidelines and safety standards request several constraints on requirement statements like well-formedness, unambiguity and formalization. Moreover, requirements should be consistent and they should be satisfied by the proposed implementation (Fig. 3.1).

The ISO standard 29148 [94] – cf. also [92, 93] – considers two main requirements engineering activities: (1) *Requirements Development* and (2) *Requirements Management.*

Requirement Development is applied during the development process on all hierarchy levels starting from development of stakeholder requirements over development of system requirements down to development of module requirements. Each cycle contains the sub activities (1) *Define Requirements* (of the considered level) and (2) *Analyze Requirements.* The top-level *Development of Stakeholder Requirements* contains a sub-activity *Elicit Stakeholder Needs*, which are transferred in the subsequent sub-activity into stakeholder requirements.

CESAR itself does not define a specific requirements engineering process – as such a process may be company- or project-specific – but CESAR provides specific methods and tools that can be used within a concrete requirements engineering process addressing the activities mentioned above. All elements are included in the CESAR *Practice Library* (CPL), which defines a general process framework and which should be instantiated for a concrete process.

During the requirements development process, stakeholders' needs are transformed into stakeholder requirements, which are further decomposed due to the hierarchical structure of the considered systems to more technical requirements. Going down in the hierarchy, requirements become more concrete and can be formulated in a more formal way. A formal or semi-formal approach supports the analysis of requirements. Due to that, formalization of requirements is a key issue of CESAR.

Within the CESAR project, a *Requirements Specification Language* (RSL), is defined, which

- Supports smooth transition from informal, through semi-formal to fully formalized requirements;

  To cover the whole range, RSL enhances existing description methods and provides different description formats including textual ones (from natural language over structured text (boiler plates) to formalized requirements (pattern-based RSL)) as well as graphical ones.

- Supports multiple viewpoints throughout the development life cycle;

  The RSL is designed for expressing functional and extra-functional properties of critical systems. RSL supports all relevant viewpoints: functional, safety, performance, timing, architecture constraints, and others. Also the combination of several viewpoints is supported.

- Provides a basis for formal analysis methods.

  Topics addressed include consistency as well as completeness of requirements, correctness of design w.r.t. requirements, impact analysis as well as requirement-based testing and other V& V activities.

Requirements analysis and validation of requirements benefits from formalized requirements. Dedicated methods and tools can be applied. Within the CESAR project, methods and tools have been investigated to support validation of multi-criteria requirements by

- Enhancing and measuring completeness of functional and extra functional requirements,
- Ensuring consistency and soundness of functional and extra functional requirements,
- Requirement-based testing for functional and extra functional requirements,
- Establishing Code of Practice for safety-critical applications.

The requirement development activities are interwoven with the management of requirements. The *Requirements Management* activity consists of the following sub activities:

1. Maintain Requirements Traceability,
2. Management requirements changes,
3. Identify non-compliances, and
4. Measure requirements.

Traceability is addressed in CESAR by defining a framework for requirements management which ensures traceability across the complete development process between requirements, modelling elements as well as derived artefacts including for example test cases. The requirements management framework supports also the establishment of a quality measurement method as for example information of the requirement analysis results can be easily accessed and used in appropriated process metrics. Changes in requirements require revalidation of the whole system. Impact analysis is a key issue to identify the relevant parts. Formalized requirements together with formal analysis methods form the basis for a novel approach for the impact analysis of requirements changes.

This chapter is organized as follows. Section 3.2 addresses the elicitation activity in requirements engineering. Section 3.3 provides details on the requirements specifications language RSL. In particular, Sect. 3.3.5.1 describes the support to derive formal requirements out of informal ones. Analysis techniques for requirements engineering in the context of the CESAR approach are given in Sect. 3.4. Requirements Management issues are addressed in Sect. 3.5. A cross-sectorial topic is product-line engineering. How this topic can be addressed in requirements engineering is discussed in Sect. 3.6.

3.2 Requirements Elicitation

The requirements elicitation process is initiated by identifying relevant stakeholders to obtain functional and non-functional requirements and to capture them in any RSL/RMM, which supports manual/semi-automatic and automatic post-processing. Domain specific requirements may be derived from a problem domain and not only from stakeholder needs. Prioritizing captured requirements is part of the requirements management. Nevertheless, it may require additional iterations of the elicitation process. Checking the captured requirements on consistency, completeness and correctness (CCC) is taken care of in the requirements analysis task of CESAR. The results of the CCC checks may re-initiate the elicitation process, e.g., caused by inconsistent requirements provided by different stakeholders. At the beginning of Sect. 3.2, stakeholder identification is addressed: methods are introduced to identify relevant stakeholders and how to obtain requirements from them. Furthermore, the interaction between different stakeholders and the impact on the elicitation process is analyzed. Different elicitation techniques, e.g., use-cases, misuse cases and quality models are addressed in Sect. 3.2.2 and their suitability for functional and/or non-functional requirements are investigated. Selection criteria to choose appropriate elicitation techniques are introduced in Sect. 3.2.3. Specifying non-ambiguous requirements in a certain domain is supported by domain ontologies. Methods and tools to obtain domain ontologies and domain-specific requirements are introduced in Sect. 3.2.4.

3.2.1 *Stakeholder Identification*

It is crucial to know who the target persons and organizations are in order to elicit the correct requirements. The activity to find those stakeholders is called stakeholder identification. Stakeholder identification is important because an inappropriate set of stakeholders is likely to lead to a wrong set of requirements. Omitting to identify a stakeholder might result in the omission of requirements; the final product will then lack a feature or will fail to adhere to a quality aspect. Eliciting requirements from stakeholders not relevant to the project might increase the development cost and time without any additional value.

3.2.1.1 Stakeholders

In order to identify stakeholders, one needs to define what precisely a stakeholder is. Glinz and Wieringa [68] define the term stakeholder in the following way: *A stakeholder is a person or organization who influences a system requirement or who is impacted by that system.*

In contrast to earlier views, this definition does not only include all kinds of users and operators who will interact with the system under development, but also the members of the development project, any person or organization who imposes regulations or constraints on the project and for example, anyone that has a financial interest in the project success. The abstraction of a group of stakeholders is called a *stakeholder role*. Roles are used for two reasons: The actual stakeholders might not be known in person at the time the elicitation is performed. Due to time and cost constraints, it might not be practical to elicit requirements from a lot of persons. Sample subjects are chosen to represent the requirements of the entire group.

Exemplary stakeholders for top-level requirements are: product and project managers, customers, market specialists, internal research experts, specialists from different domains (e.g., production, finance), regulatory authorities, experts on previous products and suppliers. At lower levels, in addition to the mentioned stakeholders, system designers and test specialists might also be involved in requirement elicitation.

3.2.1.2 Identification Methods

Preiss and Wegmann [127] apply systems science principles to the field of stakeholder identification and come to the following three conclusions: *Two systems:* A system cannot be modeled without considering the system within which it is embedded. *Two viewpoints:* The goals of a system and the means to achieve the goals must be analyzed. *Two domains of inquiry:* The system life-cycle consists of the system development and the system operation. They propose to identify stakeholders by looking for stakeholders in each of these categories.

Sharp, Finkelstein and Galal [136] define four classes of *baseline stakeholders*. *Users* are defined as persons, groups or companies who interact with the system and additionally, those who use the products of the system. *Developers* are people who are involved in the system development. *Legislators* are entities that affect the development and/or operation by means of guidelines, standards, laws or regulations. *Decision-makers* include all kinds of managers and financial controllers among the developers and users. They define three additional sets of stakeholders that are delineated by their relation to the baseline stakeholders. *Supplier stakeholders* provide information or supporting tasks to baseline stakeholders. *Client stakeholders* process or inspect baseline stakeholders' products. *Satellite stakeholders* interact with baseline stakeholders in any other way. The proposed method consists of five steps:

1. Identify the baseline stakeholders.
2. Identify the supplier stakeholder for already identified stakeholders.
3. Identify the client stakeholder for already identified stakeholders.
4. Identify the satellite stakeholder for already identified stakeholders.
5. Repeat until no more stakeholders are found.

3.2.2 Elicitation Techniques

This section describes additional common elicitation techniques which we consider to be relevant in the CESAR context. These techniques should not be thought of as excluding each other; it can be useful to combine different techniques.

3.2.2.1 Interviews

In requirements engineering the interviewer is the requirements engineer/analyst and the interviewee is a stakeholder. It is recommended to record interviews to avoid disturbing the communication flow by writing down answers. The recorded data can then be analyzed later. There are two possibilities for answers to interview questions. In *open-ended interviews* the interviewee is free to answer questions in his/her own words as detailed or succinctly as he wishes. In *closed-ended interviews* there is a given set of answers and the interviewee has to choose one of them. Structured interviews use a fixed set of questions in a fixed order. These are contrary to unstructured interviews, where the interviewer has the possibility to modify questions or to change the order as he wishes.

3.2.2.2 Group Techniques

Group technique is an alternative to single-person interviews. Here requirements are elicited from several stakeholders at once. The basic idea is that discussions and cooperation between different stakeholders will make requirements apparent that if they were elicited otherwise. Additionally, the time required for the elicitation process is lower and conflicting requirements can be resolved. However, the interaction between different stakeholders can have negative impacts (e.g. due to personal or organizational conflicts). Examples for group techniques are *focus groups*, *brainstorming sessions* and *Joint Application Design* [172].

3.2.2.3 Use Cases/Scenarios

Use cases are a representation to describe system functions. Sindre and Opdahl [138] state: "The use case construct is used to define the behavior of a system or other semantic entity without revealing the entity's internal structure. Each use case specifies a sequence of actions, including variants that the entity can perform, interacting with actors of the entity". Furthermore, a method to elicit additional requirements is the scenario walk-through. It means that a scenario is being thoroughly inspected step by step, and consequently alternative scenario courses and quality aspects/constraints are being thought through, which might generate new requirements to complement the existing ones.

3.2.2.4 Misuse Cases

Use-cases are more suitable for eliciting functional requirements rather than non-functional requirements (NFRs). The objective of misuse cases is to derive requirements regarding security, safety, reliability and other non-functional requirements A *misuse case* is a special kind of use case, describing unfavorable behavior that the owner of the system wishes not to take place. The *actor* is someone who initiates misuse cases, either intentionally or unintentionally. The actor can be a person or an external system that interacts (gets some value and produce some value) with the system being described. In this case (misuse cases), the actor concept is further specified: the actor initiates not only use cases (specifying the behavior of the system) but also misuse cases (describing unfavorable behavior).

The misuse case method is to gather and specify NFRs by providing "play and counter-play" between use and misuse cases. It is possible that the misuse cases threaten the use cases. However, additional use cases can be defined to mitigate these misuses.

3.2.2.5 Quality Models

Quality models aid the specification of quality requirements. The objective of quality models is to define and structure quality requirements in a measurable way. That is, for each relevant quality goal, one has to define specific quality factors, quality criteria, and corresponding metrics. Lamsweerde [103] states: "Quality requirements state additional, quality related properties that the functional effects of the software/system should have. Such requirements complement the 'WHAT' aspects to 'HOW WELL' aspects." Examples for quality requirements are: safety, security, reliability, usability, efficiency.

Quality models extend the taxonomy of quality requirements by adding metrics to the quality attributes. Another advantage of quality models compared to flat taxonomies is that they provide dependency relations between different quality attributes. Quality requirements have complex and nontrivial interdependencies. They can have a positive or a negative influence on each other. The impact of such influences has to be analyzed.

ISO 9126-1 Quality Model

ISO/IEC 9126 is a standard for specifying software product quality. It consists of four parts: quality model, external metrics, internal metrics, and quality in use metrics. We will concentrate only on the quality model. This model structures the software quality into characteristics such as functionality, reliability, usability, efficiency, maintainability and portability. These are refined into sub-characteristics. Sub-characteristics can be measured by metrics. Parts 2, 3 and 4 of this standard specify some example metrics. The ISO 9126 quality model provides a framework for quality requirements and quality evaluation. However, the standard gives only a high-level proposal of how quality can be specified, which is not finite and needs further detail specification. This is because each software product is individual and has different properties and metrics depending on the application domain and context for which will be used.

NFR-Method

The NFR-Method provides a systematic method based on experience-based quality models to elicit, specify and analyze non-functional requirements. The method distinguishes between quality attributes, which are structured in quality models and NFR captured in documents [44, 78]. The quality models are the models used for specifying the NFR. The requirements are the initializations of the models.

As an example for an efficiency requirement, consider the following: "The database of our new system shall handle 1,000 queries per second." This requirement constrains the quality attribute "workload of database". The value is determined based on an associated metric "number of jobs per time unit". The method consists of two stages: Quality Model Tailoring and NFR Elicitation.

Quality Model Tailoring Using questionnaires, the main quality attributes (QA) are prioritized. This task saves effort during the elicitation process. The prioritized QAs are captured in quality models, which are tailored from reference quality models (such as ISO 9126). The tailoring is done in workshops, where domain experts from the company tailor each QM, which specify the QA to the needs of the specific project. During the tailoring process dependencies and relationships between different QAs, are investigated and documented. The outcome from this stage is a quality model tailored for a specific project.

Elicitation NFR During this stage, reference checklists and templates are customized for the tailored quality model. The checklists and templates support the elicitation of NFRs by giving additional information, e.g., where the elicited information has to be documented; for which artifact (system, task) a QA is relevant, etc.

3.2.2.6 Prototypes

A prototype is an early version of a system that can be used to evaluate some aspects of the final product. It implements the features under consideration, however, may lack in specific parts. The goal of a prototype is to visualize or ascertain the system to be built at an early stage of the development process (naturally, building prototypes takes time). In presenting the generated prototype to the stakeholders and checking its functions against the stakeholder needs, the prototyping technique can be used to validate requirements, to identify missing requirements and to find errors in requirements. Prototypes are mainly used for functional requirements and user interface requirements; other quality requirements such as timing, safety or standards to be applied to the development process will often differ substantially in the final product.

3.2.2.7 Ethnographic Techniques

Ethnography is a research method to gather empirical data on human societies and cultures. In the context of requirements engineering, it means to elicit requirements by observing users in their natural work environment. This natural setting (compared to the situation in an elicitor's office) has an impact on the requirements that can be captured. Sommerville [143] mentions two kinds of requirements that can be elicited with ethnographic methods: Requirements that are derived from the way in which people actually work (as opposed to the way they think or ought to think about it), and requirements that are derived from cooperation and awareness of other people's activities.

Task observation is an ethnographic method where the requirements engineer watches a user while he/she performs the task under consideration. The observation is either direct (the requirements engineer is present during the task execution and takes notes) or indirect (the task is recorded, e.g., with a video camera and the requirements engineer observes the recordings). It is important for the observation to be as unobtrusive as possible to avoid influencing the observed task. The technique is used especially to study human-machine interfaces of existing legacy systems.

3.2.3 Elicitation Technique Selection

As can be seen from the above list, there are many elicitation techniques available. The actual technique choice often depends on the elicitor's personal preference rather than facts, possibly leading to elicitation of less or worse requirements than with another technique. In order to overcome this situation, Carrizo et al. [28] studied the adequacy of elicitation techniques. Situational attributes were collected from literature that was mentioned to have good or bad influence on elicitation with a certain technique. At first, they compiled a list of 34 attributes, then removed duplicates and attributes they considered irrelevant or not justifiable and added four new attributes they thought to be missing. This lead to a list of 16 contextual attributes they used in their study, e.g., elicitation experience and familiarity with domain. For each of the 16 attributes they define 2–3 different values, e.g. an elicitor's training in elicitation techniques can be zero, low or high. Furthermore the authors chose to study the above attributes on 15 elicitation techniques. A matrix was created from the set of contextual attributes and techniques, depicting when a requirements engineering expert recommends to use the technique under the circumstances given by the attribute or, when an empirical study shows the adequacy of the technique. In case that the authors could not find any information for an attribute-technique pair, they used their own experience and understanding for an adequacy level.

3.2.4 Domain Analysis

Domain analysis is the task to collect knowledge about the domain of the system to be built. Domain analysis is part of domain engineering/PLE. It is important for two reasons:

- The requirements engineers and stakeholders have to communicate during the elicitation process. It is thus necessary that the elicitor has a basic understanding of the problem domain.
- The requirements themselves will include terms which are domain-specific. To avoid misunderstandings between the development team and the customers these terms should be properly defined. The term definitions can be reused in other projects in the same domain.

Gathering new domain knowledge can be done in two ways: (1) Reading domain-specific documents. These can be standards specific to the domain or existing specifications from earlier projects. (2) Talking to domain experts and asking them for domain-specific terms, their definitions and relations to other terms.

3.2.4.1 Domain Ontologies

Ontologies are a common way to represent domain knowledge. Their advantage over pure glossaries is that by using relations and axioms, reasoning and consistency checks are possible to a certain extent.

How to Develop an Ontology

Noy and McGuinness [119] propose the following seven-step approach to create ontologies.

1. Determine the domain and scope of the ontology.
2. Consider reusing existing ontologies.
3. Enumerate important terms in the ontology.
4. Define the classes and the class hierarchy.
5. Define the properties of the classes.
6. Define the restricting axioms of the properties.
7. Create instances.

Ontology Tool Support

Protégé [130] is an open-source tool developed by Stanford University. It contains a graphical editor to add, edit and delete the elements of ontology, i.e., classes, instances, relations, properties, etc. Text2Onto [32] is a tool that semi-automatically creates ontology from natural language text. Text2Onto consists of several data-mining and natural language processing algorithms which can be chosen and configured by the user.

3.2.4.2 Domain Requirements

Domain requirements are requirements that derive from the problem domain itself and not from stakeholder needs. Domain requirements can be both functional requirements and non-functional requirements. Often domain requirements are requirements about safety, interfaces, or design, implementation or process constraints. Fulfilling these requirements is crucial to be able to sell products at all in certain domains. Some examples for domain requirements:

- *The system shall comply with Standard XYZ.*
- *The system shall operate at temperatures of −XY C to +YZ C.*
- *The used software tools shall have a maturity level of at least 6.*

As domain requirements are independent from the project-specific requirements, they can be easily reused from other projects in the same domain without any or with only minor modifications. Thus it is a good idea to compile domain requirements in a requirements library for easy reuse in new projects.

3.3 Requirements Definition

This section introduces the different languages developed/used in the CESAR project to express requirements. Driven by the various needs these languages range from natural language text to semi-formal textual representation (Boilerplates) and formal textual representations (Pattern). Also graphical languages are considered. At the end of the section conversion techniques are presented to create a consistent process while proceeding from informal to formal representations.

3.3.1 Guided Natural Language

Natural language is the starting point for formalization into guided natural language.

Guided natural language RSL can be achieved by using a set of requirement statements that are derived with the assistance of prescribed words from a dictionary. Such guided natural language approach retains the benefit of free text descriptions since its use does not introduce additional formal constraints on expressed requirement statements or require additional expertise training. Furthermore, requirements elicitation with the aid of prescribed words can provide the basis for semantic processing and checking of requirements.

Using guided natural language during requirements elicitation and specification can be realized by adding a number of enhancing factors. Firstly, there is the need for a mechanism to generate a dictionary of prescribed words and relations between these words (domain ontology). The second factor involves establishing the basis and appropriate technique for the semantic processing and checking of elicited or specified requirements.

The main advantage of GNL is the continuing benefit of free text expression of requirements that are qualitatively guided by an existing domain knowledge base. Suggestions on how to possibly improve the status of the requirements are also provided. A challenge with GNL approach is that while GNL can suggest on possible modalities to improve specified requirement, the accuracy of such suggestions is based on the correctness of documented knowledge base. It is also important to note that GNL is not a requirements enforcement mechanism (such as boilerplate or pattern-based RSL), and as such some of effort is still required from the analyst to abide by some external rules or pattern when specifying requirements.

3.3.2 Boilerplates

The Boilerplates for requirements specification documentation consists of a set of predefined templates – the Boilerplates. There are three main categories of Boilerplates: capability Boilerplates, functional requirement Boilerplates and constraint Boilerplates. Experiments performed at NTNU shows that also non-functional requirements can be documented using the same set of Boilerplates. The primary benefit of using boilerplate RSL is that they allow requirements to be captured in a consistent fashion i.e. reduction of spelling mistakes, poor grammar, ambiguity, etc., thus facilitating understandability, categorization and identification of the requirements. Additionally, boilerplate RSL act as an input mechanism for capturing the underlying system-related semantics behind requirements i.e. the stakeholders involved, the system capabilities and so forth. Each boilerplate has one or more attributes – places where the user can insert information pertaining to a requirement. In the example below, the attributes are enclosed in "⟨" and "⟩" brackets. Boilerplate32: The ⟨user⟩ shall be able to ⟨capability⟩

3.3.2.1 The Boilerplate Requirement Elicitation Process

The process described here assumes that new requirements are directly elicited using the boilerplates RSL. The case that existing natural language requirements are reformulated into boilerplates requirements is described in Sect. 3.3.5.1.

Two cases called the normal case and the exceptions case need to be considered. In the normal case, the requirement can be directly mapped to a boilerplate whilst, in the exception case, one or more parts of a requirement needs to be changed. When all the requirements have been specified, we may run an ontology-based check for completeness and consistency. In both cases, the requirements elicitation is a process involving the domain expert, the requirements engineer and a tester. The process is described for one requirement only.

- Normal process
  1. The domain expert states a requirement. The requirements engineer identifies an appropriate boilerplate. This is done manually or with support from a requirements engineering tool.
  2. The domain expert assigns a value to each of the boilerplate attributes. Even though not a part of the original boilerplate definition, NTNU has provided a set of definitions for event and action attributes.
  3. The tester, in cooperation with the domain expert, formulates one or more tests for the requirement. This can be an executable test or inspection of one or more artifacts, e.g. related to proof of process conformance. Upon completion of step 3, the boilerplate requirement is stored in the requirement database.
- Exception cases If no appropriate boilerplate is identified, then:
  - Reformulate the requirement

  If the requirement is too complicated for a single boilerplate, then:

  - Split the requirement into several sub-requirements and repeat the normal process for each sub-requirement. Use a combination of two or more Boilerplates.

  For all exception cases, the requirement in question must also be stored in a separate database for later analysis, possible boilerplate updates and learning.

3.3.3 Patterns

Stating requirements is one of the essential tasks in developing products. These requirements build an interface between different groups of people, i.e., customers, engineers, project managers and many more.

Typically requirements are stored as natural language text which has the disadvantage that ambiguities of the sentences can cause a huge amount of requirement changes in early as well in later development phases. To reduce the costs coming from these ambiguities formal languages are used to have a fixed semantic meaning for a requirement. But it does not only require some training to write requirements in these formal languages, it can also be hard to read them.

The pattern-based RSL fills this gap by providing an easy to learn formal language with a fixed semantic that is still readable like natural language.

Patterns consist of static text elements and attributes being filled in by the requirements engineer. Each pattern has a well defined semantic in order to ensure a consistent interpretation of the written system specification across all project participants. On the one hand this limits the possibilities of writing a requirement on the other hand it prevents misunderstandings regarding the interpretation of the sentences. To gain a set of quality requirements this limitation is necessary. However writing requirements shall still be possible in an intuitive way. Patterns allow the writing of natural sounding requirements with defined semantics while being expressive enough to formalize complex requirements.

To gain a high degree of intuitiveness the language shall consist of only a few constructs that can be easily remembered to give the requirement engineer the possibility to fully understand the RSL and choose the right pattern that fits the properties he wants to demand on the system. There exist patterns for each of the following categories:

- *Functional Patterns*: These Patterns express functional requirements of the system. This includes the relationship between events, handling of conditions and invariants and the possibility to define intervals in which the requirements or parts of them are valid.
- *Probability Patterns*: In nearly all safety critical system it is necessary to express failure or hazard probabilities. Since there are various partly redundant forms of expressing probabilities that were used quite ambiguous in common speech, these patterns guide the requirements engineer to express his needs.
- *Safety related Patterns*: Only a small part of safety related requirements can be covered with probability patterns. The major part of the requirements outlines relationships between system components. These patterns enable the specification of single point of failures or other failure and hazard dependencies between different components.
- *Timing Patterns*: These patterns can be used to describe real-time behavior of systems. This includes periodic and aperiodic activations, jitter or delay.
- *Architecture Patterns*: These patterns specify the existence of architecture elements like components, events or connections between components.

- *Mapping Patterns*: These patterns can be used to express requirements on the mapping from the functional design perspective to the physical design perspective.

Patterns are recorded in a form with optional parts that are not necessarily instantiated. A functional pattern describing the causality between two events does look like this:

whenever `request` **occurs** `response` **occurs during** `[0 ms,10 ms]`.

Phrases in square brackets are considered optional, bold elements are static keywords of the pattern and italic printed elements represent attributes that have to be filled out by the requirements engineer. When filling the attributes there are no general restrictions on the names but they should be descriptive. The names of the attributes can be used later to be mapped to a given architecture or they can be used to generate the architecture. When specifying requirements there is no strict typing like int or real but there are categories that specify what is described by the different attributes. To not complicate the language too much and to be also expressive enough the set of supported attributes has to be carefully chosen:

- *Event*: Events represent the activity in the system. An event can be for example a received signal, a user input by a pressed button, a computational result or the change of a value. Events occur at one point of time and have no duration in time.
- *Condition*: A condition is a logic and/or arithmetic expression based on variables and the status of events. Conditions can be used in two different ways in patterns. On the one hand they can be used as pseudo events to trigger an action if the condition becomes true. On the other hand the condition can be the system state that has to hold for a specified time. The values the condition is based on may change with time only.
- *Interval*: Intervals describe a continuous fragment of time which can have relative and quantitative time measures as delimiters or events. Intervals can be *open* $]x, y[$ or *closed* $[x, y]$ or a combination of them.
- *Jitter* Jitter of an event is the variation in the point of time of occurrence in a real time context. In this document jitter is considered as always positive, i.e., the event is delayed.
- *Component* Components refer to entities that are able to handle events or condition variables. Typically these components refer to an actual architecture, but in early design stages, where the system has not been designed completely there are also possible component names allowed. These identifiers can either be used to generate the missing architecture or be mapped to the design in later development stages.

Table 3.1 contains some example Pattern requirements related to a doors management system of an aircraft.

Table 3.1 Some example pattern requirements

Requirement
`whenever openingAttempt under (!onGroundSignal) occurs unlock does not occurs during[0 ms,openingAttempt[.`
`whenever unlock occurs open occurs during [0 ms,100 ms].`
`always(onGround== onGroundSignal).`

These Requirements can be analyzed using various pattern-based tools. Analyzes for Consistency and Completeness of functional and safety requirements are described in Sect. 3.4.3. Management of impacts through requirement changes in a system specified with patterns is explained in the Sect. 3.5.2.

More Info:

The Pattern-based RSL is described with all its details in the Architecture Modelling Document found on the OFFIS website under: http://www.offis.de/cesar

3.3.4 *Graphical Specification*

3.3.4.1 Introduction

As described in the introduction of this chapter, Requirement Specification Languages may differ from each other depending on the degree of formality of requirements specification, and depending on the way of representation of requirements. Graphical Requirements Specification Languages will represent requirements in a graphical way. The specification of requirements using a graphical RSL may be informal, semi-formal, or formal. In this section we will focus on semi-formal graphical RSL. Semi-formal thereby means that the language has formal and precise syntax, while the semantics are imprecise and allow different interpretations. This is for example the case for some modelling languages such as SysML. SysML (Systems Modelling Language) is a general purpose, visual, descriptive modelling language for systems design. It has been submitted to the OMG for adoption in March 2006. It is based on UML 2.0 but some UML diagrams have been modified from and suppressed, and two diagrams have been added – the Requirements Diagram and the Parametric Diagram.

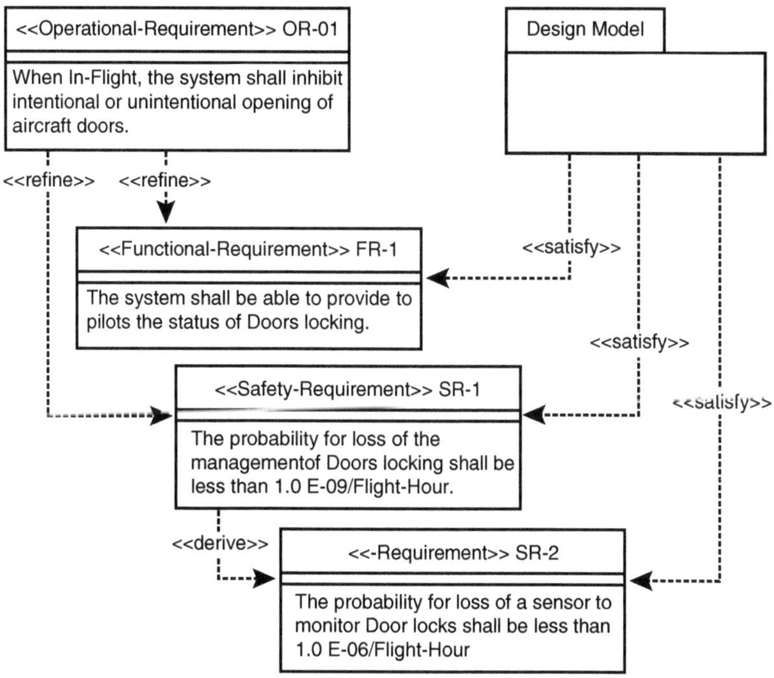

Fig. 3.2 Example SysML Requirements Diagram

3.3.4.2 Requirements Diagram

The SysML Requirements Diagram has been added to UML 2.0 in order to provide a diagram that can be used to represent the relationship between Requirements. Figure 3.2 provides an example of this diagram.

The Requirements Diagram is adapted to provide a structural view of requirements. It allows to document textual requirements, meta-data for these requirements (e.g. type of requirement) and dependencies between requirements and from design solutions to requirements in a graphical manner, without being tied to a specific methodology. Furthermore, it provides the possibility to have textual requirements and design elements represented in the same model, which can help to manage traceability between both worlds.

A challenge for the use of the requirement diagram is related to scalability: It is difficult to manage and to review requirements diagrams with hundreds or thousands of requirements. Some SysML tools counter this challenge by providing additional tabular views for requirements that can be also used to edit requirements. Only a relevant subset of requirements is represented using the requirements diagram.

3.3.4.3 Operational Scenarios Using SysML

The SysML-based graphical RSL is a graphical specification language that uses elements of SysML to describe requirements. Concretely, SysML Use Case Diagrams, Activity Diagrams, and Sequence Diagrams are used for representing requirements.

- Use Case diagrams are used to represent the main intended use of the system (respectively the top-level functions and features of a system)
- Then activity or sequence diagrams are used to describe detailed operational scenarios for each use case.

Operational scenarios are combining functional requirements that have sequential relationships to each other into scenarios. Operational scenarios are typically invoked by an actor that performs an operation. In the following, the operational scenario describes how a system shall react to this operation. Operational scenarios describe the nominal reaction of the system to this operation as well as alternative reactions. In addition to the actor and the System under Development (SuD), external systems may be represented on operational scenarios. However system components shall not be indicated at this stage as the definition of system components is a design activity. Operational Scenarios may be defined at different occasions. They can be defined by requirements engineers in parallel to atomic functional requirements as part of requirements elicitation activities. They can also be defined by system designers as part of an operational and functional analysis, using textual functional requirements as input. Within CESAR, the definition of operational scenarios is thus being addressed by both SP2 and SP3. In the following, the application of the SysML-based graphical RSL for the definition of operational scenarios is demonstrated on an example that has been derived from functional requirements of the simplified Doors Management System (DMS) case study (see Sect. 9.2.2). At first, we have transferred individual functional requirements for the DMS example into an operational scenario in tabular format (see Table 3.2).

Use Case Diagrams

The SysML-based graphical RSL provides two main diagrams that can be used to represent this operational scenario in a graphical way: Activity Diagrams and Sequence Diagrams. In addition, it provides a Use Case Diagram which can be used to give a global view about all main use cases of the System under Development. The principal elements of Use Case Diagrams when used for Requirements Specification are:

- Use Case: Used to represent the main intended use of the system. For each Use Cases a detailed description can be provided, resulting in an "Operational Scenario".
- Actors: Used to indicate Human Actors and external systems

Table 3.2 Tabular description of an operational scenario for the DMS example

Activation condition of the operational scenario for the use case "set speed":		
Pilot requests to pressurize the cabin in order to prepare take off of the aircraft	1	
Nominal reaction of the system:	1	
	1	The DMS checks whether all Doors are closed, latched and locked
All Doors are closed, latched and locked	2	The DMS pressurizes the cabin
	3	The DMS sends a message to the Pilot that cabin is pressurized, and that he can proceed with the take-off procedure
	4	The Pilot receives the message
	5	End of nominal scenario
Alternate scenario 1:		
On step 2: The Doors are not fully closed, latched and locked	2b	The DMS inhibits cabin pressurization
	3b	The DMS sends a message to the Pilot that cabin is NOT pressurized, and that the pilot has to check the Doors

- System Boundary: Used to group all System internal Use Cases
- Association: Used to indicate the actors involved in a Use Case
- Relation: Indicates the relationship between use cases. The two main types of relationships are "include" and "extend". Note that Use Case Diagrams do not provide sequential relationships between use cases. A detailed description on how Use Cases can be elicited is provided by the deliverable D_SP2_R3.2.

Activity Diagrams

In a subsequent step, for every use case a detailed description is provided. For the SysML-based RSL this detailed description can be described in form of Activity Diagrams or Sequence Diagrams. Figure 3.3 exemplifies the use of Activity Diagrams for the representation of an Operational Scenario for the Use Case "Manage Door Operation" of the DMS case study. The same information is provided as in the tabular description of this operational scenario shown in Table 3.2. The main elements that are used when representing Operational Scenarios via Activity Diagrams are as followed:

- Swimlanes: They are used to allocate all other elements of the Operational Scenario to the Actors or the System under Development.
- Activation State: Used to indicate the start of an operational scenario
- Termination State: Used to indicate the end of an operational scenario

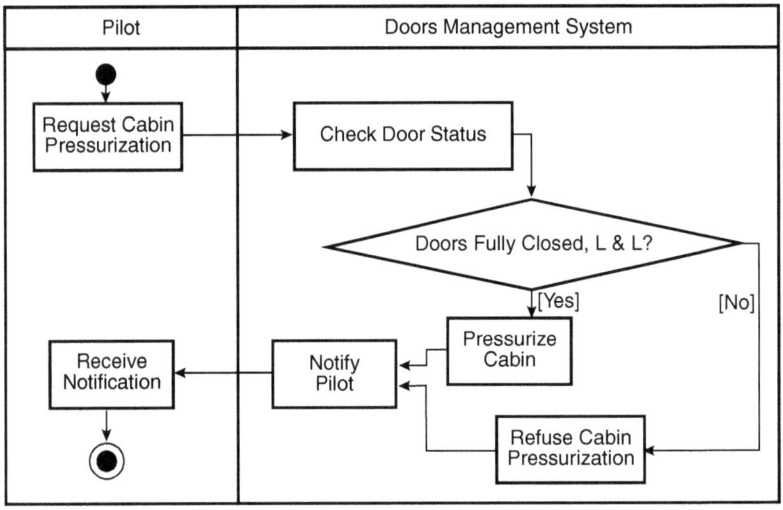

Fig. 3.3 SysML Activity Diagram for the representation of an operational scenario of the DMS example

- Activities: Used to represent one function or operation within an operational scenario
- Decision Nodes: Are used when several subsequent steps are possible within the operational scenario, depending on whether certain conditions are given or not. With the use of decision nodes it is possible to indicate alternative paths through the operational scenario.
- Transitions: They connect functions, operations and decision nodes and are used to indicate the functional and operational flow (the order in which functions/operations are invoked).

Sequence Diagrams

Sequence Diagrams are an alternative way to represent functional requirements and operational scenarios. The principal elements of Sequence Diagrams when used for Requirements are as followed:

- Condition Bar: Indicates the condition that needs to be fulfilled at the start of the scenario
- Lifeline: Represents the actors or the System under Development
- Message: Messages are sent from Actors to the System under Development
- Function: Represents a function of the System under Development
- Response Message: Response Messages are sent from the System under Development to the Actors

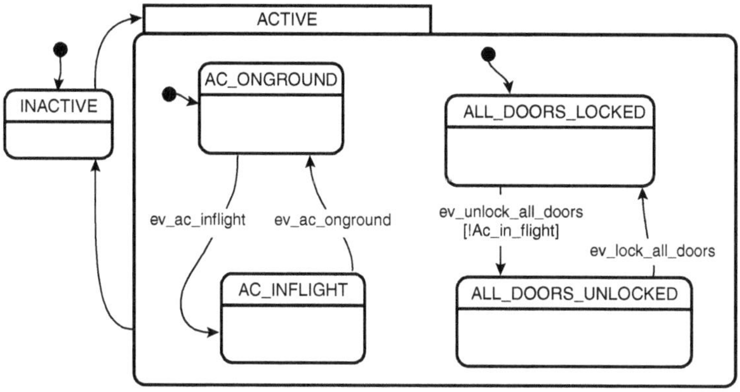

Fig. 3.4 SysML State Machine for the representation of the states/transitions of the DMS example

- Alternative Bar: Indicate Alternative paths within the operational scenario
- Reference Bar: Refer to another Sequence Diagram that represents a different scenario.

For the SysML-based graphical RSL we do not impose the use of Activity or Sequence Diagrams for the representation of operational scenarios. Both have their advantages or disadvantages. From real life experience we have seen that Sequence Diagrams tend to be easier to describe and to read than Activity Diagrams when no or very few alternative paths have to be indicated. They quickly seem to become unreadable when an operational scenario includes many alternative paths. One way to handle such cases is to use reference bars in order to describe alternative paths by additional sequence diagrams.

3.3.4.4 SysML State-Machine Diagrams for the Representation of Requirements

Like Activity Diagrams, State Machine Diagrams are a class of behavioural models. But the purpose of State Machine Diagrams is to focus on the states of the SuD and the transitions from one to another, whereas Activity Diagrams picture the operational scenarios of using the SuD. Both diagrams are of course closely related. Figure 3.4 illustrates the use of a State Machine to describe the behaviour of the SuD in the example of the DMS. Events (such as "ev_unlock_all_doors", triggering a transition in the State Machine Diagram) have corresponding actions in operational scenarios (such as "Passenger tries to unlock the door" in the Activity Diagram).

SysML State Machine Diagrams are not always necessary when representing requirements graphically. They are mainly useful for expressing requirements which refer to variables or states of the SuD, for instance: the aircraft shall not be "in flight" and with "all doors unlocked" at the same time. In the DMS example shown in Fig. 3.4, note that a guard "!Ac_in_flight" (not in flight) indeed prevents the doors

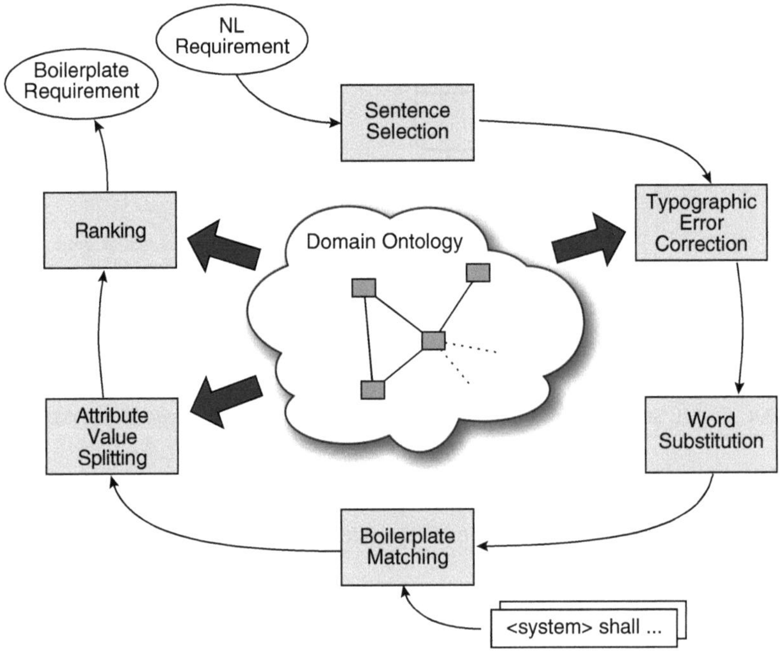

Fig. 3.5 Six-step semi-automated conversion process

from becoming unlocked when the aircraft is in flight. But it also makes obvious that apparently no protection has been put to avoid a transition from "AC_ONGROUND" to "AC_INFLIGHT" when the doors are unlocked. Here we may have to add another guard or make sure that the triggering event "ev_ac_inflight" cannot be sent whenever the doors are unlocked, maybe by writing a new requirement that was missing before. This exemplifies the use of State Machine Diagrams within the SysML-based graphical RSL to improve the completeness of the requirements.

3.3.5 Conversion Techniques

3.3.5.1 From Natural Language to Boilerplates

Figure 3.5 shows the semi-automated process for converting natural language requirements into boilerplate requirements as implemented by DODT.

The process was first described in [61]. The process inputs are choices made by the requirements engineer, a set of boilerplates, and the domain ontology. The goal of the process is automate as much of the mechanical transformations for requirements that are already similar to boilerplates requirements. For the requirement "The passenger shall be able to open the door." and the boilerplate ⟨user⟩ shall be able to ⟨action⟩ it is immediately clear to a human that the original requirement can be trivially transformed into a boilerplate requirement by assigning "The passenger" to the attribute ⟨user⟩ and "open the door" to ⟨action⟩. DODT tries to automatically translate such requirements. This way the requirements engineer can focus on the more complicated requirements which cannot be handled by a tool.

Sentence Selection

A boilerplate requirement represents a single sentence. Natural requirements which consist of several sentences are split into several requirements. The tool uses the GATE sentence splitter to detect sentence beginnings and ends. If more than one sentence is detected, the requirements engineer is asked to select the sentence to transform into the boilerplates representation. In the evaluation we found that sometimes only the first sentence is an actual requirement and that the following sentences are comments, explanations or rationales. Such information should be clearly separated from the requirement statement, e.g., using a requirement attribute.

Typographic Error Correction

To correct typographic errors in the natural language requirement, the domain ontology concepts are used as a dictionary. A POS (part-of-speech) tagger is used to detect nouns in the original requirement. For any sequence of nouns (detected by a POS tagger) which is not contained in the domain ontology, the tool computes the Damerau-Levenshtein distance (the number of single letter edits to get from one expression to another) to each of the domain ontology entities. If there are domain ontology entities similar to the noun sequence, i.e., below a threshold for the distance, the requirements engineer is asked to select from a list of possible corrections.

Word Substitution

The goal of this step is to transform a requirement in such a way that matching boilerplates can be found in the next step. User-defined substitution rules are applied to replace expressions of the original requirement with synonymous fixed syntax elements. The substitutions we used in our evaluation are to replace occurrences of "must", "should" and "will" with the word "shall" in order to match our set of boilerplates based upon "shall" and to replace "in case that" with "if". The actual transformations that should be applied here strongly depend on the original natural language requirements and on the used boilerplates. Therefore the tool allows the requirements engineer to configure the substitution rules. The default substitution rules of DODT are listed in the following table:

From	To
Must	Shall
Will	Shall
Should	Shall
In case that	If
For at least	For a period of at least
Shall {be no more than, be no greater than, not be higher than}	Shall be less than
Probability for	Probability of

Boilerplate Matching

For each boilerplate, the tool searches all possible ways the boilerplate matches the requirement. For the matching only fixed syntax elements are considered. They match if identical words or punctuation exist in the requirement. The text between fixed syntax elements will be assigned to the attributes in the next step. A boilerplate can match at several starting positions in a requirement and also in different ways for the same starting position, e.g., for the boilerplate "if ⟨event⟩," each comma following one occurrence of "if" will be considered a different match. After all matches have been found, conflicts between matches are computed. Two matches conflict if they span overlapping parts of the requirement. Then all combinations of match sets that are maximal – in the sense that no further non-conflicting matches can be added – are computed. For this we use a simple enumeration of conflicting matches as the number of conflicting boilerplate matches is low in a requirement (the maximum was 10 in our evaluation using the Doors Managements System requirements).

Attribute Value Splitting

In this step, text occurring between fixed syntax elements is assigned to boilerplate attributes. When there are two or more adjacent attributes, a decision has to be made about where the text should be split. The tool iterates over all possible ways (word boundaries) to split the text and computes ratings. The rating is composed of a syntactic part and a semantic part. The syntactic part checks if the splitting agrees with the syntax tree of the requirement by increasing the rating if the children of a syntax tree node are assigned to the same attribute. The semantic part checks if the expressions assigned to an attribute are semantically sound by using the domain ontology links from concepts to boilerplate attributes. Cardinals semantically fit to the ⟨quantity⟩ attribute. The splitting with the best rating is then used.

Ranking

The Boilerplate Matching step typically generates several solutions on how to map the original natural language requirement to the boilerplate representation. It is the responsibility of the requirements engineer to choose one of the proposed solutions. However, in order to support the requirements engineer with this decision, the tool computes a ranking of the solutions. This ordering is computed from the percentage of the requirement text that could be transformed, the number of words that could be matched with fixed syntax elements and the quality of the matching between boilerplate attributes and attribute values. ⟨user⟩ would be considered a better match than ⟨system⟩ if the corresponding attribute value contains "passenger" (linked to ⟨user⟩ in the domain ontology).

Final Adjustments

After the six steps described above the requirements engineer can adjust the requirement before saving it. Parts of the requirement which could not be transformed are displayed by the tool. Step Boilerplate Matching creates a maximal matching which in some cases splits a requirement into too many boilerplates. This is a deliberate design choice as removing an extra boilerplate requires much less effort than adding a missing one. The tool allows the requirements engineer to remove a boilerplate with a single click and will transfer the text of the boilerplate to neighbor boilerplates automatically. Refinement links are automatically created to connect the original natural language requirement with the new one in order to provide requirements traceability.

Caveats

As mentioned above, the semi-automated process tries to find the maximum number of matching boilerplates. This is done on purpose as the removal is very easy (simply click the "X" icon in the top-right corner of a boilerplate to remove it). Especially the boilerplates for the prepositions (for ⟨entity⟩, etc.) are often matched in a nonsensical way. Those boilerplates should only be used if the preposition applies to the main verb of the requirement, not to some other part. Example: The DMS shall be able to control two to ten doors. For this requirement DODT suggest the boilerplate representation "⟨The DMS⟩ shall be able to ⟨control two⟩ to ⟨ten doors⟩". Removing the second boilerplate and agreeing to the question whether to preserve the attribute value will result in the reasonable translation "⟨The DMS⟩ shall be able to ⟨control two to ten doors⟩".

3.3.5.2 From Boilerplates to Patterns

The requirement engineer may specify patterns directly by entering the patterns in an editor. Even if this is an appropriate approach in many situations, the user needs support in the formalization of requirements that are specified in natural language or in boilerplates.

Table 3.3 Functional patterns

Functional patterns		
Key words	Occur	Pattern F1, F2
	Always	F3
	Hold(s)	Pattern F2
Hints	Two or more events/actions	F1
	One event/action + equation	F2
Related boilerplate categories	Function(needs break down)	Pattern category "functional"
	Safety(for unwanted events)	Pattern F1 "does not occur"
	Timing/sequence	Pattern F1
	Timing/delay	Pattern F1
	Timing/duration	Pattern category "functional"

In this section, we described how information that is extracted from natural language and Boilerplate requirements can be used to suggest patterns to the user that are considered suitable for the formalization of a given requirement. The focus is on definitions that can be implemented in a tool. The patterns that are available in the Pattern-based RSL are structured in categories like functional, real-time, safety, and so on. For each of the categories is the following indicators are presented:

- Key words: Words that occur in the requirement and indicate that a pattern of the category should be suggested. This mechanism can be used both for natural language and Boilerplate requirements.
- Hints: General indicators for a pattern or a pattern category. These indicators are derived from ontology information that is stored with the Boilerplate requirements in the most cases.
- Related Boilerplate categories: The user may assign a category to Boilerplates during the specification process. The category field may be filled with any free text, but there are some standard categories that can be used to suggest patterns of a corresponding pattern category.

Table 3.3 displays an example how functional Patterns are processes:

3.4 Requirements Analysis

The main reason to justify the effort in formalizing requirements, a defined traceability to system artifact and complex requirements management cycles is the increased analyzability. It is the aim to find errors in the specification as early as possible to prevent malfunctioning implementations. In this section the typical analysis goals are presented as an consolidation from requirements from safety standards from different domains. This analysis goals are represented by the terms *consistency*, *completeness* and *correctness*. A short definition and excerpts from the different analyzed standards are presented in Sect. 3.4.1. Two basic approaches are

Table 3.4 Standards used for CCC criteria definition

Aeronautics	ARP 4754, DO 178B, DO 254, ABD 0100
Space	ECSS-E-ST-10C, ECSS-E-ST-40C, ECSS-Q-ST-80C
Automotive	ISO 26262, ISO/TS 16949, ISO/IEC 15504
Railway	EN 50126, EN 50129, EN 50159, EN 50128

used in CESAR for analyzing requirements: An ontology-based approach operating on domain ontologies and the contract-based approach using the pattern-based RSL. These approaches are introduced in Sect. 3.4.2. The actual analyzes are described in Sect. 3.4.3.

3.4.1 Consistency, Completenesss and Correctness

The notions of Completeness, Correctness and Consistency (CCC) based on the industrial needs shall be defined. This CCC criteria have been elicited from various standards from aeronautics, space, automotive, railways domains (Table 3.4):

3.4.1.1 Completeness

Short Definition

The completeness for a set of requirements about a system is the quality of this set of requirements which indicates whether every needs and constraints for the system have been identified and taken into account.

Two complementary approaches can be taken to study completeness:

- Make sure that all categories of requirements (by categories we understand environment requirements, functional requirements, maintenance requirements, ...) are covered. This is represented below by the use of viewpoints.
- Make sure that all sources of a set of requirements are covered at lower level of a system. We can consider that the set of requirements is generated from two sources: the first one external to the system to be specified itself (stakeholder needs, higher level requirements), and the second one being all additional input brought to the system (assumptions based on lessons learnt, internal development constraints, ...).

Figure 3.6 proposes an overall vision of the notion of completeness for requirements.

Viewing this picture and seeing the word "all" in every bubble, we can identify the raised problem of completeness as: which are the means we can use to how can we guarantee that we did indeed identify all requirements?

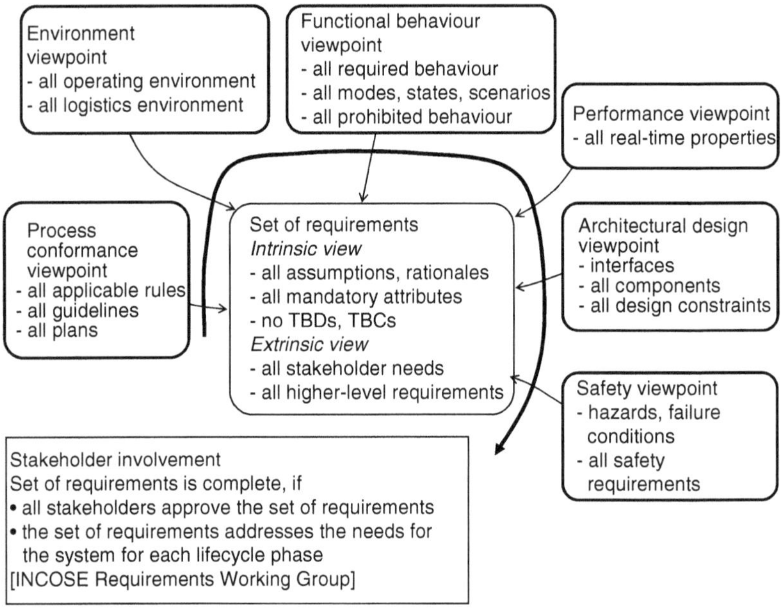

Fig. 3.6 Overview of requirement completeness

This picture helps since it shows some kind of refinement (we started from a general completeness of a specification or a set of requirements to reach completeness per categories). However, it is not the final answer as we still need to justify that we have identified all requirements/conditions in these categories.

Some CESAR technical items (processes, methods and tools) are able to help requirements engineers to improve the completeness of their specification. These technical items will be described in following sections.

3.4.1.2 Correctness

Short Definition

The correctness for a set of requirements about a system is the quality of this set of requirements which indicates whether the requirements are in accordance with the elicited needs and constraints of the system, and also whether the set of requirements is in accordance with quality and formalization rules for requirements.

Figure 3.7 proposes an overall vision of the refinement of the notion of correctness of requirements. This classification can be considered as arbitrary and might thus not be the best suited. However, the main intent is to group the requirements of the several standards.

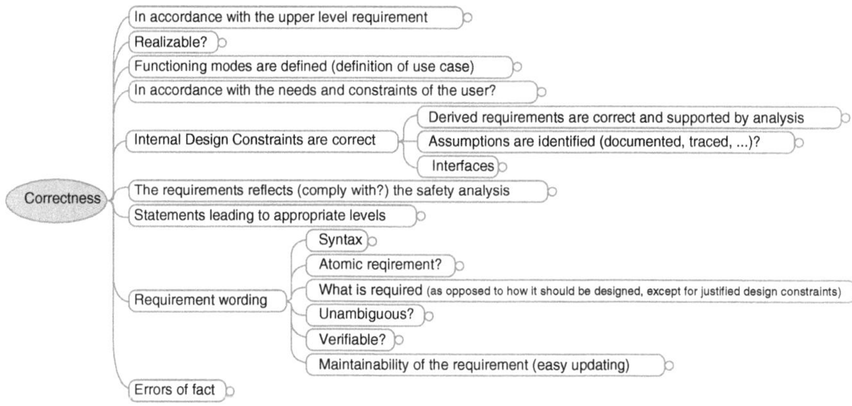

Fig. 3.7 Overview of correctness criteria

Some CESAR technical items (processes, methods and tools) are able to help requirements engineers to improve the correctness of their specification. These technical items will be described in following sections.

3.4.1.3 Consistency

Short Definition

The consistency for a set of requirements about a system is the quality of this set of requirements which indicates whether the requirements are necessary and pertinent, do not conflict with other requirements of the project and allow clear understanding of the system to produce.

The notion of consistency of a set of requirements can be refined by following criteria:

- Requirement does not conflict with other and with higher requirements? The aim of this criterion is to check if a requirement is clear and correct when compared to other requirements. That means that the requirements do not contradict each other.
- Redundancy of requirements?
- Necessary? The aim of this criterion is to check that the requirement is necessary for the set of requirement to be complete. In other terms, is this a combination of other requirements or it is redundant with another one? Another case when this requirement might not be necessary is if it does not bring any added value to the set of requirements.
- Combination? The aim of this criterion is to avoid having multiple requirements with the same purpose. Requirements expressed differently can have the same

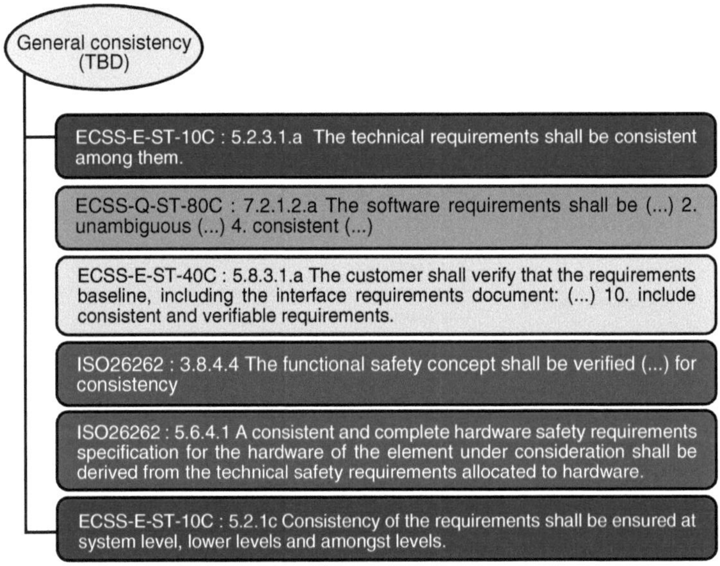

Fig. 3.8 General consistency mapping with standards

purpose. In this case, it is better to combine them into one requirement to avoid redundancy and complexity. On the other hand, some requirements can have sense only considered together. These requirements must also be combined to avoid incompleteness.

- General consistency check (Fig. 3.8)?

3.4.1.4 Compatibility

Short Definition

The compatibility for a set of requirements about a system is the quality of this set of requirements which indicates whether the requirements are aligned with project constraints as physical architecture or reused constraints.

One last notion about quality of a set of requirements is the compatibility between requirements and project constraints as physical architecture (Fig. 3.9).

The aim of the compatibility criterion is to ensure that compatibility exists between systems or subsystems.

Compatibility or interoperability is the capability of a product or system, whose interfaces are fully disclosed, to interact and function together with other products or systems, without any access or implementation restrictions.

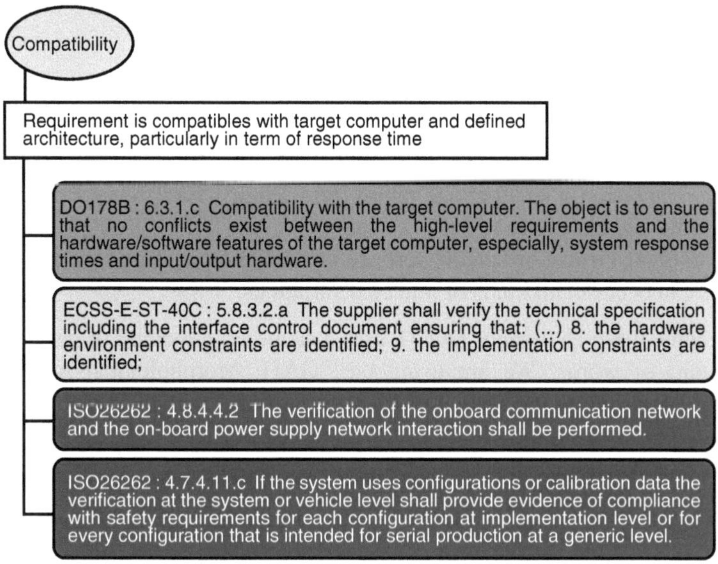

Fig. 3.9 Compatibility mapping with standards

These criteria ensure:

- Interoperability between Software and Hardware Components
- Interoperability of Communication networks
- Interoperability of Power supply networks
- Hardware and Software environment and implementation constraints are defined.

3.4.2 Approaches for Requirements Analysis

Two different ways to approach the CCC criterias have been developed within the CESAR project. The ontology-based approach focuses on semi-formal analyzes based on a domain ontology, the contract-based approach describes a way to specify the requirements of a system in assumptions and promises pairs.

3.4.2.1 Ontology-Based Approach

The basic idea of ontology-based analysis is to compare the information found in requirements with the domain ontology's information. Generally the ontology is expected to be the correct one because it should have been validate by domain experts and should be more stable than a set of requirements. However, in some cases it still might be necessary to update the ontology.

Requirement Statement Analysis

In order to analyze requirements further we must be able to to map the words of the requirements to the concepts of the domain ontology. This is not as trivial as it may sound because simple string matching will perform poorly:

- Nouns can occur in singular and plural form. English has irregular plural forms, e.g., child and children.
- Concepts can consist of several words (compound nouns), e.g., "armed position".
- Verbs can occur in all kinds of inflected variations.

To overcome these challenges DODT uses a stemmer, a method to get the lexical root (stem) of words. We are using the "GATE Morphological Analyzer" as stemmer. Words not found in the ontology are further classified by a part-of-speech (POS) tagger. Part-of-speech is a classification of words into one of several categories of syntactical role in a sentence: noun, several kinds of verbs, adjective, preposition, determiner, etc.

Having the mapping of words to concepts we can run several kind of analyses. They are described in Sect. 3.4.3.1.

3.4.2.2 Contract-Based Approach for Requirements

The problem with the SPEEDS approach to contract-based design [129] is that the design flow must be adapted to the use of contracts. This has the advantage of putting in forefront both the definition of contracts as part of requirements capture, and then the management of contracts and their verification at design phase. This, however, also impairs acceptance as industrial users are very careful at drastically changing their design flow.

Some Important Aspects of Requirements

By having a wider body of partners, both industrial and academic, CESAR SP2 broadens the scope of studies on requirements capture and management. Consequently, it reduces the risk of bias when performing research on this topic. From the preliminary work performed in CESAR regarding requirements, the following broad aspects can be identified regarding them:

- *Domain specific aspects*: the issue here is to keep control over the dictionary and wording used in requirements capture. This is achieved by making use of *ontologies* or by restricting the language or notations used, e.g., with the help of *boiler plates*. Handling these aspects well is key to PLM (Product Line Management) and other data base related activities where system parts are managed. While this helps keeping control over the complexity of requirements documents, it generally gives little support for automation of or event assistance for requirements management and then verification at design phase.

- *Behavioural aspects*: this term should be taken in a broad sense. By this we mean behaviours contributing to the functions of the system, behaviours constitutive of its performance (non-functional), as well as technical characteristics of its hardware (e.g., the IMA or AUTOSAR platform used in the design).
- *Stakeholders' aspects*: in large systems design, requirements are defined and then taken care of by a large number of different teams, possibly in different entities or companies. It is thus important to trace who has produced a given requirement. This is generally part of requirements capture. There are, however, issues related to stakeholders which do not seem to have attracted the attention they deserve. It is understood that requirements apply to the system under design. But this is not sufficient for systems designed under large and widely distributed supplier chains. It must be clear as part of the requirement to which stakeholder a requirement applies. Frequently, part of a requirement acts as an assumption under which the rest of the requirement should be satisfied. If stakeholders' responsibilities are put as part of requirements, then limits to this responsibility should also be made clear.

Requirements Should Better Identify Roles and Responsibilities of Stakeholders

This is a first step toward having requirements (or a dedicated subset of them) serving as a technical support for sub-contracts in the supplier chain. Recommendations toward enabling this are listed now:

- *RSL*: nothing specific is required. It might, however, be advisable giving provision for specifying roles and responsibilities as part of ontologies; particularly so if notions are domain specific.
- *RMM*: a *stakeholder* attribute is provisioned for requirements. This attribute puts a relation between a requirement and a stakeholder. This generally means that the requirement is assigned to this stakeholder. This, however, needs to be refined, for some requirements in which boundary conditions play a role:
  - A requirement may comprise more than one *field;* for example, one field indicates the *obligation* and another field indicates environmental conditions under which the requirement holds (the *assumptions*); the latter may be split into several fields if needed.
  - Each field is given a *stakeholder* attribute, assigning the field to this stakeholder. In the simple case of two fields, obligations are assigned to a stakeholder (as usual) whereas assumptions are assigned to "the environment" to refer to an unspecified stakeholder.
  - In case some requirements are considered as technical support for legal sub-contracts, a special *contract-item* attribute may be useful. It is, however, not clear that this must be firm part of RMM.
  - Alternatively the same could be obtained by having, as part of RMM, a generic mechanism for enriching the meta-model with additional attributes. This generic mechanism would then also serve for other purposes.

- *Traceability links*: the key point about having roles and responsibilities better defined in RMM is that this would facilitate the checking of important attributes of traceability links, such as "verify", "is consistent", "refines", etc. when formal approaches are considered, this approach would better secure separate design of different sub-systems, as explained in the SPEEDS White Paper [144] on SPEEDS approach.

Consequences for Requirements Management and Traceability Links

In Sect. 3.5.1, several *traceability relations* are listed and discussed. Some of them are central in distributed and concurrent development in multi-tier supplier chains. We explain this for a few of them:

- *Satisfies, Verifies*: these two links express that a design conforms a (set of) requirement(s), either because V&V actions have substantiated it, or because formal verification was performed, or because the designer asserted so. The change is that conformance has to be checked modulo assumptions: the design should meet its obligations provided that boundary conditions are met by the environment.
- *Refines*: indicates that a new set of requirements can safely be substituted for the original set of requirements (the former *refines* the latter) in that every design conforming the former also conforms the latter. *Refinement strengthens obligations but relaxes assumptions.* This is where the new features become important.

Of course, while doing this, it is understood that not specifying roles and assumptions for a requirement or a set thereof amounts to saying that the whole requirement is put to an obligation of the stakeholder it has been assigned to, under any environmental conditions. Following such a discipline would reduce ambiguities regarding "implicit" assumptions under which designs are sometimes performed without the integrator being aware of it. Following such a discipline is possible regardless of the availability of a formal support consisting of engines that can validate/invalidate the above links. Of course, further benefit can be obtained by using such formal approaches and tools. Advances of research during the last decade on the topic of *interface theories* and *contract theories* will soon enable this.

3.4.3 Analyses

3.4.3.1 Ontology-Based CCC Analyses

Based on the definitions in Sect. 3.4.2.1, we define the following seven analyses. They are based on work originally by Kaiya and Saeki [97], which was later extended by Stlhane and Omoronyia [148]. We use the following definitions for our analyses:

$$
\begin{aligned}
R &\ldots \text{Set of requirements}\\
O = \langle C, L\rangle &\ldots \text{Ontology}\\
C &\ldots \text{Set of ontology concepts}\\
C_o \subseteq C &\ldots \text{Set of concepts marked as being obsolete}\\
T &\ldots \text{Set of ontology link types}\\
N &\ldots \text{Set of nouns occurring in the requirements but not in the ontology}\\
L \subseteq C \times C \times T &\ldots \text{Set of ontology links}\\
c_r : R \rightarrow \mathcal{P}(C) &\ldots \text{Maps a requirement to the concepts that occur in it}\\
n_r : R \rightarrow \mathcal{P}(\mathbb{R}) &\ldots \text{Maps a requirement to the numbers that occur in it}\\
\delta_p : C \times C \rightarrow \mathbb{N} &\ldots \text{Shortest path between two ontology concepts}
\end{aligned}
$$

For equivalence axioms we define a transitive closure for a set of concepts:

$$
\begin{aligned}
cl_{eq}(cs) &= cs \cup \{c \mid (c', c, \text{equivalence}) \in L \wedge c' \in cl_{eq}(cs)\}\\
c_{req}(r) &= cl_{eq}(c_r(r))
\end{aligned}
$$

Further we define the set of all concepts used in the requirements including equivalent concepts:

$$
C_r = \bigcup_{r \in R} cl_{eq}(c_r(r))
$$

- *Completeness*: The completeness analysis tries to detect missing requirements. It analyses the set of used concepts in the requirements and finds links to other concepts which should be defined in the requirement specification as well but are not. We define the set of concepts that can be reached using one link by the concepts used in the requirements:

  $$
  C_{com} = \{c \mid (c', c, t) \in L \wedge c' \in C_r \wedge t \in T \setminus \{\text{subclass-of, contradict}\}\}
  $$

  The completeness analysis checks if all concepts in this set are covered by the requirements.

  Example: If we have the domain knowledge that a door needs to have a door sensor, we will link them with a relation in the domain ontology. Now if the requirement set contains a requirement containing door but none containing door sensor, this lack be found by the completeness analysis.

We define the metric m_{com} to measure the completeness of a requirements set. A value of one means the requirements are complete, a value approaching zero means almost all concepts have links to other uncovered concepts.

$$m_{com} = \frac{|C_r|}{|C_r \cup C_{com}|}$$

- *Inconsistency*: The inconsistency analysis detects contradictions between two requirements. Two kinds of inconsistencies are detected:
  - Requirements using concepts which are explicitly linked by a contradiction axiom. Some concepts, most of them related to quality factors, can be said to be inconsistent in the sense that a trade-off needs to be found between them, e.g., throughput and cost. The analysis simple finds pairs of requirements using concepts that are linked by a contradict axiom. This does not necessarily mean that an error exists but it can be an indication that two requirements need to be inspected closely. The set of requirements linked by a contradict axiom is defined by:

$$R_{inc} = \{r \mid \exists\, r', c, c : c \in c_r(r) \wedge c' \in c_r(r') \wedge (c, c', \text{contradict}) \in L\}$$

  - Requirements which use the same set of concepts but a single differing quantity. This kind of inconsistency analysis tries to find requirements that are numerically contradicting. We look for pairs of requirements that use the same concepts and each one exactly one number which is different from the other requirement one's. The set of requirements having single differing quantities is defined by:

$$R_{inc2} = \{r \mid \exists\, r' : c_{req}(r) = c_{req}(r') \wedge dq(n_r(r), n_r(r'))\}$$
$$dq(ns, ns') = |ns| = 1 \wedge |ns'| = 1 \wedge ns \neq ns'$$

  Example: This analysis would find an inconsistency between the requirements "The aircraft shall have 8 doors." and "The aircraft shall have 12 doors."

  We define the metric m_{inc} to measure the (combined) inconsistency in a requirements set. A value of zero means no requirement is inconsistent, a value of one means all requirements are inconsistent.

$$m_{inc} = \frac{|R_{inc} \cup R_{inc2}|}{|R|}$$

- *Ambiguity*: The ambiguity analysis detects requirements using terms that could be replaced with more concrete sub-classes of a concept. The intent is to find places where invalid over-generalizations are used. The set of requirements containing ambiguous concepts is defined by:

$$C_{amb} = \{c \mid c \in C_r \wedge \exists\, c' : (c', c, \text{subclass-of}) \in L\}$$

Example: If the domain ontology has two concepts cargo door and passenger door defined as subclasses of the concept door, then the analysis will detect usages of plain door and ask whether one of the two subclasses was meant instead.

We define the metric m_{amb} to measure the ambiguity in a requirements set. A value of zero means no concept is ambiguous, a value of one means all concepts are ambiguous.

$$m_{amb} = \frac{|C_{amb}|}{|C_r|}$$

- *Noise*: The noise analysis finds usages of nouns which are not defined in the domain ontology. Such words will generally fall into one of the following categories:

1. Outside of domain: The requirement states something which has no relevance to the domain and system to be built. The requirement should be rephrased.
2. Common word: The word meaning is commonly known and there is no ambiguity about its meaning among stakeholders.
3. Incorrect word: The word is incorrectly used but the correct one exists in the domain ontology. In the simplest case it is a typographic error.
4. Missing in domain ontology: The domain ontology is missing a vital term which is required for the specification. In this case the domain ontology needs to be enhanced.

DODT supports ignoring words known to WordNet [173] (to avoid false positives due to case 2).

We define the metric m_{noi} to measure the noise in a requirements set. A value of zero means there are no noisy nouns, a value of one means there are only noisy nouns.

$$m_{noi} = \frac{|N|}{|N| + |C_r|}$$

- *Opacity*: The opacity detects the usage of unrelated concepts in a requirement. We define the relatedness of two concepts by their distance in the ontology, the minimum number of links needed to travel from one concept to another one. The intention of this analysis is to catch errors due to unintended usage of an incorrect word. In our experiments the limit of four to consider two concepts to be unrelated works well. The set of requirements containing such opacity is defined by:

$$R_{opa} = \{r \mid \exists\, c, c' : \{c, c'\} \subseteq c_r(r) \wedge \delta_p(c, c') > 4\}$$

Example: Let us consider the (assumed) incorrect requirement "The Doors Management System shall be able to measure the flight altitude." If the number of links in an aircraft ontology that contains the concepts Doors Management System and flight altitude is more than four, the two concepts will be reported as unrelated by this analysis.

We define the metric m_{opa} to measure the opacity of a requirements set. A value of zero means there are no requirements containing opacity, a value of one means there are only opaque requirements.

$$m_{opa} = \frac{|R_{opa}|}{|R|}$$

- *Redundancy*: The redundancy analysis detects redundant requirements by finding requirements that use the same set of ontology concepts and numbers. This analysis is useful two find duplicate requirements, especially when you have thousands of requirements. By comparing concept sets rather than strings the analysis works even in the presence of minor typing differences or reordered sentences. The set of redundant requirements is defined by:

$$R_{red} = \{r \mid \exists\, r' : c_{req}(r) = c_{req}(r') \wedge n_r(r) = n_r(r')\}$$

We define the metric m_{red} to measure the redundancy in a requirements set. A value of zero means there are no redundant requirements, a value of one means there are only redundant requirements.

$$m_{red} = \frac{|R_{red}|}{|R|}$$

- *Obsoleteness*: DODT supports marking certain domain ontology concepts as being obsolete which means they should not be used in requirements. The obsoleteness analysis detects those requirements. An important use case is when a company uses two or more terms for one phenomenon (synonyms). In order to keep the requirements consistent and to avoid confusion about whether two names are refer to the same entity or not, one concept can be set to be obsolete and DODT is then able to automatically use the non-obsolete term in requirements. The set of requirements containing an obsolete concept is defined by:

$$R_{obs} = \{r \mid c_r(r) \cap C_o \neq \emptyset\}$$

We define the metric m_{red} to measure the obsoleteness in a requirements set. A value of zero means there are no requirements containing obsolete concepts, a value of one means that all requirements contain obsolete concepts.

$$m_{obs} = \frac{|C_{obs}|}{|C_r|}$$

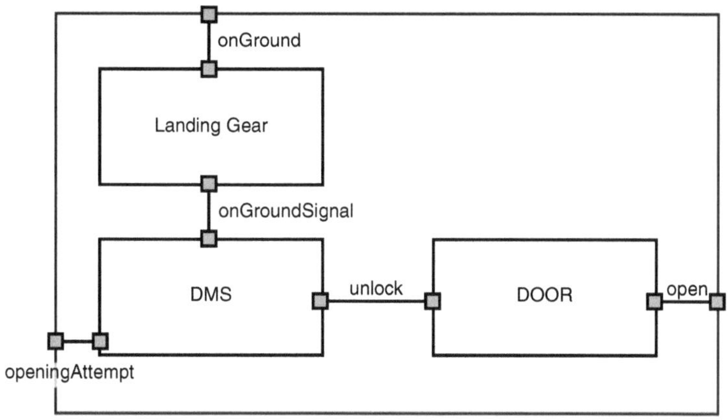

Fig. 3.10 Architecture of the doors management system

3.4.3.2 Functional Completeness

During the development of a system requirements get split up and assigned to components and subcomponents. It is important to know if the specified behavior of the requirements assigned to the sub-components cover the behavior that is intended by the top level requirement. This is a typical question of completeness of requirements.

Failures violating the completeness of the specification typically result from two sources:

- It is not clear which sub-component takes over which parts of the behavior described by the top level requirement. This results in wrong interfaces and wrong behavior of the subcomponents that may not interact in the desired way.
- There are different assumptions on the environment and thus the components may not cover the same operational situations, environmental conditions or expected inputs.

Contracts allow to state the assumptions needed to prove a correct integration of sub-components while Patterns enable to state the assertions in a formal but still readable way.

In the following example we demonstrate how the entailment analysis can be used to detect a design failure caused by missing environmental information:

The system depicted in Fig. 3.10 is a simplified doors management system which shall ensure that the doors of an aircraft may only be opened if the plane is on ground. The stated requirements are listed in Table 3.5.

Running the entailment analysis tool, the stated requirements are not dominating the Top-Level Requirements.

The result from the counterexample is depicted in Fig. 3.11.

Table 3.5 Pattern requirements of the system with refinement links

Component	Requirement
Top	`whenever openingAttempt under (onGround==1) occurs open occurs during [0 ms,200 ms].`
Top	`whenever open occurs onGround holds during[0 ms, 0 ms].`
Dms	`whenever openingAttempt under (onGroundSignal) occurs unlock occurs during[0 ms,100 ms].`
Dms	`whenever start occurs unlock does not occur during [0 ms, openingAttempt[.` `whenever unlock occurs unlock does not occurs during [0 ms, openingAttempt[.`
Dms	`whenever openingAttempt under (!onGroundSignal) occurs unlock does not occurs during[0 ms,openingAttempt[.`
Door	`whenever unlock occurs open occurs during [0 ms,100 ms].`
Door	`whenever start occurs open does not occur during [0 ms, unlock[.` `whenever open occurs open does not occur during [0 ms, unlock].`
LandingGear	`always(onGround== onGroundSignal).`

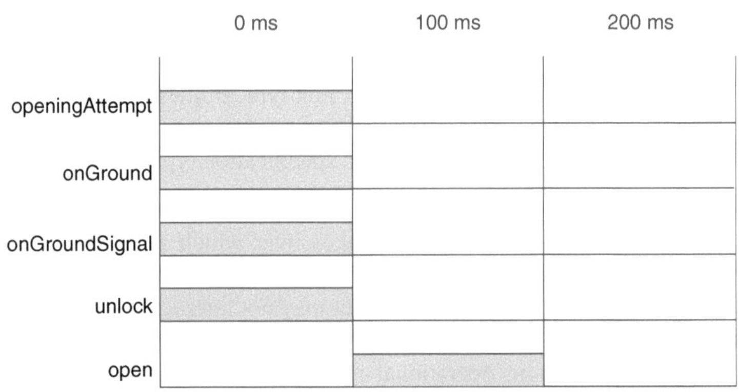

Fig. 3.11 Trace of the erroneous behaviour

At the beginning of the trace the plane is on ground and an opening Attempt occurs. The landing gear indicates the status with a onGroundSignal. The DMS sends an unlock command to the door. At the same time the plane takes off and the door will be opened while the plane is in the air. This is not a permitted operation. The problem is the capability to detect the ground status reliably. Since it is always possible that the OpeningAttempt occurs shortly before take-off the detection has to be improved. The change to fix the problem is an extension of the landing gear system to detect if the gear is in an expanding state. If the opening attempt occurs while the landing gear is already expanding the attempt shall not result in an unlock command.

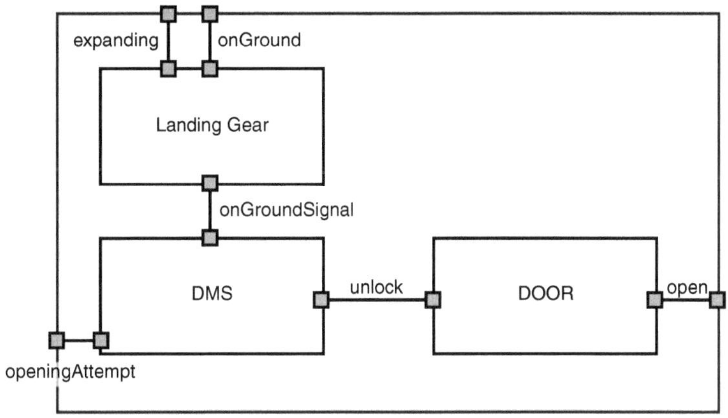

Fig. 3.12 Changed architecture with added expanding port

Table 3.6 Changed requirements to specify behavior of expanding port

Component	Requirement
Top	`whenever openingAttempt under ((onGround==1) && (expanding==0))occurs open occurs during [0 ms,200 ms].`
LandingGear	`always((onGround && !expanding)==onGroundSignal).`

The changed architecture is depicted in Fig. 3.12.

Furthermore, some requirements of the DMS need to be changed to correctly specify the new behavior. These requirements are depicted in Table 3.6.

Nevertheless, an additional run of the entailment analysis fails again, since there was an implicit assumption, that the plane will reside for a longer time (longer than 200 ms) in the expanding state. The approach requires all these assumptions to be stated properly, since implementation highly depend upon this information. After adding the assumption (see Table 3.7), the entailment of the requirements is given.

3.4.3.3 Functional Consistency Analysis

This analysis identifies inconsistent requirements stated in the pattern based RSL (see Sect. 3.3.3). There are two kind of inconsistencies, syntactical inconsistency and semantical inconsistency.

Syntactical consistency criteria:

- Are the names of events and continuous variables consistently used?
- Are all intervals consistent? When intervals are specified by absolute times, it is checked, whether the start is before the end. Example: `whenever trigger occurs action occurs during [7 ms, 5 ms]`

Table 3.7 Requirements with added assumption

<table>
<tr><th>Component</th><th>Requirement</th></tr>
<tr><td>Top</td><td>Assumption:
whenever tr(expanding) occurs ((expanding==1)
&&(onGround==1)) holds during [0 ms, 100 ms].
always((expanding==1) -> (in[tr(onGround),fs(onGround)])
|| (in[start, fs(onGround)])
Promise:
whenever openingAttempt under ((onGround==1) &&
(expanding==0))occurs open occurs during [0 ms,200 ms].</td></tr>
<tr><td>Top</td><td>Assumption:
whenever tr(expanding) occurs ((expanding==1)
&&(onGround==1)) holds during [0 ms, 100 ms].
always((expanding==1) -> (in[tr(onGround),fs(onGround)])
|| (in[start, fs(onGround)])
Promise:
whenever open occurs onGround holds during[0 ms, 0 ms].</td></tr>
</table>

The semantical check is performed for a set of requirements and detects failures in syntactical correct requirements.

Example:

`RQ1: whenever trigger occurs (variable == 50) holds during [0s, 5s]`

`RQ2: whenever trigger occurs (variable != 50) holds during [0s, 7s]`

RQ1 and RQ2 cannot be fulfilled at the same time if the event trigger occurs. Beside this obvious specification problem, inconsistencies can be caused by multiple requirements in a transitive way. To be able to find also this complex inconsistencies a model checking approach is used for this analysis.

3.4.3.4 Safety Consistency and Completeness Analysis

Using functional requirements for early system verification has already been shown in the previous sections. Analyzing safety requirements with respect to consistency and completeness is also crucial for reducing costs and time-to-market. An example safety contract to formalize fault containment properties is the following:

- **Assumption**: `none of failure-sets f1, f2, f2, f3 occurs`
- **Guarantee**: `none of failure-sets f0 occurs`

The safety pattern used in the assumption and guarantee describes scenarios that are characterized by sets of failure-conditions, which are not allowed to occur. Informally the above contract specifies the required fault containment properties, it states that any combination of failures that do not involve (f1 AND f2) OR (f2 AND f3) will never lead to a situation where failure-condition f0 can occur. When a safety specification is formalized it is important to distinguish between the assumptions under which a safety concept has to hold and the promise what a component – that later will implement this specification – shall do to keep the system safe. This principle enables the supplier to build a system without having to communicate with the integrator on an informally ambiguous way. For example, a failure rate for failure modes on a component can only be met by an implementation if there is knowledge about the failure rates of propagated failure modes on the input Ports of the component. Same applies for argumentation not taking failure rates into account: The non existence of a failure mode on a port can only be shown under the assumption that only a known number of faults can occur at the same time.

In order to express the relationships between the failure modes and the counter measures, thus implementing a technical safety concept, a formalism is needed that allows the statement of the assumptions as well as the promises in a semantically well defined and unambiguous way. The language used in the examples below is based on patterns that consist of a fixed part and variable attributes. Only few patterns are needed to specify error propagation and counter measure functionality. To be able to express temporal dependencies between failure modes in patterns a special kind of mode variable has been introduced that can be used within any pattern associated with the respective component. These modes are a tool to simplify the specification language; they have no relation to implementable system modes. A good example is a state undetected in which a failure has already occurred but has not been detected yet. Since the failure conditions can be quite complex (value failure, timing failure, unexpected occurrence, etc.. . .) also the failure condition itself shall be expressed with the specification language.

We use the formal pattern-based RSL for expressing both assumptions and guarantees.

The approach allows specifying a safety concept in terms of failure modes, failure rates, their propagation, and the usage of counter measures expressed in assumptions and promises. The safety patterns of the pattern based RSL are used for this specification. This method allows verifying decomposition and integration of safety concepts. The safety concept can be seen as requirements on safety that do not want to force a special implementation but requires a defined behaviour regarding failure propagation. Typical requirements are the non-existence of a single-point-of-failure. In particular the safety-modes used for stating temporal properties between patterns do not have a direct relation to the implementation. On the other hand the safety specification can include already partial details about countermeasures like voting or validity checks to realize a required fault containment. Expressing such elements is in particular important for verifying if the solution that has been created by a supplier still fits into the overall safety concept. Countermeasures can be seen as a gateway between functional behaviour and safety argumentation. There are two main scenarios where completeness and consistency of a safety specification needs to be checked:

- On the one hand if the OEM refines a system to distribute the sub-parts to one or more suppliers he needs to prove that the refinement still satisfies the upper level safety goals.

- On the other hand a supplier can offer a solution (that could also have different assumptions as actually needed in the development process) that refines the OEMs view on the system. In this case it is important to prove that this externally developed component fit into the already existing component structure and the top-level safety goals are still satisfied.

The CESAR Meta Model implements the design-by-contract paradigm enabling modelling of complex heterogeneous systems and formal specification of requirements. Both together, the framework and its associated analysis methods, provide the instruments for a parallel development based on an early virtual integration of components. This virtual integration avoids problems and incompatibilities during the later integration phases, and hence speeds up development and reduces costs and time to market. A typical and repeating step in the design process of component-based development is the decomposition/refinement of components.

The Starting point for the component-based design is typically a set of requirements and the idea for an abstract high-level component, which should realize these requirements. Also a preliminary interface of the abstract component can be derived from the requirements. The methodology envisages that the requirements the component should satisfy are captured in terms of contracts CRi and applied to the rich component R modeled for the abstract component. In the actual decomposition step, both the high-level component and its requirements are broken down. This results in a set of sub-components, each intended to realize parts of the high-level requirements.

In the CESAR Meta Model, the decomposition of a rich component R also leads to a set of sub-components Si. But again, the decomposed requirements are formalized by contracts $CSij$ and applied to the respective rich components. This decomposition step is a fault-prone step, because deriving the correct requirements for the sub-components is very difficult. Also faults in this step are often detected very late in the development, primarily during the integration process, and cause a lot of rework. Thus the contract-based approach tackles this problem and provides analysis tools and techniques to perform a virtual integration already at this stage of development. Based on the contracts, in particular two kinds of analyses are part of the virtual integration:

- *Consistency Analysis*: This analysis verifies whether the assumptions and promises of a set of contracts allow an implementation that fulfills the requirements.
- *Completeness Analysis*: This kind of analysis composes the contracts of a set of interconnected components and then verifies whether this assembly is covers the contracts of their parent component.

In the case of completeness, one can say that the contracts of the sub-components imply the contracts of their parent component. Both analyses together enable the developer to ensure that the decomposition step was correct, in the sense that the defined sub-components will work together and satisfy the requirements of their parent component, provided that the sub-components satisfy their own contracts.

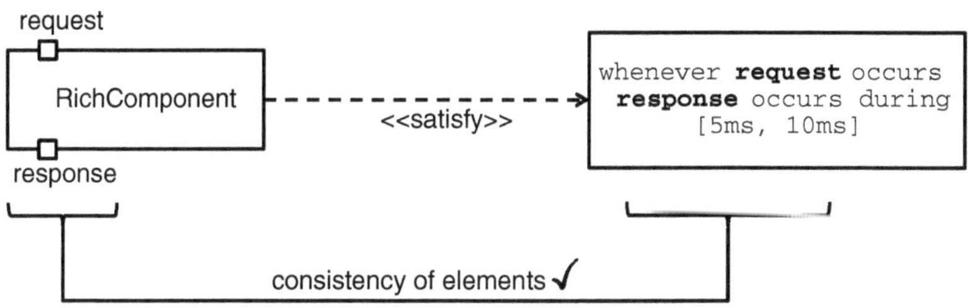

Fig. 3.13 Idea of the architectural consistency analysis

After the incremental verification and validation step, all derived sub-components are sufficiently characterized and can be designed independently. The developer now has the alternatives to iterate the decomposition step again, implement the sub-components or select an existing implementation from a library. The developer must ensure that any implementation that is provided, either newly developed or selected from a library, satisfies the sub-components contracts.

3.4.3.5 Architectural Consistency Analysis

The Architectural Consistency Analysis is used to check if the attributes like events of formalized requirements are consistent with the components that shall satisfy these requirements.

All data that refers to the architecture is extracted from the requirements. This includes ports, events and continuous variables. This data is compared to the architecture and mismatches are reported.

In the following example a requirement is expressed by a functional pattern:

`whenever request occurs response occurs during [5ms, 10ms]`

In this pattern the events "request" and "response" are referenced, so the corresponding component (see Fig. 3.13) is required to have ports/flows of kind event with these names.

This very simple analysis can be used prior to more advanced analysis to check if these analyses are possible.

Before the Architectural Consistency Analysis can be started all requirements in the current folder have to be assigned to components via Satisfy links (see Sect. 3.5.1). Requirements without Satisfy links are ignored by the analysis and a warning is added to the results.

3.4.3.6 HazId with Boilerplates

Introduction to Safety Analysis

Our present focus for safety analysis in CESAR is the requirements engineering phase. In this phase the most important participants are the customers with their domain knowledge and the developer or requirements engineer with his software development knowledge. It is not always practical to involve a safety expert at this stage of development. Thus, to guarantee that both development knowledge and usage knowledge is used to its fullest potential in the safety analysis, we need a method that is efficient and simple to use.

In CESAR we have considered four methods for safety analysis: Hazard and Operability analysis (HazOp), Preliminary Hazard Analysis (PHA), Failure Mode Effect Analysis (FMEA) and a combination of HazOp and FMEA originating with Motorola [161]. We have decided not to use HazOp since this requires a quite elaborate process. PHA (Preliminary Hazard Analysis) is at the other end of the scale in that it contains no prescribed process at all. FMEA (Failure Mode Effect Analysis) is a good choice since the failure modes offer some guidance for the safety analysis. However, we may not have any defined system components in the early stages and a traditional FMEA may thus be difficult to execute.

A large amount of companies already use some form of FMEA throughout the whole development process – e.g., in railroad systems development, where FMEA is the only method that is highly recommended for hazard identification [131]. For this reason, we have decided to focus on FMEA as the preferred method for safety analysis. Using FMEA will also allow us to use several ideas from the Motorola approach [161]. For the rest of this chapter we will use the term HazId (Hazard Identification), which in our case is an FMEA table, initiated with generic failure modes.

On Generic Failure Modes

A failure mode is a way in which a component can fail. A generic failure mode is a term used to capture a high-level description of ways to fail. Generic failure modes are used in several industrial domains. The failure modes will vary, depending on industry type. An example for mechanical industry can be found in [167], while the OREDA handbook [139] is an example from the offshore industry. In addition to variation over domain, there is also a large variation according to level of details. Some industrial areas and companies use one set of failure modes for all types of components, e.g. [79], while OREDA [139], SRC [150] and NUREG [110] use different sets of generic failure modes for different component types. Both OREDA and SRC include failure mode rate data – OREDA as failures per 106h (approximately 100 years) and SRC as the percentage of all reported failures for each generic failure mode. In addition, some of the publications on generic failure modes also split the failure modes into groups – e.g., "Catastrophic" or "Critical", "Degraded" and "Incipient" [3, 139]. The number of generic failure modes used varies. OREDA typically uses 10–12 failure modes for a component while for instance Airbus in many cases use only two [1, 135].

Most of what is published on generic failure modes concerns hardware – e.g. CPUs, sensors and actuators – while a few publications focus mainly on software – e.g., [56, 110], which are quite detailed and [70], which is at a high level.

In addition, [30, 76, 111] also describe generic software related failure modes at different levels of detail. In [63], ALA uses three quite generic failure modes while Airbus uses four generic functional failure modes.

It is a problem common to all the generic failure modes referred to above that they do not use the concept of time – e.g., failure modes like too early and too late. As will be seen in the section on generic failure modes for the DODT tool, it is easy to add these terms as failure modes for any generic failure mode set. In our opinion, they should be added to the failure modes for control systems and data networks.

Generic Failure Modes for DODT

The generic failure modes are intended as default failure modes. Several of the CESAR partners already have their own generic failure modes. It is thus important that the DODT tool allows the users to insert their own failure modes – generic or specific. The failure modes set should also be updated as we get more experience with using the DODT tool. We will, as a first approach, suggest two sets of generic failure modes for the DODT tool.

The first set is taken from [95], adding the generic failure mode "incorrect" from [111]. Thus, we have the following failure modes: Omission, Commission, Stuck, Incorrect, Too early and Too late.

The second set of generic failure modes for the DODT tool is taken from Airbus [1] and has only two generic failure modes: *Loss* (e.g., no output, no action) and *Erroneous* (e.g., wrong output, wrong action). As above, we should also add here one or more generic failure modes related to timing. To make the chosen terms more concrete and thus work better for non-experts, the failure modes should be specialized according to their use, e.g.:

- Sensors: No output, Wrong output
- Actuators: No action, Wrong action
- Computer system: Function not present when required present, either partly or in full – omission, function present when not required – commission, function incorrect, system reacting too late, annunciated loss of function – discovered by a control unit

HazId Based on Components

The DODT tool uses the Motorola approach for hazard identification, combined with one of the two generic failure mode sets suggested in the previous section. The user can select either the general failure mode set containing "omission, commission, stuck and incorrect", or the more concrete set containing "no output or action, wrong output or action". In the table below, it is possible to relate an element – e.g., a pump – to a specific requirement. As an example, we have inserted the requirement R4 for a steam boiler:

Table 3.8 DODT automatically fills in element and failure modes

Req. id	Element	Failure mode	Failure effect	Cause or mechanism	Recommended actions
R4	Water_level_contol_unit	Omission			
		Commission			
		Incorrect			
		Too late			
		Annunciated			
	Pump	No action			
		Wrong action			

R4: Water_level_Control_unit **shall** control water_level **from** pump

In this requirement the DODT tool is able to identify two components: the Water_level_Control_unit (a PLC) and a pump (an actuator). What is shown in the table is what the DODT tool can fill in automatically. The rest must come from a safety expert, a domain expert or from a reuse repository. Table 3.8 below is a minimum version. It is, however, straight forward to extend it with columns for occurrence probability and consequence.

HazId Based on Requirements

A boilerplate requirement will in the general case consist of three parts:

- Pre-condition (PR): one or more operational conditions that need to be true in order to move on to the action part of the requirement. This is usually contained in the prefix part of the boilerplate.
- Active part (A): one or more actions performed or abilities made available if the pre-condition is true. This is always found in the main part of the boilerplate.
- Action-condition (AC): one or more operational conditions or restrictions that applies to the action or the way the action is performed. This is usually found in the suffix part of the boilerplate. Each of these parts can fail.
- Pre-conditions (PR): active part is: activated even though condition is false, not activated even though condition is true, activated too early, activated too late.
- Action part (A): one or more actions (action/entity pairs or action/entity/equipment triples) are not performed, e.g., due to equipment failure, only partly performed. This might be dropped when we are doing early analyses, performed in a wrong way.
- Action-conditions (AC): not performed according to the action condition, e.g., too late or too slow.

Below we show a small example, again taken from the steam boiler requirements. The requirement is first shown in full and then as it is split into pre-condition, actions and post-conditions by the DODT tool. For the two requirements used in the example the DODT tool is able to identify (1) a pre-condition, (2) an action

Table 3.9 DODT automatically fills in failure conditions

If water level **greater than** max water level **the** control system **shall** stop feeding pump				
If PR1 the control system **shall** A1				
Failure condition		Effect of failure	Type	Remark
PR1 true	A1 not done			
	A1 done incorrectly			
	Something else is done			
	A1 done too late			
PR1 false	A1 done			

A – e.g., stop feeding pump – and (3) an action A realized using some equipment E – e.g., control water level using a feeding pump. What is shown in the table is what the DODT tool can fill in automatically.

Note that the failure condition is generic, e.g., an action is not done, while the effect of the failure is dependent on the specific product and the environment in which it is operating. Thus, the table can be initiated with generic failure conditions, while the failures' effect must be filled in by the analyst. Table 3.9 is a minimum version. It is, however, straight forward to extend it with columns for occurrence probability and consequence if needed.

3.5 Requirements Management

The management of requirements covers many aspects ranging from the representation and traceability of requirements to complex tasks like impact analysis and measurement activities.

The concepts for representing requirements and the different types of traceability links between requirements and other model artefacts like V&V cases and components are presented in Sect. 3.5.1. Based upon the presented requirements model typical change management will be considered in Sect. 3.5.2. The presented impact analysis will allow to estimate the effects of changes to the requirements of an system. The measurement methods for requirements (Sect. 3.5.3) will allow to evaluate the quality and the state of the system under development based on the requirements and performed verification activities.

3.5.1 Management and Traceability Concepts

CESAR defines the Requirements Management Meta-Model (RMM) as a conceptual meta-model for the requirements domain. The RMM contains the concepts for managing and storing requirements. It is intended to form a part of the RTP

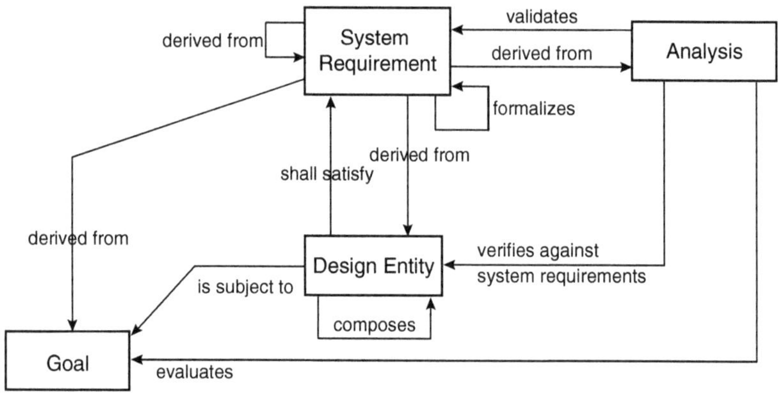

Fig. 3.14 Traceability concepts for system requirements, design and analysis

specification which includes the elements relevant for handling requirements as well as other elements for components and system verification amongst other things. An important aspect is the intra-traceability between requirements, i.e., that links between the requirements can be created to express their relation e.g. which requirement was derived from other requirements. General properties of the requirements itself can be defined in the RMM. Good examples are the categories of requirements like "system requirement" or "process requirement", their rationale and association with stakeholders.

The actual form in which a requirement is expressed is not part of the RMM. A requirement can be stored in many different forms, e.g., as natural language text, as formal patterns or even as a video file. The definition of the language and the semantics of the requirements is the purpose of the Requirements Specification Language (RSL). Languages capable of describing requirements have their own conceptual elements. For example, timed automata would have the concepts of states, transitions, clocks and guards. Boilerplates could have different concept: related to the ontology being utilised.

The RMM defines requirements and goals as well as their relationship to analyses and design. The traceability concepts of the RMM reflect existing standards like SysML [121], EAST-ADL [154], meta-models from the MeMVaTEx project [2] or the traceability reference model by Jarke and Ramesh [132]. Regarding the traceability concepts there are only few differences between these meta-models. In fact many approaches are related to each other. The MeMVaTEx requirements meta-model uses traceability links from SysML and the abstraction layers from EAST-ADL. The EAST-ADL traceability links are very similar to SysML. Figure 3.14 provides an overview of the RMM traceability concepts. These concepts are formalization, derivation, satisfaction, evaluation and decomposition. They will be explained in the following sections.

Figure 3.15 illustrates an example that shows two abstraction levels of the system where each of the abstraction is described hierarchically. The figure shows examples

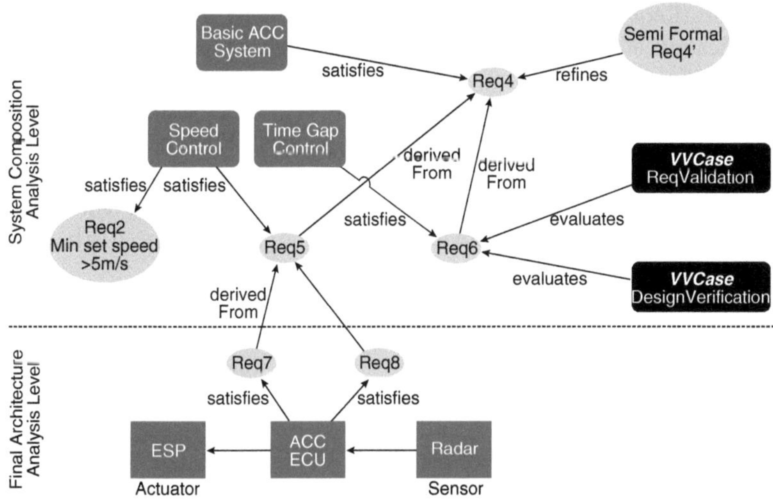

Fig. 3.15 Example: Multi-design-level traceability of system requirements, design and analysis

of vertical traceability along the design hierarchy in a development process as well as horizontal traceability between requirements of these two abstraction levels. Both kinds of traceability are supported to enable traceability along the whole system development process. For simplicity reasons, each link is indicated as a simple arrow rather than objects of association classes as indicated in the main figure of this section. Derived links exist between requirements at different levels in the system hierarchy (Req5 and Req6 are derived from Req4) and between requirements of different abstraction levels (e.g., Req7 and Req8 are derived from Req5). Furthermore, the example shows that Req4 represented using semi-formal techniques like activity diagrams refines the textual requirement Req4. It adds more detail to the textual requirement. Each requirement is satisfied by one component. There are also two V&V cases as analysis artefacts, one validating a requirement and another verifying the requirement against the design.

3.5.1.1 Refinement/Formalization

System requirements and goals can be related to another in terms that one specification refines the other one. The refinement of a system requirement adds detail to its specification while preserving the semantics of the abstraction. Typically, system requirements are refined for the purpose of formalization of the described properties and of the intended behaviour of a design entity. The result of applying a refinement operation to a specification item (say requirement AR1) is a new item (say RR1), such that RR1 has more detail than AR1, but RR1 has the same intent and meaning as what AR1 specifies. A refining specification is written in a formal RSL (Requirement Specification Language). This allows formal validation and verification of properties that are described in a system requirement with a formal specification.

In a meta-model like SysML, a Refine link represents a specific dependency between requirements on different level of abstractions where one or multiple target requirements adds further detail to the source requirement. The existence of the resulting requirements is completely defined by the superior source requirement. If the latter is changed, the resulting requirements of the refinement need to be changed as well.

3.5.1.2 Refinement/Derivation

System Requirements and goals can be derived from other requirements, design decisions, goals, analysis results or recommendations. Typical design decisions are composition, realization and deployment of design entities as presented in SPES Architecture Modeling [39]. This typically involves structural refinement of design entities. Along with these design decisions, requirements have to be derived for introduced and related design entities.

When deriving requirements, they can be related to another in terms that a set of system requirements entails one system requirement. An entailment relationship between system requirements exists when a design entity with a specification given by a system requirement is refined into a composition of other design entities with related specifications. The composed specifications shall entail the specification of the context specification. An analysis of the system requirements that are related via entailment shows whether satisfying design entities can be composed or not.

3.5.1.3 Satisfaction

A system requirement can be satisfied by design entities. When a design entity is related to a system requirement, this indicates that the design entity with its properties and its behaviour shall satisfy the system requirement. Therefore, the dynamics of the implementation of a design entity shall meet the intended properties and the behaviour that are described by the system requirement. This concept is inspired by SysML [121] and by EAST-ADL [154]. The satisfaction concept reflects contract-based specification, because assumed and guaranteed properties can be expressed by system requirements that shall be satisfied by design entities. The concept of contracts assigned to design components is described in the HRC meta-model specification of the SPEEDS project [129]. Satisfaction must be verified by analysis methods that show whether a design entity meets a system requirement or not.

In a meta-model like SysML, a Satisfy-link between components (components of different abstraction layers, source code) and system requirements indicates that the artefacts shall satisfy the requirements they are linked to. The justification of this link is done by verification means. A change of a requirement may have an impact an assigned component. Therefore, a change of a requirement typically leads to a complete re-verification whether the component still satisfies the requirement.

3.5.1.4 Evaluation

An analysis validates system requirements and verifies that a design entity fulfils a system requirement. Verified fulfilment of a system requirement means that the respective design entity claims the properties described by the system requirement. For given metrics, a goal can be evaluated in an analysis and the degree of fulfilment by a design entity can be determined. An analysis is typically defined in a V&V case.

In a meta-model like SysML [121] such a link between a V&V case and a requirement indicates that the V&V case connected to the requirement shall verify/validate the requirement. The actual verification/validation of the requirement within the V&V case is a question of the analysis methodology to be applied. Nonetheless, a change to a requirement may have an impact on the results of the verification/validation and therefore lead to a complete re-verification/validation.

3.5.1.5 Composition

When system design structure is refined there are design entities that are composed of others. That means that one design entity can consist of other design entities which are its parts. Being parts one design entity is used in the environment of other design entities. They are linked to another by using interconnections. The composition of design entity as interconnected parts defines how the composed design entity is internally built and how it internally behaves. Therefore, the composition of behaviour of different design entities results in the overall behaviour of the composed design entity. Composition in system design leads to a hierarchy of structured design entities.

Interconnections such as connectors define the relationship of the interfaces of design entities, which are composed in the context of another design entity. They are delegating when an interface of the surrounding context design entity is connected. Connectors declare how the interaction items of the connected interfaces are linked to another. They typically define how services are invoked and how data is exchanged. Connectors can be different starting from simple links up to complex protocol links. Simple links define the identity of connected interfaces. Complex protocol links define causality between connected interfaces. Therefore, connectors define the actual control and data flow between the interfaces of design entities.

The specification of a design entity shall be satisfied by the dynamics of its implementation. In a very clean specification even the specification of the design entity parts should entail the specification of the composed design entity taking the composition structure into account. Knowing only the specifications of the composed design entities but not their implementations provides a gray-box view on the composed design entity.

3.5.2 *Change Management and Impact Analysis*

Changes in the requirements of a safety critical embedded system typically require the revalidation of huge parts of the design. Especially in late development phases this revalidation effort can be extremely expensive. We describe a approach for automated impact analysis based on requirements. We demonstrate how formalized requirements can be used to limit the costs of revalidation by proving that parts of the system cannot be impacted by a change. The systems engineer is guided through the change management process by a systematic inspection of the relevant components and gets suggestions where compensation techniques could be applied.

An impact analysis needs to access every part of the system starting from initial functional models with high-level requirements down to software or hardware models with their respective requirements. A couple of different tools are used including IBM DOORS or Requisite Pro as requirements engineering tools and modelling formats like EAST-ADL [154], AUTOSAR [12] and SIMULINK [114], only to name a few. To obtain a unified view on all of these system artefacts we are using the CMM-API [17], which provides a system representation based on the CESAR-Meta-Model (CMM) [15]. Although the presented approach is applicable to any kind of requirements, the highest degree of automation can be reached by using formalized requirements.

In addition to the automation of the impact analysis we discuss a technique to stop typical change propagation by rephrasing requirements. This approach is based on common over-approximation in system requirements due to the use of library components (COTS) that are purchased from different suppliers. This technique allows the compensation of a change by rewriting some requirements and revalidation of a single component.

3.5.2.1 Related CESAR Technology

The CESAR project has developed a couple of technologies allowing to efficiently analyse the impact of changed requirements. One technology is the Cesar Meta Model with its capability to express components (see Chap. 4) and requirements (see Sect. 3.5.1) as well as the traceability between these artefacts. For impact analysis we restrict to the most basic artefacts:

- *Rich Component*: A rich component describes a structural unit having ports to communicate with other Rich Components. Rich Components can be encapsulated. Their intended behaviour is described through attached System Requirements.
- *System Requirement*: A condition or capability the system shall fulfil in order to satisfy the stakeholders needs.
- *V&V Case*: Specifies a validation or verification activity and the result whether the linked component complies with the required properties in the V&V case.

The method used in these activities is not limited and can vary from unit tests to model checking or consistency checks.

The different requirement specification languages (RSLs) (see Sect. 3.3) allow formal representations of the requirements. For the impact analysis we focus on the formal pattern based RSL (see Sect. 3.3.3).

Furthermore, the entailment analysis is an essential element of the impact analysis enabling us to determine if a requirement is correctly split up in sub-requirements. The impact analysis uses these technologies to combine and extend them in a single process.

3.5.2.2 Using Automated Impact Analysis for Avoiding Revalidation

The impact analysis is a task inside a change management process as required by modern safety standards like the ISO 26262 [152]. After analysing different use-cases in the CESAR project, we assume a very generic change process in six steps:

1. *Baseline V x.y*: The requirements are in a consistent state, all V&V cases are executed successfully. There exist V&V cases requesting correct requirement decomposition for each component.
2. *Change Request Creation*: The change request has been created, initially one requirement is changed in the request itself, the original model has not been modified.
3. *Impact Analysis*: The impacted requirements and V&V cases are determined, the change request grows in the proceeding of the impact analysis.
4. *Accept/Reject Change Request*: After the impact analysis run has completed the system engineer/product manager decides if the change is reasonable and the change may be implemented.
5. *Revalidation of Implementation*: The changed parts of the system need to be revalidated if the changes are made in a consistent way.
6. *Baseline V x.(y+1)*: After successful revalidation of the changed requirements a new baseline has been reached.

The change request itself contains the initially changed requirement, and a pointer to the requirement, which should be replaced. The change request also manages the requirements and verification activities that are potentially affected by the change, the "suspect" elements. These are the elements that still need to be checked for consistency after the introduced change. In Sect. 3.5.2.2, the marking of suspect requirements and V&V cases is described in detail. Through the analysis of suspect elements, other requirements or verification activities may be identified that need to adapted to implement the changed requirement. This propagation of changes is described in Sect. 3.5.2.2. A technique to stop this propagation is introduced in Sect. 3.5.2.2.

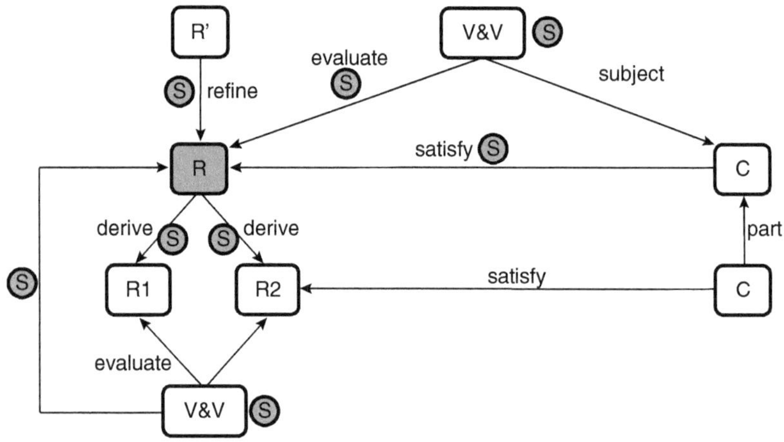

Fig. 3.16 The highlighted requirement R gets changed, suspect components are marked with an "s"

Suspect Elements

A suspect system artefact might be affected by a change, but an investigation is still pending. Only elements directly linked to a changed requirement or V&V case may be marked as suspect.

Figure 3.16 includes a requirement R (highlighted in grey) that gets changed. R has been refined from R′. There is a V&V case that verifies if component C fulfils the specification (entailment case). Furthermore R has two derived requirements R1 and R2. An additional entailment V&V case verifies if the decomposition of R in R1 and R2 is correct.

All the elements that now get marked as suspicious are tagged with the grey "s". This includes basically all links related to the Requirement and the V&V cases that point to the requirement. The following listing explains why the elements got marked and how the suspicious flag can be removed. When possible resolving strategies are executed is summarized in Sect. 3.5.2.3 dealing with the overall impact analysis process.

- *Suspect V&V case*: After the change of a requirement the V&V case and all of its previous results are no longer valid since the test-results still belong to the old requirement. Furthermore it might be necessary to even modify the verification method itself. This is typical for manually created unit tests.
- *Suspect Evaluate Link*: In addition to the V&V case, the evaluate link can become suspicious. The changes in the requirements can result in a reassignment of the V&V cases to the requirements. The evaluate link is marked suspect to check if the V&V case is still meant to be assigned to this V&V case. This reassignment is typical if the way requirements are split-up is changed.

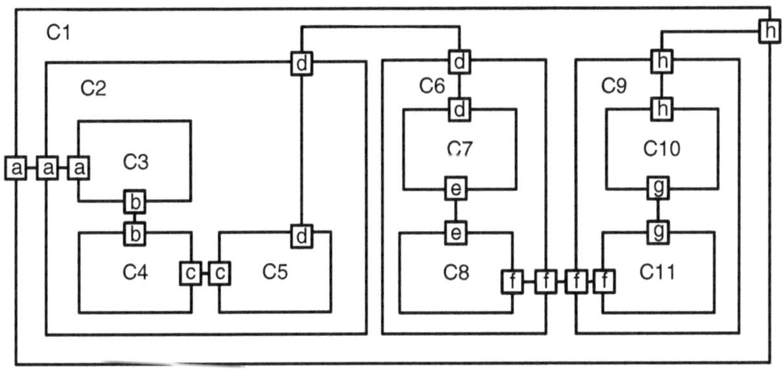

Fig. 3.17 Architecture of the example system. A signal a is the input, is processed by multiple components and h represents the output of component $C1$

- *Suspect refine links*: It has to be verified that the refinement is still valid, i.e., that the requirements R' and R still express the same facts. This is easy to verify if R and R' are both available in a formal representation. In case of informal requirements this step has to be done manually.
- *Suspect derive links*: Through the change of requirement R it has to be verified if $R1$ and $R2$ still cover R. In case of formalized requirements an entailment check can show the correctness of the derived requirements.
- *Suspect Satisfy Link*: Similar to the suspect evaluate link it has to be checked if the requirement still belongs to the correct component.

Marking an element as suspect results in the need to verify if the element is affected by the change. Depending on the result of this verification more requirements and test cases need to be modified which leads to a growing change request. This is called *change propagation*.

Propagation of Changes Through the Requirements Model

Currently we focus on a subset of the elements from the CMM. We restrict the links to the satisfy link, which connects requirements to components and the evaluate link, which connects V&V cases to requirements. We denote the requirements R that are satisfied by a component C by $sat_R(C)$, the V&V cases assigned to a requirement r by $eval_V(r)$ and the requirements connected to the V&V case v by $eval_R(v)$. The handling of derive links can be reduced to the handling of satisfy links. The entailment analysis is a function $ent(r, R)- > 0, 1$ where r is a top level requirements and R a set of derived requirements. The result indicates of the top level functionality is fulfilled by the derived requirements (see Sect. 3.4.3.2).

The following example (see Fig. 3.17) demonstrates the basic principles of change-propagation .

The example system consists of 11 components ($C1$–$C11$) that are connected through ports a–h. The composition of the components is directly represented by nested boxes. The initial change is performed to one of the requirements from $sat_R(C6)$. All V&V cases $eval_V(sat_R(C6))$ connected to that component are marked suspect. Furthermore it is necessary to check if the entailment relation is still valid for both directions, towards the upper level and towards to lower level (if there are not entailment V&V cases anyway):

$$ent(sat_R(C1), \{sat_R(C2), sat_R(C6), sat_R(C9)\}) \tag{3.1}$$

$$ent(sat_R(C6), \{sat_R(C7), sat_R(C8)\}) \tag{3.2}$$

Depending on the result of the re-run of these entailment analyses more requirements may need to be added to the change request. If relation (1) fails the components $C2$, $C9$ or $C1$ need to be modified to be compatible with the change again. It is a decision of the engineer if it is more suitable to change a top level requirement or change parts inside of the system.

If $sat_R(C2)$ are changed they also have to be added to the change request and $eval_V(sat_R(C2))$ have to be marked suspect. The entailment relation

$$ent(sat_R(C2), \{sat_R(C3), sat_R(C4), sat_R(C5)\})$$

needs to be checked also and further changes to $C3$, $C4$ or $C5$ might be necessary. Furthermore the entailment relation (2) needs to be considered and might result in changes in the components $C7$ and $C8$.

Note that the components $C9$, $C10$ and $C11$ do not need to be considered for revalidation although the component are connected through ports. Our proposed impact analysis allows to cut off complete subtrees from the potential affected requirements. Only the marked suspect V&V cases need to be re-executed and/or modified.

Nevertheless, in the worst case a lot of requirement are affected by the change and huge parts of the system needs to be reverified. Additional measures are needed to stop a change from propagating.

Stopping Change Propagation by O-Shifting

A failing entailment test caused by the change of a requirement indicates a wrong decomposition of requirements. An additional change is needed to compensate this inconsistency. This further change leads to a propagation behaviour of the change through the system (see Sect. 3.5.2.2).

In order to avoid this propagation a novel technique, which we call "O-Shift", can be applied. O-Shifting allows to identify if a change can be compensated by modifying a set of other specifications without violating any test and verification results.

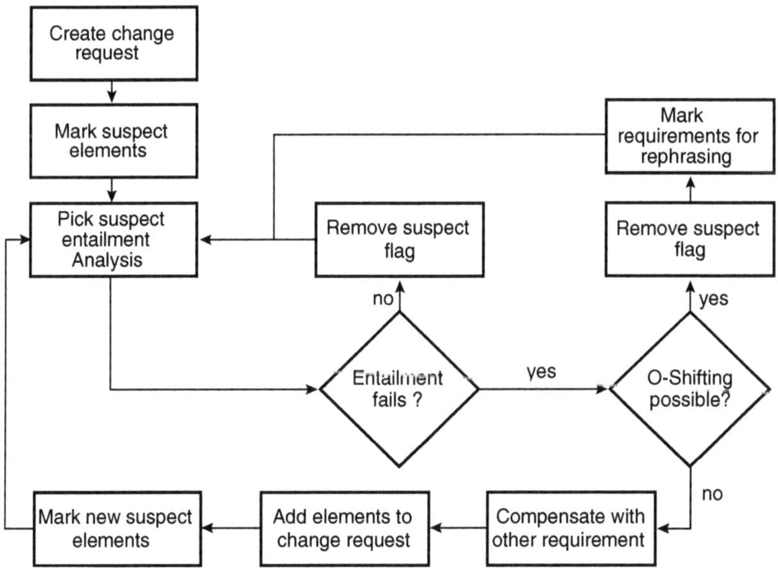

Fig. 3.18 Basic impact analysis process

More Info:

A description and formalization of the O-Shifting approach is currently submitted for review and will be published soon on: http://www.offis.de/cesar

3.5.2.3 Process for Impact Analysis

Figure 3.18 depicts the basic impact analysis process.

After the creation of a change request the system artefacts that are directly related to the initially changed requirements are marked for later inspection.

Now a loop starts with picking a suspect entailment V&V case and running it again to determine if the change affects the composition of other requirements. In case the entailment analysis fails it can be checked via O-shifting if the propagation can still be stopped. If O-shifting is also not possible then the initially suspect marked requirements is added to the change request and the system artefacts connected with the requirement are also marked suspect.

3.5.3 Measurement of Requirements

The INCOSE Systems Engineering Measurement Primer [151] points out that measurement is a process for obtaining the vital insight into the progress, products, and/or processes of the project or system being developed. This insight helps the decision maker to make more informed decisions, identifying deviations from plans earlier, thus allowing mid-course corrective actions when it is least expensive to make them. There are many existing metrics, but there are still challenges in realizing such a measurement process in an efficient way. For example, which are the correct metrics and how do we efficiently calibrate a company wide metric definition to a single project. Each project has its own tailored development process that leads to other information demands and metrics. Quality relevant data must be interpreted depending on defined process goals. For example, different verification methods are demanded by safety standards to create enough confidence that the system works as intended. Depending on the system's ASIL, there will be different process steps and a different way of measuring quality. The system requirements remain the same; the process context adds a new dimension.

This section describes a process oriented quality measurement method to dress these challenge and is structured as follows: Sect. 3.5.3.1 introduces the main idea of this approach. Section 3.5.3.2 describes briefly the model based quality measurement framework to implement the idea. The chapter closes with an example in Sect. 3.5.3.3.

3.5.3.1 Main Idea

A development accompanying quality measurement that integrates and aggregates quality data for project monitoring, demands the following: the definition of target quality in terms of quality goals and their comparison with current development data, measured by quality metrics. Both terms are defined as follows:

Definition (Quality Metric): A quality metric measures the as-is value for a defined product quality characteristic at a dedicated point in time in development.
Definition (Quality Goal): A quality goal defines a target value for a quality metric. This target must be fulfilled at a dedicated point in time. Quality goals serve as completion criteria for tasks and milestones.

Although requirements are one necessary input for the target quality, they are not sufficient on their own. To capture the needed process context and enable efficient calibration of quality measurement to one project, quality metrics and goals must be defined in a process oriented manner. Figure 3.19 illustrates the central idea behind the methodology: A process consists of multiple tasks. Each of them has a dedicated view on the system model or one of its parts. A task is atomic and is applied one

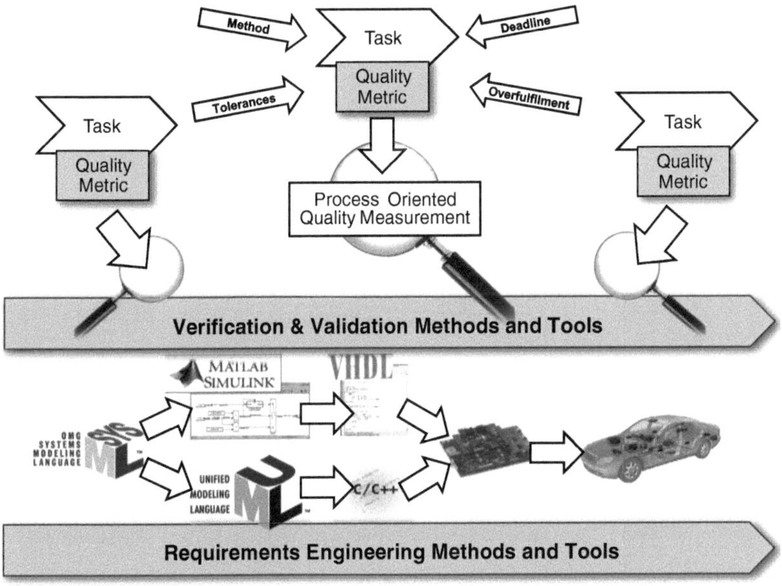

Fig. 3.19 Process oriented quality monitoring – the idea

or multiple times in one or more process definitions. Quality definition starts from tasks. A quality metric evaluates one quality characteristic like fulfilment degree of requirements. A process oriented quality definition considers information like deadlines, tolerances, over fulfilment and even the applied verification method that is possibly demanded by a safety standard. Results are annotated to tasks and are directly usable for project control.

3.5.3.2 Quality Monitoring Framework

A modelling framework as a common semantic basis realizes an integrated view on the current quality status of a project and supports project planning and monitoring across different levels of the entire V-cycle. Figure 3.20 sketches the two parts of the framework: a process meta-model that allows the definition of quality metrics and goals to assess current quality and therefore the progress of different process activities and a system meta-model to represent all product related artefacts. The integration between these two perspectives allows a process oriented quality assessment based on current development data.

Furthermore, there is a connection to development tools. As this is mainly the responsibility of the RTP infrastructure, this will not be detailed here.

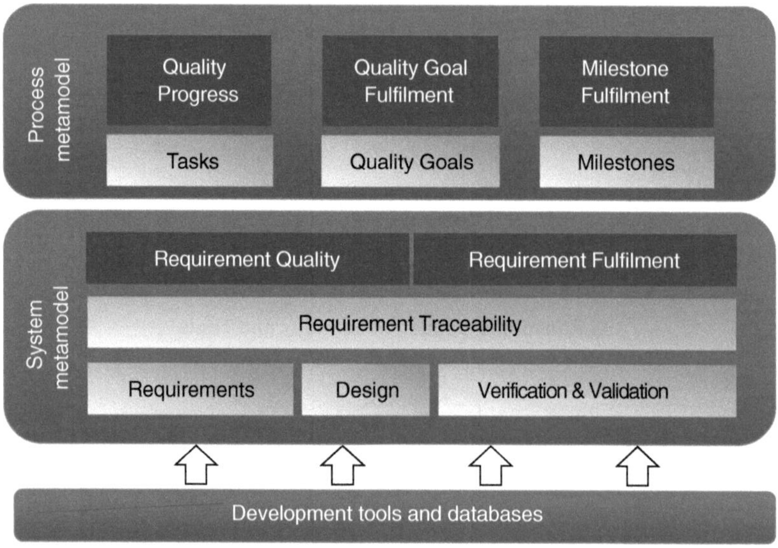

Fig. 3.20 High level overview modelling framework

System Meta-Model

To support a requirement-based quality evaluation, existing requirement traceability concepts and unified access to data like requirements, components V&V definitions, their results and their traceability is needed. Here, the approach relies on the core part of the CMM definition and its implementation in the RTP infrastructure. Using these system related information, it is possible to assess requirements, their quality and fulfilment and derive assessments of process specific progress metrics from them and finally represent the results to detect early deviations from the initially defined plan.

Process Meta-Model

Once the basis for requirement quality and fulfilment assessment is created, the process model part of the modelling framework is presented. The product oriented quality evaluations need to be interpreted from a process point of view. In CESAR, the Software & Systems Engineering Meta-Model (short SPEM [122]), is used for process modelling. SPEM offers a conceptual framework to describe development methods and processes. It distinguishes between the definition of methods and processes. The same idea can be applied to the definition of quality metrics. An extension of the MethodContent package in SPEM with generic measurement

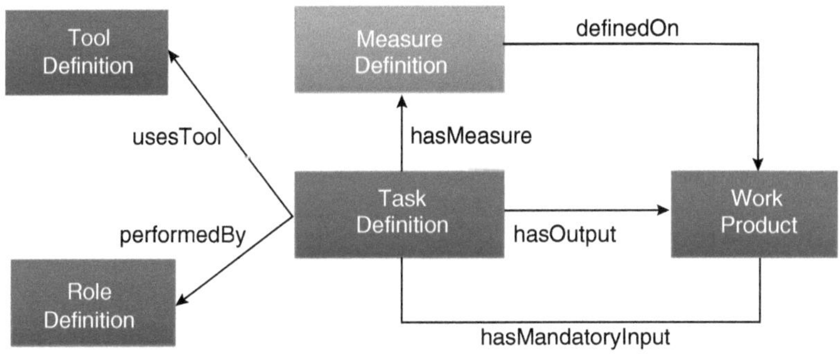

Fig. 3.21 Method content extension in SPEM

concepts (see Fig. 3.21) as described in [71] will allow the definition of such measures independently from a concrete process and their reuse in many process instances.

A measure definition can range from a simple BaseMeasure just querying the requirement coverage, up to more complex indicators. Therefore, measures are defined in a hierarchical manner. The dependencies between indicators and measurable parameters, as well as the aggregation functions are described using the meta-model elements. The following measure concepts exist (see also [71]):

1. *BaseMeasure*: Base measures can be read directly from the model via a defined MeasurementMethod.
2. *DerivedMeasure*: A derived measure aggregates multiple base measures. The aggregation function is defined via a MeasurementFunction having other measures as inputs.
3. *Indicator*: An indicator allows the integration of advanced AnalysisModels like classifiers (e.g. Bayes or Fuzzy Classification).

As for the *MethodContent* package, also the *Process with Methods* package is extended to allow the reuse of quality metrics in multiple process definitions. When defining a process, each activity can have one or more assigned quality metric. Any *WorkBreakdownElement* can define goals based on these defined metrics. The combination of quality metrics and goals gives project managers a wide range of planning possibilities. Figure 3.22 gives a simplified illustration of concepts and relations.

By defining these SPEM extensions it will be possible to define reusable measures and company-wide quality measurement plans. The extended framework is not restricted to a specific development phase. It is adaptable to different contexts and stakeholders in the supply chain.

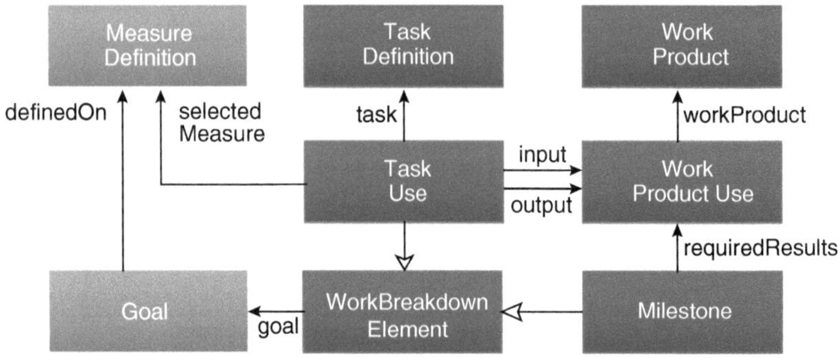

Fig. 3.22 SPEM extensions to describe quality goals

Integration of System and Process Perspective

The process-oriented definition of quality metrics like requirement fulfilment or requirement quality must be integrated with the product perspective to define and execute these metrics against concrete development data.

Integration is needed on two levels. On the one side, we need a conceptual integration. All measures shall be defined based on the system meta-model, its concepts and properties. The measurement definition itself shall be generic and assessable for every work product instance the measure is defined for. One example could be a simple measure that counts the percentage of fulfilled requirements in a verification phase. The measure needs to be defined in way to allow its evaluation against any requirements model.

Furthermore, integration on instance level is necessary. A process (including its measure definition) can be instantiated multiple times. For example there could be a definition of verification steps that needs to be performed. This process description could be instantiated for different components in a project plan. On a process level it talks about the same work products, but the concrete component represented by the system meta-model is different and needs to be linked to the work product in the process instance to allow development accompanied assessments of defined measures. In the following, both integration issues are discussed:

Conceptual Level – Obviously, the main concepts that need to be aligned are the work product descriptions in the SPEM and the system meta-model concepts. As we aim to create a loose coupling to minimize dependencies between these two meta-models, a query-based approach (e.g., through using the Object Constraint Language) is used. This is exemplified in Fig. 3.23.

The example shows a Work Product "Requirement Model" and a defined metric "Number of fulfilled requirements". Basically, the query is used to identify at run-time which SystemRequirement instances belong to the work product "Requirement Model". Each query needs one or more context objects, the query runs against. In the example, it is the concept of RichComponent as a process definition is instantiated for a system, a sub-system or a system component (see integration on "Instance Level").

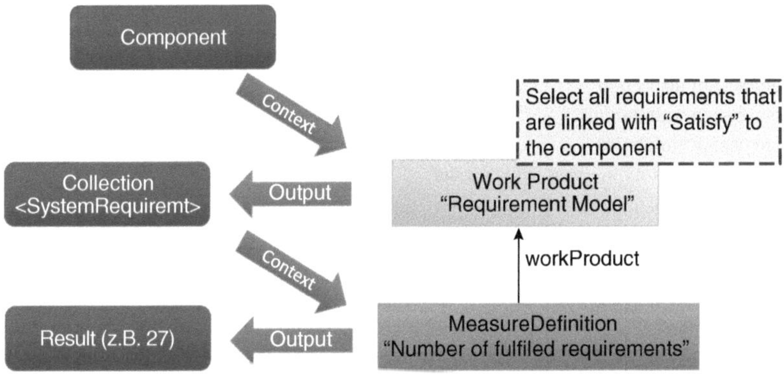

Fig. 3.23 Conceptional integration example

The work product "Requirement Model" is defined as a collection of SystemRequirement instances that are linked via a "Satisfy" link to a RichComponent instance the process is instantiated for. The result is a collection of SystemRequirement objects. The MeasureDefinition works with the same principle, it needs one or more context objects it will be executed against. Usually, it is the result of a query that is defined for the work product. In the example, the MeasureDefinition runs against a collection of SystemRequirement objects and calculates the fulfilment degree of each requirement to come to a result (e.g., 27 as indicated).

Instance Level – Now, work products and metrics can be defined based on the system meta-model definition. But the determination of concrete instances, the queries are executed against is still missing. For this purpose, defined process and metric definitions are instantiated for a concrete project and its system under development. Result is a calibrated process and metric definition that contains just the needed process activities, e.g., the ones that are demanded from a safety standard depending on the ASIL of the system and its components.

3.5.3.3 Example

This section describes the usage of the developed methodology for functional safety management [152]. The method will support managers as well as developers in checking which activities demanded by a standard are already fulfilled.

The example concentrates on the System Design Phase from the standard ISO 26262 with the following three goals:

1. *ISO 26262 compliant process model*: The first goal was to develop an ISO 26262 compliant process model for the system design phase as the basis for quality monitoring.

Table 3.10 ISO 26262 system design verification and their ASIL recommendation

Methods	A	B	C	D
System design inspection	+	++	++	++
System design walkthrough	++	+	o	o
System design simulation	+	+	++	++
System design prototyping and vehicle tests	+	+	++	++
Deductive analysis	o	o	+	++
Inductive analysis	++	++	++	++

2. *Requirement quality and fulfilment metrics*: Central parts of the ISO 26262 deal with verification and validation of safety requirements. Quality metrics were defined and annotated to the defined process model.
3. *Demonstrate the calibration to a single project*: The ISO 26262 reference process contains optional parts that depend on the ASIL level of the system. It is shown how tailoring to a specific project looks like.

Process Definition Foundation for a process oriented quality monitoring according to ISO 26262 is the reference process that is recommended by the standard itself. The ISO 26262 contains requirements and recommendations for the product (e.g. probabilities for random hardware failures) as well as for the development process. From the latter, a development process was defined for part 4.

All variable process parts and methods for these two phases were modelled as tasks. Depending on the systems' ASIL classification, the recommendation to apply a certain method in the development process differs. Besides the system verification shown in Table 3.10 this is typical characteristic in different parts of the standard. For process definition, we used the following interpretation:

1. (++) – The method is highly recommended → Mandatory use in development process
2. (+) – The method is recommended → Optional use in development process. Decision during process instantiation.
3. (o) – The method has no recommendation for or against its use → Not used in the ASIL specific process definition.

Using the given interpretation, we created process descriptions for ASIL A, B and C as shown in Fig. 3.24. With increasing ASIL level, the needed steps to formalize requirements as well as the needed steps to verify the system differ.

Process-oriented Quality Definition Requirements play a central part in the ISO 26262, especially in the system design phase. Therefore, a requirement-based quality definition was applied in this example. In the following, we will describe its application.

Requirement Quality One of the primary goals in system design is the development of a complete and consistent set of requirements for the system under

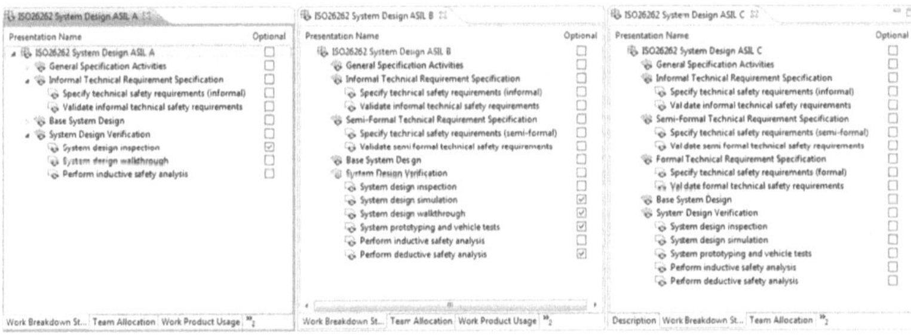

Fig. 3.24 ISO 26262 process definitions

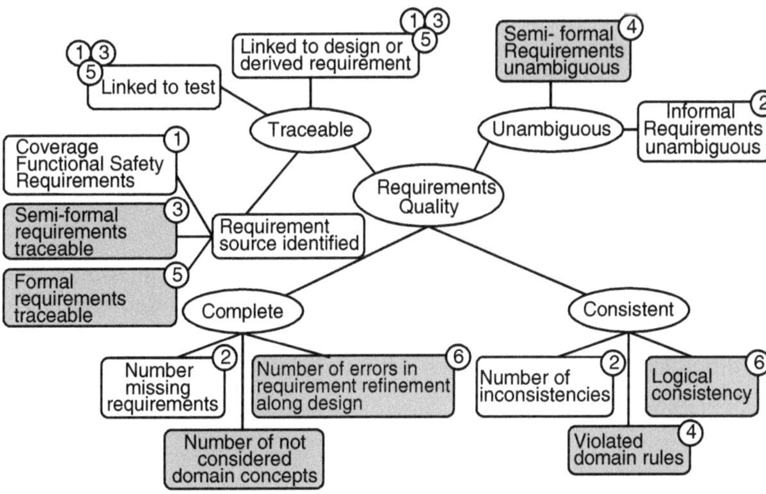

Fig. 3.25 An extract from the requirement quality model

development. To follow such a process, we defined metrics to measure requirement quality and annotated them to the defined process model. The definition is based on ISO 26262 Chap. 8 [152], where quality criteria for requirements were defined.

The model shown in Fig. 3.25 describes four quality characteristics, i.e. Traceable, Unambiguous, Complete and Consistent. These criteria again are decomposed into measurable quality attributes, where blue attributes are mandatory and green attributes are optional. The number associated to each quality attribute shows its connection to a task (see Fig. 3.25).

The optional quality attributes will only be considered when requirement formalization steps are performed. By assigning these quality attributes to tasks, the quality model is calibrated once we instantiate the process. Example: For an ASIL A system, we will not formalize requirements, so the respective tasks will not be part of an instantiated process and neither will be the quality attributes (Fig. 3.26).

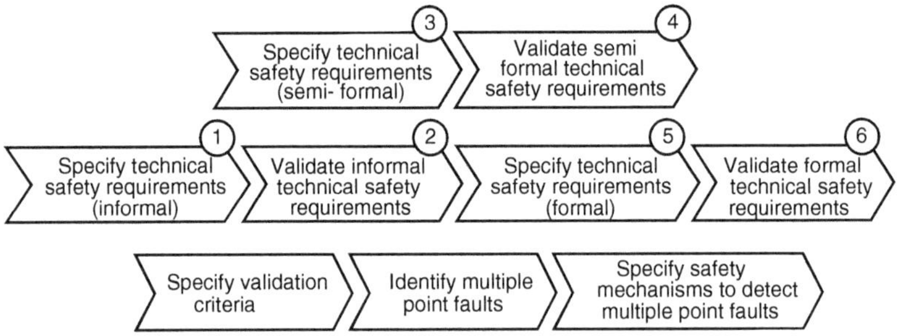

Fig. 3.26 Associations from quality metrics to tasks

Requirement Fulfilment For the following system verification tasks, we defined requirement fulfilment metrics:

1. System design inspection
2. System design walkthrough
3. System design simulation
4. System design prototyping and vehicle tests

For each task this quality metric calculates the ratio between all technical safety requirements and the number of fulfilled safety requirements. Requirement fulfilment is calculated based on results of associated tests. The metric implementation for the different verification activities just differs in one parameter, the demanded verification method.

Project Plan Creation

The defined ISO 26262 system design process is instantiated for an example dataset. The main process instantiation criterion is the ASIL classification of the system and its components.

Figure 3.27 shows an instantiation for parts of an ASIL C system architecture with two subcomponents classified with ASIL A and B. In the beginning, the architecture itself is developed. This includes specification and validation of technical safety requirements as well as architecture development. When a first architecture is completed further system design tasks like requirement formalization and design verification can be instantiated according to ASIL classification of architecture components.

This process instantiation results in a tailored quality definition for each component.

Quality Assessment

Defined requirements, validation and verification results are frequently synchronized with the repository. For any point in time, it is possible to assess development

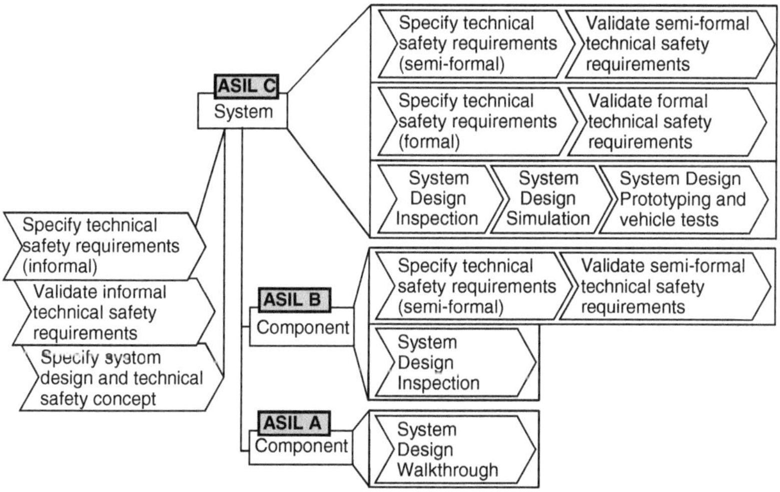

Fig. 3.27 Process instantiation for requirement definition and system verification

status per component and process task based on development data and compare the results with planned data.

Figure 3.28 shows an extract of the "model world". Depending on the process context requirement fulfilment is assessed for the two tasks "Verify System Design through Simulation"? and "Verify System Design through Prototyping"?. Both tasks verify "Technical Safety Requirements" (short TSR) and produce verification results, but they use different verification methods. This process-specific parameter is taken into account by means of the following metric implementation:

1. *Derived Measure "Sum Requirement Fulfilment (Simulation)"*: self.requirementFulfillment("Simulation") -> sum()
2. *Base Measure "Requirement Count"*: self -> size()
3. *Base Measure "Requirement Fulfilment Degree (through Simulation)"*: Sum Requirement Fulfilment/Number of requirements

The metric of the other task is implemented in the same way with "Prototype" as the parameter for the "requirementFulfillment method". The developed tool assesses the metrics against all requirements that are connected to the root component of the system architecture (see formalization of TSR). In the example there are two.

Requirement fulfilment shown by simulation is 0.75 (75 %) as three of the four VVCases with the type "Simulation"? are passed. The verification by prototyping is still pending. Requirement fulfilment is 0 %. Both quality metrics are used as indicators for the progress of the tasks.

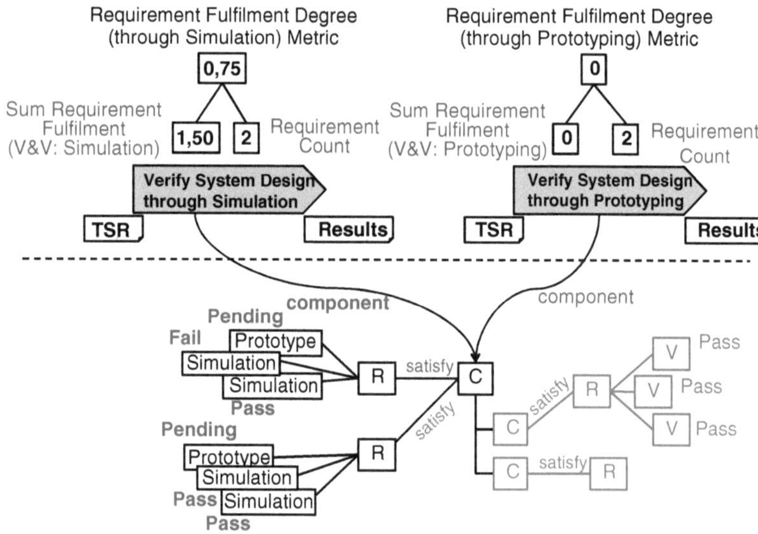

Fig. 3.28 Quality assessment from a view of two verification tasks

3.6 Requirements Engineering and Variability Modelling

This section concerns itself with variability with respect to requirements. We give an introduction to the general concepts surrounding variability and requirements and then explain the general notions behind architecture exploration and product line engineering. Following this we describe some of the supporting tools and methodologies for variability management in respect of requirements.

3.6.1 Requirements of Variability and Variability of Requirements

A reasonable requirement for my mobile phone is that I should be able to call from Norway and from most other places in Europe. For this requirement it is sufficient that the mobile supports the GSM 900 network or the GSM 1800 network. Thus a phone that only supports GSM 900 or a phone that only supports GSM 1800, both satisfy my requirement.

Is this a requirement of variability? Is this a requirement of a product line (of mobile phones)? Are those two questions really the same question? The two questions are not the same. My requirement does not call for any variability, but there are various ways to satisfy the requirement. Therefore, it may be seen as a requirement to a product line. A product line is characterized by the fact that its products vary, but they also share common features which logically binds them together in a "line" or "family". Mobile phones come in such product lines. In principle a product line could consist of a phone on GSM 900 and a phone on GSM 1800.

The reason why there are both GSM 900 and GSM 1800 networks is because the GSM900 did not have enough capacity to handle the growing mobile traffic. Therefore, the telecom operators wanted to distribute the traffic over both frequencies. This is why the network operators wanted people to get phones that could switch between GSM900 and GSM1800. Thus the network operator requirement was that each phone has certain variability. Our simple GSM900 phone does not satisfy the network operator requirement. Even though the distinction is subtle the mobile phone can only be a GSM900 phone or a GSM1800 phone at one particular point in time. A phone does not support both networks at the same time.

The variability as such is exactly the same in both our situations, namely "GSM 900 or GSM 1800 but not both at the same time". Still in practice there was a considerable difference between the two requirements. The distinction that first comes to mind is that this is a matter of binding time. In the first situation we talked about variability at design time: either you designed a GSM 900 or a GSM 1800 phone. In the last situation the binding is continuously at runtime and we often call this adaptability (or adaptivity) because it is about every product instance having the capability to adapt to its environment. Thus adaptivity requirements are requirements of variability, while design-time variability is variability of requirements. The binding distinction is not captured by the variability notations and one may ask whether the binding time is completely orthogonal to abstract variability.

In CESAR situations related to embedded systems the binding time distinction is often significant. Embedded systems often have very limited resources and it is not always the case that all the functionality can reside on one physical product. For pure hardware systems this is obvious. The variability must be resolved at design-time or load-time. This means that some of the important decisions are linked to the binding-time of variability, and there is a need for means to describe that decision.

One may even say that the decision of when to bind variability is associated closely with what is perceived as "one product". A customer never buys a product line. A customer buys a product. In fact a customer buys an individual product instance. A vendor may offer a product line of products. A vendor distinguishes between product instance and product (type). A product may be adaptable or configurable and it may result from (partially) resolving the variability of a product line. We may say that a product is a partial resolution of a product line where the product retains some adaptivity. In complex cases some of this adaptivity is called configurability and the configuring is done by specialists often from the vendor.

Let us return to requirements and variability. Let us assume that we have a variability description that represents some requirements to a product line. How are we going to interpret whether the requirements are met? This is related to refinement and of satisfying requirements. Let us assume that our product line requirement is defined by the simple alternative GSM900 or GSM1800. Is this requirement met if all the phones can only connect to GSM900? The opposite position would be that the offered product line must offer at least the variability defined by the variability model. This would in our simple case mean that the product line had to offer not only GSM900 phones, but also GSM1800 phones. In addition the phones could have supplementary variability relating to e.g., color or localization. From a formal refinement viewpoint there is obviously a significant difference between these two interpretations, but in practice we know that product line vendors make market-oriented decisions to limit their offer without admitting to have violated any product line requirements.

We have here talked about variability modelling in conjunction with product lines, but this is not the only way variability modelling can be used. In our mobile phone case, the operators wanted the phones to switch between the networks depending on the network capacities. From the perspective of the individual phone that has this switching capability their decision of which network to use at any point in time is not perceived as selecting a product. Rather the phone is optimizing some quality criterion, in this case, the capacity. The phone explores the available architectures and chooses the one architecture (or configuration) that best satisfies the quality criteria. This is what we call Architectural Exploration. A product line is about deliberately resolving the variability to select a product.

3.6.2 Product Line Requirements Engineering

Product lines are a concept that has been widely exploited in the industry to exploit commonalities and the variability among a product family to reduce costs and time-to-market. In this sense, the common part needs to be produced only once, and the evolution of the product as well as its production can be easily controlled by means of managing the variations, exploiting the systematic reuse for the software, as in the assembly lines. Systematic Software Reuse is appropriate for organizations developing similar projects leading to a family of software products or applications; this is the software product line.

To sum up the main benefits of defining a software product line are the following:

- Homogenization of produced programs
  - Code standardization and normalization
  - Homogenization for different development teams
- Improvement of Product Maintenance
  - Centralization of the product maintenance activities (Avoiding uncontrolled evolution)
  - Solutions are automatically generated based on Normalized and Tested artefacts

- Improvement of quality of the final artefacts
  - High percentage of already tested code
- Reduction of the development time
  - In production & testing & development
  - Avoiding repetitive activities and human errors
- Increment of Customer satisfaction
  - More Robust solutions from the customer point of view
  - Reduction of time-to-market

Software Product Line (SPL) practices are the most widespread systematic software reuse principle, focusing on the development of a software product family defining a common configuration and variable features of the software systems. Related to requirements specification, the main goal will be the identification of the commonalities among the families and the individual products. There are several customers and stakeholders involved. Therefore, a well-established requirements engineering process is essential for any product line practice.

In the SPL approach, the problem space is defined in terms of the product line's commonalities and variability, which reflects the desired range of applications or product variants, in the product line domain and their relationships and dependencies. The assets of the product line or platform and their relation to the problem space are described by the solution space. Under this definition, two main processes are defined: domain and application engineering.

Domain Engineering is the process of SPL engineering in which the commonality and the variability of the product line are created. Some of the key artefacts of a product line are its requirements, its architecture and its components. However, the reusable core assets also include many other artefacts; for example test cases, budgets, schedules, specifications, etc. All of these different types of artefacts are developed to be easily (re)useable within the product line.

Application Engineering is the process of SPL engineering in which the applications of the product line are built by reusing artefacts. The main input to this activity, besides reusable core assets, is requirements for the new product. During the application requirements activity, product requirements are analysed and related to the existing product line requirements. The resulting product requirements will typically include both instants of existing product line requirements and new product specific requirements.

In the context of product line requirements, several types of variability can be defined. As an example of the expression of this variability:

- Mandatory or optional, for all or only some products in a product line.
- Dependent on other requirements to make sense in a system.
- Mutually exclusive respect to other requirements.
- Variations in their detail or configuration

3.6.3 Guidelines

The definition of product line engineering is driven by three main phases:

1. *Product Management*: the identification of the main costs and economic implications of applying a product line. Basically, this means that the return of investment (ROI) of the application for a product line shall be studied and analysed before its introduction within the company product life cycle.
2. *Domain Engineering*: Identification and analysis of the commonalities and variability in the current products
3. *Product Engineering*: the specific mechanism to use the product line to get a product.

The first phase implies that if the application of a product line is not profitable in the medium-long term, then it is better not to apply it. So, the best way to analyse its profitability is through the study of the return of investment. Assuming that the company wants to establish a SPL with potential return on investment, the next phase (Domain engineering process) shall be applied, and this phase will be mainly focused on the requirement elicitation process. This requirement elicitation process is the most important one, and several steps shall be followed.

(a) Selection of a small set of the whole products of the company to do a first attempt to establish a product line
(b) Collection of all the possible information, documents, and even interview with the experts in order to clarify any possible doubt about the systems, components, etc. And how they are related.
    (i) At this point, it is a good practice to have periodical interviews with the system engineers, and the product managers
(c) Analysis of the common requirements & the variability –
    (i) It is good practice to start with a small example from scratch, trying to get the requirements from a non-real customer. This will provide a view of possible non-consistencies among some requirements, or even detecting dependencies between requirements, that are not trivial to see at a first glance.
        1. Definition of the common/variable architecture in terms of the tool specifications.
        2. Definition of the dependencies among the elements of the architecture according to the tool specifications

All the previous steps will be translated into decisions (business decisions) related with particular configurations of the final product. This is translated into the decision model within the PLUM tool. For each new product from that family, a new instance form the decision model will be obtained (please, refer to the previous example to see how this is done in practice). All the decisions taken will be controlled through the dependencies previously identified. This leads towards to non human errors in the configurations of the new product, and to reduction of the time-to-market, since the configuration is quicker.

The use of the PLUM tool automates the implementation of the SPL within the product life cycle within a company. The main benefits of applying a SPL are: the standardization of the product configuration, less human errors, and reduction of the time-to-market. The use of the PLUM tool benefits the definition of the several roles within the company (domain engineer, application engineer and executive engineer) related with the product definition in the different stages of the product life-cycle.

Chapter 4
Architecture Exploration

Magnus Persson, Stéphane Bonnet, DeJiu Chen, Martin Törngren, Sergey Zverlov, Matthias Biehl, Odile Laurent, Andreas Mitschke, and Marcus Shawky

In this chapter, we will explain and concretise the concept of *architecture exploration* as an approach to architecture design of distributed embedded safety-critical systems. Architecture design involves the mapping of behaviour to structure, or functions to components. Alternative behaviours, structures and mappings are possible, and it is the subject of the architecture design to make a well-informed choice about which alternative(s) are preferable. Architecture design can be seen as an iterative process, as illustrated by Fig. 4.1.

The idea of architecture exploration has previously been successfully employed methodically in other contexts, such as electronic design automation, but the usage has so far been limited in domains such as aerospace, automotive and automation. This is partly due to the different scope of these methodologies, which focus mainly on design problems within the EDA community, i.e., special cases such as system-on-chip design [66, 102].

Architecture exploration involves the development of several different models of architectural design alternatives in order to perform full or partial design optimization according to some selected design criteria. There are a number of different ways in which architecture exploration could be exploited, for example

M. Persson (✉) · D. Chen · M. Törngren · M. Biehl
Kungliga Tekniska Högskolan, Stockholm, Sweden
e-mail: magnus.persson@md.kth.se

S. Bonnet · M. Shawky
Centre National de la Recherche Scientifique, Paris, France

S. Zverlov
KTH/TU München

O. Laurent
AIRBUS Operations SAS, Toulouse, France

A. Mitschke
EADS Deutschland GmbH, Munich, Germany

A. Rajan and T. Wahl (eds.), *CESAR - Cost-efficient Methods and Processes for Safety-relevant Embedded Systems*, DOI 10.1007/978-3-7091-1387-5_4,

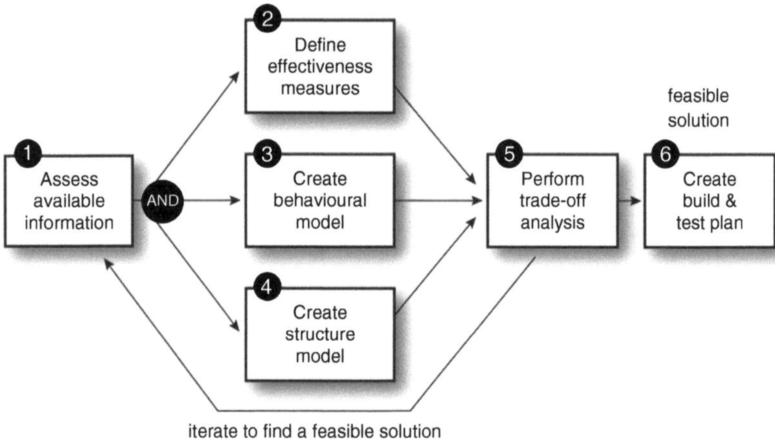

Fig. 4.1 Illustration of key steps and basic concepts in architecture design (Adapted from [120])

for design of a new product software and hardware architecture or for incremental additions to a product line architecture. Such scenarios are described in Sect. 4.1.

The core contributions of this chapter are as follows:

- A discussion and elaboration of concepts and challenges in architecture exploration.
- A methodology for architecture exploration that is applicable to safety-critical systems in general at different levels (system, subsystem, node).
- A discussion on industrial adoption and tool selection.

During architecture exploration, the architectural solution is developed iteratively. As illustrated in Fig. 4.2, at times, parallel development and evaluation work may be carried out on several design alternatives concurrently. A prerequisite to apply the approach is the availability of system information, design goals, design constraints and analysis techniques.

Architecture exploration could be applied on architecture designs at different levels of abstraction, as long as there are sufficient amounts of information available to perform architecture analysis. As such, this means that the methodology could both be applied at early stages, using "back-of-an-envelope" rough analysis methods, as well as at later design stages, with very detailed analysis models. It is however foreseen that the methodology will mainly be applied during the system design stages, i.e., the left parts of the V.

Architecture exploration does not as a concept in itself imply an automated environment, and can potentially be done completely manually by a team of engineers. However, to be able to assess different alternatives against each other efficiently, a significant amount of analysis and evaluation of the architecture alternatives are necessary, and automation becomes indispensable as soon as more than a small number of variants are to be compared.

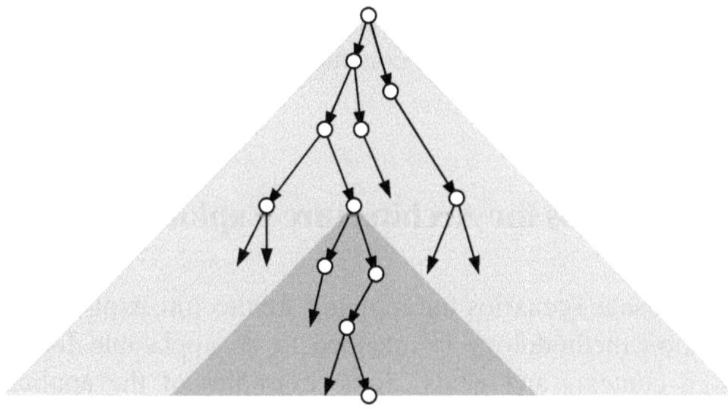

Fig. 4.2 Illustration of design space exploration – *arrows* indicate possible design choices, *circles* design choices which have been further evaluated. Every design choice further delimits the possible future design space, as indicated by the different *grey triangles* in the background

In general, some goals will be easy to analyse, such as the weight, whereas others will be difficult to assess; either because of non-linear relations from design parameters to the qualities, or just because the qualities are hard to assess objectively, such as maintainability. Moreover, many of the goals will be conflicting and the resulting trade-offs and design choices may be far from trivial to make. When changing some aspect of the system design it is typical that this will benefit some qualities whereas have a negative impact on others. For example, if a decision is made to distribute more functionality over additional processing platforms this may be done to improve reliability but will then also increase weight and hardware cost. As a consequence, it is non-trivial to come up with a notion of optimality and trade-offs will have to be dealt with.

These properties and limitations of the approach in the current design environment are clearly illustrated by one of the CESAR generic pilot applications; the simplified doors and slides control system (SDSCS), presented in Sect. 9.2.2. The pilot application covers several design goals and constraints that need to be fulfilled in order to find a usable and acceptable architecture proposal: cost, weight, safety requirements, performance, reliability, etc. In order to calculate the weight of the SDSCS, the weight of the system components (sensors, cables, network components, computers actuators) simply needs to be summed up. Calculating the reliability of the system is more difficult: To assess for example the failure probability of the key functions that the SDSCS fulfils, it is not sufficient to know the failure rates of each individual component. In addition, it must be identified whether components are redundant (meaning that they fulfil the same functionality), and whether the components are connected in a redundant way, meaning that there are several completely independent passes of how information is passed through the components.

In the remainder of this chapter we introduce usage scenarios for architecture exploration in Sect. 4.1. Section 4.2 introduces basic concepts and Sect. 4.3 describes the architecture exploration methodology. Section 4.4 provides a brief review of current types of tools and provides criteria in selecting tools to support the methodology. Sections 4.5 and 4.6 conclude with a discussion of the results and suggestions for further work.

4.1 Usage Scenarios for Architecture Exploration

There are several usage scenarios for applying architecture exploration. The architecture exploration methodology is intended to be applicable for a number of different design contexts and goals. Some examples of the application of the architecture exploration are as follows:

- Optimization ("trimming") of an architecture with respect to one or several criteria (e.g. cost, performance). For example, an existing automotive architecture could be optimised with regards to any or several of the following: the number of ECUs, the amount of cabling, the overall cost of the electronics system, or the overall weight of the system. This could be done continuously as new technology becomes available, making improvements of the optimization goal(s) feasible. The design variability that is available in this example would be changing the amount of included features (e.g. maybe remove the more complex ones), exchanging components for alternative ones, reallocation of software among the ECUs, and possible different variants of network layout (including complete removal of nodes that have no software allocated to them).
- Incremental changes of an existing product, e.g. by adding a new feature or modifying an existing one. A typical case is for the automotive sector: with each new model year, typically a few new features are added to the car. A common approach is to do so merely by adding new components to the existing platform. This is however a suboptimal approach, since changing the existing platform (exchanging components, reallocating software etc.) may imply better overall performance. Essentially, the incremental change can be seen as a special case of the first example, however where a new feature is introduced to kick-start the work.
- Performing sensitivity and/or robustness analysis of the selected architecture, e.g. with respect to errors of design parameters or estimation errors of resource usage, performance or other extra-functional properties. A good example of application of the methodology on this scenario would be a preliminary hardware and software architecture design at an early stage, where many of the extra-functional characteristics are only known at an estimated (budget) level. Such preliminary designs easily break due to misestimation of the necessary resources. Such misestimation may have small impacts (e.g. if the misestimations are for a resource which is not a system bottleneck) or it may have a big impact if it implies overuse of a scarce resource. Using architecture exploration, varying the estimates/budgets, it is feasible to find out which design parameters are most sensitive to misestimation/budget overruns, and which will not have a big impact on the overall system qualities.

- Evaluation of different design concepts when producing a new product entirely from scratch. Here, different concepts would be designed down to sufficient detail level to make relevant extra-functional analyses. Through that, pros and cons of different principal design concepts may be evaluated against each other. The design space would in this case be a small number of concrete design ideas, for example different network topologies for an automotive network.
- Changing part of the system platform, e.g. a network or the processing platform. An example would be if an existing hardware platform has been discontinued. Generating a new hardware platform meeting the same design constraints would be made feasible.
- Decision on design parameters that have not yet been fixed, e.g. the exact variant of processor (cache memory, number of cores, clock speed). This could be based on e.g. performance analysis and cost/benefit trade-offs. The design variability would be the list of processor variants available from the chip manufacturer.
- Uncovering relationships between quality attributes (metrics) and estimate their correlation in the current design. By varying the different input design parameters, it is possible to use architecture exploration to find unknown dependencies, conflicts and other relationships between the different design parameters of a design problem or architecture. By knowing these constraints between the design parameters and the wanted design qualities.
- To perform validation of one or several architecture concepts. If there is no design variability, i.e., the design space has cardinality equal to one, this means that the process collapses down to architecture validation. Architecture validation is used to check if a certain architecture proposal meets its intended design requirements and design goals or not. See further in Sect. 4.3.10.1.

Architectural decisions are not only central to the development at hand, but also very important to document. The decisions capture part of the rationale for the development, and may thus be very important to understand for system maintenance and evolution. System architects should avoid focusing on a single criterion when assessing alternative architecture solutions. Furthermore, they need to take into account that local optimization (e.g., of a subsystem) does not necessarily lead to the best global solution. While the purpose of applying architecture exploration in these different scenarios differ, they all have a common theme: the designer is systematically evaluating alternative designs in order to find a design that is acceptable in the design context.

The main assumption to be met for architectural exploration to be performed is the availability of a model (in some sense of the word) of the system under design to be available for analysis. The model may be expressed formally, using a formal or semi-formal language (examples of the latter are UML and SysML), or simply expressed in natural language. In the latter case, it is only possible to perform architecture exploration manually. Possible design variants should also need to be described.

In the field of hardware development, especially ASICs and FPGAs, the topic of co-design is a common challenge, often, several different implementation alternatives are possible – ranging from more or less parallel hardware solutions to pure software solutions. Depending on the design trade-offs, different solutions may be optimal in different usage scenarios.

4.2 Basic Concepts

To systematically perform architecture evaluation, there is a need for exploration and validation techniques. These involve techniques for analysing individual properties, such as reliability and cost, and techniques for performing trade-offs.

Architecture design is supported by the use of reference architectures, styles and heuristics that can be used to build the initial architecture to be explored. Quantitative metrics together with modularization techniques can be used to automate part of the design space exploration. Architecture design could be performed mainly using quantitative approaches or using qualitative approaches. For more complex systems, due to the trade-offs involved, qualitative approaches dominate and in practice a combination will be used.

4.2.1 A Multi-view Approach to Architecture Exploration

Architecture exploration is closely related to a multi-view approach to modelling (as discussed previously in Sect. 2.3.2), since the architecture exploration typically has to consider several different properties of the proposed architecture that are typically not strongly coupled to each other.

In the setting of architecture exploration, a viewpoint – as a set of rules used to construct and analyse a specific view – should at least include the following information:

- The models and languages used to represent the view
- The metrics and concerns it handles (a single viewpoint can be used for closely related metrics). Even though not a part of the IEEE1471 standard [86], it seems quite clear that metrics for evaluation of the architecture can quite easily be considered part of some suitable view, e.g. reliability metrics with regards to critical failures can be used as a part of the safety view. This would inherently also create a multi-metrics and multi-criteria modelling environment, since metrics from different views will be used to model the same system.
- The analysis methods to be used to compute the metrics and/or analyse the view according to the exploration objectives. While some concerns are directly addressed by the descriptive part of the view (i.e., if the concern is 'what are the functions of the system', a functional model by itself is an acceptable answer), some need to be derived from the architectural description. For instance, the probability of occurrence of a specific failure condition is derived from the

functional hazard assessment of the system, the associated fault trees and cut sets, and the specific failure probabilities of software and hardware components. Although the view itself would just be a list of probabilities for each failure condition, the viewpoint also has to define both the necessary analysis methods and the associated analysis models.
- Tools that are to be used to perform the aforementioned analyses. While not explicitly stated in the IEEE 1471 [86] standard, which only refers to analysis methods, a set of common, reusable tools used to build the view according to those methods may also be part of a viewpoint. It is of particular importance in a multi-view architecture exploration process, as applying those tools to the architectural description is central to the process concretisation.

4.2.2 Metrics, Criteria, Goals and Constraints

As a key part of the system development process, requirements are formulated for the system under design. Some of these requirements will be quantitative requirements on extra-functional aspects of the system, for example on the needed performance, real-time properties or reliability. Each such characteristic may have one or more different concrete metrics, i.e., a concrete way of measuring the property.

The metrics are of different nature. Some are technical (for example related to performance) and some are strategic (for example allocation constraints that are due to organizational or contractual issues). Moreover, the metrics can be easy or difficult to quantify and compare, for example reliability vs. extensibility.

The value of the metrics for a certain system may be computed in different ways. The most common include:

- Analysis results (based on formalised analysis, e.g., cost calculations, safety analysis methods such as FMEA, model checking).
- Statistics, e.g., from simulation or partial test runs.
- Judgement from experienced engineers, such as estimates or assessments.

Design criteria may be formed over the metrics, in order to evaluate the architecture in question. Such assessment criteria may be more or less formal (see Chap. 3 on requirements engineering, which proposes RSL (Requirements Specification Language) & RMM (Requirement Meta-Model) as an approach to formalization of requirements). Assessment criteria differ in nature depending on their usage. First and foremost, some metrics are better if minimised, others if as close to a certain value as possible, yet other should be maximised. Further, for some metrics, hard limits are deployed, making them into absolute *design constraints* (e.g., the reliability level of a safety-critical system), whereas other rather are used as *design goals* and only make sense relative to the results from other architecture options (e.g. overall component cost) [41]. Hence, and to adjust for scaling effects, it is common to convert metrics into some type of common scale.

The metrics may also need adjustment. As an example, sometimes further improvement does not give any (or a very small) additional advantage. In a certain application, maybe increasing the processing speed of the hardware above a certain threshold will not bring any additional overall performance improvement. If the hardware processing speed is used as one of the evaluation criteria for the hardware platform, the metric should be adjusted in such a way that design solutions that increase the processing speed over the required level does not bring any improvement of the metric.

4.2.3 Selection of Metrics and Criteria

The choice of viewpoints has a big influence on the development process. Given a viewpoint, the next step would be to decide on metrics, design goals and criteria. Since there will be an infinite number of aspects that have a more or less vital impact on the overall suitability of the system under design, it is important that the choice of viewpoints and metrics are relevant to the system under design. If not, the system may be optimised for the wrong setting due to the wrong design characteristics being measured ("what you measure is what you get!").

The criteria need to be based on the requirements that apply to other system under development, e.g., based on relevant real-time constraints from the environment. Due to missing system understanding and available time, this may not always be feasible, so simplified constraints that may be partly representative may sometimes by used instead.

When defining the metrics and criteria for a specific evaluation to be performed, it is important to avoid a single-sided assessment of architectural options. It is desirable to use an appropriate criteria space, making sure that an appropriate overall quality of the architecture is maintained. Standards for quality requirements, e.g. ISO/IEC 9126 [90], may provide guidance.

4.2.4 Design Space

Architecture exploration is characterised by the design space that is available for exploration. Considering the development of a new product from scratch, a designer may have many options at hand, including the choice of suitable behaviour, choice of hardware elements and structure, and many solutions for mapping behaviour to the structure.

In practical settings, there will often be many constraints due to already-made design decisions. For example, a fixed network structure will set limitations on the feasible allocation of software onto a network due to bandwidth and delay constraints. Some functions may already be associated with suppliers that deliver integrated components, implying complete absence of allocation freedom.

In many cases, either because the available design space is too large or even infinitely big, it is not possible to exhaustively explore the entire design space. In those cases, by necessity, in practice only a smaller search space will be possible to actually explore. Those search spaces may be chosen in different ways [102].

4.2.5 *Trade-Offs*

Architecture design, due to its multi-criteria nature, by necessity includes trade-offs. End system qualities, such as reliability and performance, depend on shared design parameters, e.g., the number of computers, the allocation of functions etc. Decisions about these design parameters will affect multiple qualities. It is the normal case that no single proposal can be judged to be *the* best one. Rather, different solutions tend to have different advantages and disadvantages in comparison to each other, depending on which stakeholder's point of view is taken. A common case is to have a trade-off between different design properties, e.g., cost vs. performance, such that there may be a fast but expensive system that is suitable in some contexts, and a cheap but slow that is suitable in others. Such a solution, which is the best from at least one viewpoint, is said to be a dominating one, or alternatively: called Pareto-optimal and lie on the so called Pareto border.

Depending on which view is deemed most important, the opinion on which system architecture is most suitable will hence differ. Changing one aspect of the design will typically improve some qualities but may well reduce others. One key challenge in architecture design is thus to resolve these trade-offs.

4.2.5.1 Quantitative Trade-Off Analysis

Of course it is feasible to perform ad-hoc-based trade-off analysis between different design alternatives, but there are also methods with mathematical basis, as outlined by e.g. citeShell2003SynthesisOptimalSystems, introducing mathematical trade-off functions as a means to combine different partial metrics into an overall one. It can be noted that such an approach is equivalent to introducing additional design goals acting as "tie-breakers" among several solutions where there is no single clearly better solution.

Most such mathematical based "tie-breakers" are based on the idea of calculating one joint scoring function based on all relevant metrics. There are several ways to do so. The references [13, 41] provide a short but incomplete list: linear combination, product combination, exponential combination, sum minus product, compromise combination, certainty factors. Of these, the linear combination is most commonly chosen, of which special cases are averages and weighted sums. These are commonly known from e.g., sports ratings and school grades. The other types of combination functions are less commonly used but more interesting from a theoretical standpoint: they all represent means to accomplish trade-offs between different design goals.

Table 4.1 Comparison of combination functions (scale is 1–5, 5 is best). The best option for each combination function is highlighted

Option	Metric 1	Metric 2	Metric 3	Sum	Weighted average (1,2,2)	Product
A	4	4	4	12	4	**64**
B	5	2.5	5	**12.5**	4	62.5
C	3	4	5	11	**4.2**	60

Which combination factor is chosen has a significant impact on which system architecture will ultimately be chosen, especially when the systems are relatively comparable, see Table 4.1 for an illustration of such effects.

The optimisation problem of architecture design now formally is simple: just optimise the relevant compound metric over all possible architectures. In practice, this is not quite as simple. First and foremost, due to state-space explosion, it is normally impractical to test all variants. Further, it is seldom feasible to enumerate design variants automatically; architecture exploration typically is a manual or possibly semi-automated process. Finally, the choice of trade-off function in itself artificially limits the available design space, possibly excluding interesting design solutions. An alternative would be to use multi-objective optimization methods, but that is a significantly harder mathematical problem to tackle.

4.3 Methodology

Architecture design concerns the definition of the fundamental organization of a system. The word "fundamental" derives from the fact that this system organization determines key system properties, such as performance and cost. IEEE defines the term "system architecture" as follows:

> The fundamental organization of a system, embedded in its components, their relationships to each other and the environment, and the principles governing its design and evolution [86]

Architectural design is supported by design patterns, norms, templates and other guidance (e.g., reference architectures). Key design tasks include:

1. Given a certain context and requirements at this level – decide which viewpoints to use and hence to determine metrics, design criteria and design goals to guide the architecture design.
2. Determine/develop system behaviour and structural candidates if not available earlier
3. Perform a mapping of behaviour to the structure – providing architectural candidates which are evaluated against the metrics.
4. Continue the refinement process at a more detailed level if relevant or provide specifications for the continued design.

The process of architecture design is characterised by its recurring and "fractal" nature. Given a top-down approach, the mapping of requirements and behaviours to structural artefacts could for example be seen to take place at the system level (e.g., complete aircraft), at the subsystem level (e.g., propulsion control), at the processing node level, and later even at the chip level.

Whenever alternative behaviours, structures or mapping solutions are possible, it is also the task of architecture design to elicit and describe the design options, and to provide the selection of which option(s) is preferred. While many design alternatives can often be identified and resolved based on stakeholder consensus, heuristics, or design patterns, an explicit support for architecture exploration and evaluation would allow a more systematic investigation of alternatives.

In the following we describe the architecture exploration methodology developed in CESAR.

4.3.1 Architecture Exploration: An Approach to Architecture Design

Architecture exploration is a design approach that provides evaluation of architectural design alternatives in order to achieve full or partial system optimization. It allows a designer to systematically examine alternative design options, and thereby complements the usage of heuristics and domain experiences for justifying design optimization. One of the CESAR project outcomes is such a methodology, as given in Fig. 4.3.

The methodology identifies the key work tasks and artefacts that are of particular concern in regard to architecture exploration in distributed embedded systems:

- Preparation, preparing the work and producing a choice of viewpoints, including tool chain and metrics, design constraints and design goals to be used, starting from informal requirements, use cases, scenarios and similar. See Sect. 4.3.2.
- Baseline architecture design, producing a baseline architecture based on design patterns, norms, templates, existing components and known operational and environmental constraints. See Sect. 4.3.3.
- Design Variability Modelling, producing a design variability model describing the design space under consideration. See Sect. 4.3.5.
- Selection of Search Space, which selects a subset of the design space for actual evaluation (since exhaustive search normally is not feasible). See Sect. 4.3.6.
- Architecture Analysis, providing concrete analysis results for each architecture instantiations that are evaluated. See Sect. 4.3.7.
- Trade-Off Analysis and Decision, where final design decisions are taken based on trade-off between different design goals. See Sect. 4.3.8.
- Documentation, where the outcome of the architecture exploration is documented for future use. See Sect. 4.3.9.

These key tasks and artefacts could be instantiated and configured in different ways. depending on the exact design environment that is used. For example, different companies and industrial domains are likely to use different tool environments.

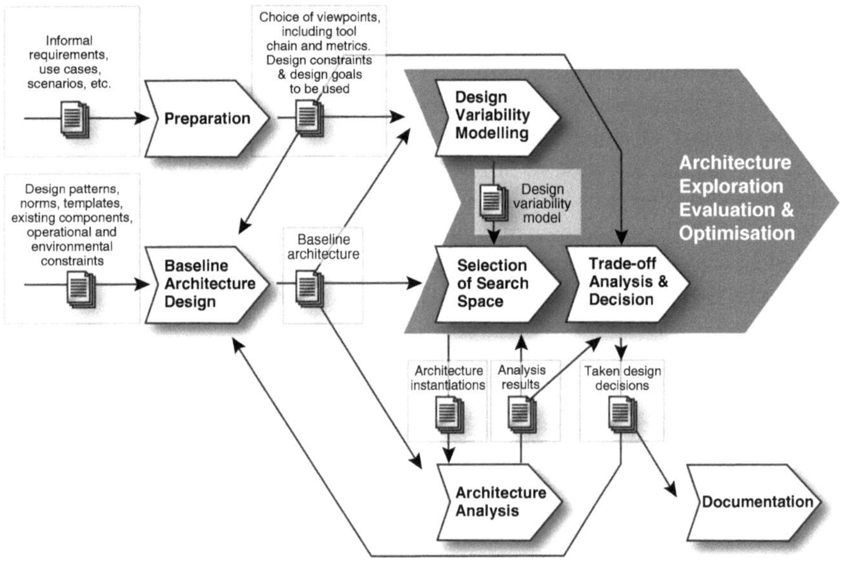

Fig. 4.3 An overview of the architecture exploration process as described in Sect. 4.3.1. The *blocks* represent activities, while the *arrows* between them represent the data flow between the activities

As there is no dependency on using a specific tool environment to use the methodolgy, this means the actually used tools and the methodology need to be adapted to each other before usage. The methodology emphasises necessary dependencies of work tasks and artefacts. For example, the implied workflow is not strictly sequential; the artefacts are not all mandatory, and the process may very well be applied in an iterative manner such that the same work task is revisited multiple times, either with an increasing level of design detail, or due to backtracking some taken design decisions, before its completion. The main point of iterative architecture exploration is to be able to deduce changes to the baseline architecture design (i.e., design decisions) getting it closer to satisfying all the design criteria (constraints and goals) until all are satisfactorily met.

The process inherently employs the multi-view/multi-criteria approach in that it makes it possible to take several viewpoints into account, and allows modelling a system at different detail levels.

Many architectural decisions are taken on the basis of best practices and application domain experience. Resolving design variations in safety- or mission-critical systems in addition requires substantiated decisions with explicit evidences supporting the rationale. It is the task of designers to derive the metrics and criteria for architecture evaluation, to initiate the related analysis, verification and validation tasks (e.g., simulation and function testing), and to assess the analysis, verification and validation feedback.

4.3.1.1 Different Approaches in Applying the Methodology

There are many different ways to apply the methodology, which will have different implications on what type of models are suitable, how the design variability[1] is best described, and other aspects on how to implement architecture exploration. Some of these concerns are briefly described below.

Architecture exploration can be applied both within a certain design abstraction level (to resolve certain design questions or to produce a more in-detail design), or across two (typically downwards, e.g. to allocate software components onto a physical architecture). We distinguish these two as horizontal and vertical architecture exploration, respectively.

Depending on the support for product family, component- and platform-based reuse, two fundamental approaches to the identification of design variability for a baseline architecture can be observed:

- Fan-in: starting from some externally a priori developed functions/components, investigating their conformity and usage in an architecture design (possibly with variability defined) in regard to some functional, extra-functional, and technical criteria of the design.
- Fan-out: starting from an architecture design (with or without variability), searching for candidate functions/components that meet some given functional, extrafunctional, and technical constraints from the design.

For both cases, variability modelling provides the fundamental support for describing the dependencies of alternative solutions as well as being a prerequisite for the related rationale for the eventual design decision. As an important basis for the design space definition, variability models augment the baseline architecture with information about the related functional and technical options.

In practice, these two approaches may be mixed. Some part of the design or some subset of the allowed variation may be approached with one of the approaches, whereas the other approach is applied for other parts of the design.

4.3.2 *Preparation*

In order to start the architecture exploration process, it needs to be prepared. Preparation can be seen as part of the requirements engineering efforts in which

[1]The term *variability* is used within the product line community to describe variations within product lines. We here extend the term to also cover proposed changes to the architecture, that possibly won't end up in the final design, during the system design stage, e.g. in order to perform architecture exploration. When we want to explicitly distinguish between the two types of variability, we talk about *design variability* in architecture exploration and *product line variability*.

requirements are analysed, refined and formalised to make them usable for supporting architecture exploration.

The viewpoints to be used for the specific system under design need to be chosen, including choice of relevant metrics and design criteria. By choosing viewpoints, indirectly also choices of tools are performed, including the choice of how the tools are to be set-up into tool chains. The choice of viewpoints and metrics also implicitly gives a quality attribute model, i.e., a (mathematical) model giving the relations between the choice of subsystem components and other design parameters to overall system qualities, including extra-functional ones such as timing and reliability. The criteria and metrics are often of different nature. Some are technical (for example related to performance) and some are strategic (for example allocation constraints that are due to organizational or contractual issues). Moreover, the metrics can be easy or difficult to compare, e.g. reliability vs. extensibility. It is therefore a challenge in architecture design to refine those qualitative objectives with concrete scenarios or other quantitative metrics. The proper choice of metrics is a vital prerequisite for a successful overall architecture exploration process.

The choice of metrics should be directly related to the system requirements and stakeholder concerns. However, system qualities may be difficult to measure directly. Therefore, often indirect measures are used (one example is to use the number of physical connectors in automotive cable harnesses as indicators of reduced reliability, especially those used in a harsh environment).

In line with the proposed multi-view approach to architecture exploration, choosing viewpoints is one of the cornerstones of the architecture exploration process, as that choice dictates the implementation of most of the subsequent steps.

4.3.3 Baseline Architecture Design

As a starting point, there will always be some design decisions which are typically not formally evaluated, but taken more or less ad-hoc, e.g., based on domain knowledge, design patterns, standards or informal discussion, possibly coupled with outcomes from previous iterations of the architecture exploration process. We call the result from such decisions the baseline architecture design. The reason that the design decisions in the baseline architecture may not have been fully formally evaluated may be that they were already given in terms of input, e.g., an already existing product line and/or legacy parts that are difficult to change due to poor documentation or lack of formal evaluation. The baseline architecture design then corresponds to documenting and (more or less) formalising what this baseline is. As a special case, the baseline would be empty – corresponding to a new design built with the architecture exploration methodology from scratch. In this case, the baseline architecture design would consist of constructing basic structural alternatives, component choice alternatives, etc. This corresponds closely to the approach of platform-based design, where a set of possible platforms are created.

The decisions within the baseline architecture are not expected to change during the continued design work (unless the design concept turns out to be infeasible and rework becomes necessary). The baseline architecture will be kept constant – it can be called the "anti-variable" part of the architecture. Variability may be put into the baseline architecture, either to describe unresolved design issues or to describe resulting variability (e.g. product lines, customization possibilities, etc.). See discussion in later Sect. 4.3.5.

The granularity used for the baseline architecture models may differ significantly between different organizations and development projects; a small high-level model may be used or a very detailed one.

4.3.4 Architecture Exploration, Evaluation and Optimization

There are certain tasks within architecture exploration that are more closely coupled to each other, as indicated in Fig. 4.2. These are design variability definition (Sect. 4.3.5), selection of search space (Sect. 4.3.6) and trade-off analysis and decision (Sect. 4.3.8). Since they are closely coupled with each other, they may be handled jointly by many tools, indicating the need for grouping them closer together in the figure.

Many tools may approach more than one of these three sub-worktasks in an integrated manner, possibly also adding architecture analysis (Sect. 4.3.7). By integrating them, it is feasible to customise a tool for a certain selection of design problems, but integration with tools covering other viewpoints is made more complex. This is made especially obvious considering the fact that many analysis tools are very much specialised towards a single (or a small group of) design viewpoints, e.g., design of the wiring harness only, and few tools are general enough to be able to contribute throughout all design aspects.

4.3.5 Design Variability Modelling

A prerequisite for architecture exploration is the system design space definition [104], i.e., the set of alternative architectures given in terms of design parameters and their possible variability. These need to be (more or less) formally described in order to make it possible to explore the architecture design space. The degree of formality depends on the involved stakeholders, related work tasks and degree of automation involved in the design process. Formalization is however a fundamental necessity for ensuring that correctness of a system implementation can be objectively assessed, possibly automatically by letting a computer do the evaluation. By capturing the design variants (and through the previous choice of viewpoints and metrics their impact on system attributes), the design space for a certain system architecture is spanned.

There are different approaches to variability modelling, as discussed in Sect. 2.6.2. One approach to create a design variability model is to use explicit variability points, i.e., explicit model elements that describe alternative implementation solutions. Using such a variability description, the design space simply becomes all possible combinations of ways to resolve each individual variability point. CVM and PLUM are such approaches. In other approaches, variability points may be more implicit – such as is the case with a model fragment replacement approach, as in CVL.

The baseline architecture, together with the design variability definition and the list of design choices in a particular case, form the necessary basis for instantiation of one of the architecture variants. Together with the choice of viewpoints and metrics from the preparation task (i.e., the so-called quality attribute model), these instances can be used to evaluate specific architecture alternatives from different viewpoints.

4.3.6 Selection of Search Space

In some cases, the architect will only need to evaluate a small number of combinations, but generally, the total size of the design space could be very large (or even infinitely large). Several design variation points may combine with each other combinatorially, creating a design space too large for exhaustive search. In such cases, for practical reasons, it may become necessary to only evaluate a certain subset of the available architecture alternatives. Such selection may be made by using one of the following principles [102]:

- Exploration by hand, i.e., the variants evaluated are selected manually by the architect
- Exhaustive search, i.e., all possible variants are checked
- Reduction to a single objective, i.e., the multi-objective design goals are reduced to a simplified trade-off function, e.g. a weighted arithmetic mean of similarly scaled metrics
- Black-box randomised search – random sampling of the design space
- Problem-dependent approaches – knowledge about the problem structure (e.g. design space pruning based on parameter independence, restriction to promising regions of design space, etc.) is used to reduce the search workload. This group of approaches includes e.g. heuristic algorithms in general, genetic algorithms and different deductive approaches, e.g. by application of heuristics and "coupling tightness" between different metrics. Also, the approach of *design of experiments* [118], which to our knowledge has not been applied within architecture exploration, could be a relevant problem-dependant approach.

The selection of a relevant subset of the design space to perform searches on is sufficient to produce architecture instances for evaluation. In some cases, the selection is very simple (e.g., simply taking the entire design space or a few manually selected instances); in other cases, complex search algorithms will be used. The issue is further complicated by the fact that the order of resolving different variation points might greatly influence the overall amount of work to be performed. For example, during design of an automotive system, the choice of engine fuel (petrol, diesel or other) will have a big influence on other design decisions, e.g., a particle filter will not be needed for a petrol car.

4.3.7 Architecture Analysis

Architecture analysis provides support for predicting the satisfactions of functional and extra-functional requirements by concrete architecture variants, and thereby means for characterising and comparing different solutions. For the analysis, it is necessary to develop/provide analysable models to capture the relevant properties of the design according to the adopted analysis methods and tools (e.g. component behaviours, end-to-end timing, system safety or reliability).

Each architecture analysis tool typically provides support for evaluation of one architecture instance from a single viewpoint. For example: timing analysis requires models that capture the triggering, periodicities, execution time of software in functional chains; whereas one type of safety analysis requires models that capture the failure modes, errors and error propagation. There are however exceptions; either tools that support concurrent analysis of several architecture instances by including variability in the analysis model, or tools that support concurrent analysis of several different viewpoints of the same architecture instance.

Depending on the choice of representation of the baseline architecture, architecture analysis can be applied either directly on the baseline architecture or to architecture instances where design space variability has been resolved. The choice of approach will vary depending on the type of baseline design used and the general approach to architecture exploration. The former approach requires the baseline architecture to constitute a correct (i.e., a complete, analysable) architecture model, whereas the latter allows the baseline architecture to be incomplete, and to give the remaining parts of the design as part of the variability binding for each specific architecture alternative.

4.3.8 Trade-Off Analysis and Decision

Trade-off analysis aids designers in establishing a suitable design choice among candidate architectures. It is sometimes claimed that such a design choice should be optimal; however, such a statement is not necessarily a suitable way to formulate the problem of making a design choice. Firstly, in a multi-criteria approach, there seldom is a single solution that, from all viewpoints, is a global optimum. Secondly, in practical settings, it is seldom necessary to find *the* best solution; rather often a solution close to that level is sufficient.

Trade-off analysis comes in several flavours; it may be more or less formal; there are many options, ranging from simple round-the-table discussions, over illustrative graphing techniques, to full-blown mathematical optimization.

As an analytical method for decision making, quantitative trade-off analysis allows the comparison of alternative design solutions in regards to multiple metrics (also known as for example figures-of-merits, FoMs, if using a standardised scale). With a given architectural preference, all the related design variability can now be bound by making a design decision (also referred to as "delta architecture", i.e., the implicated architectural changes from the design decision), which forms the main feedback to the overall architecture design task from the trade-off analysis. These design decisions are used in two main ways: they form part of the subsequent version of the baseline architecture (or the final architecture to be used) if iterative architecture exploration is performed, and secondly they are made part of the system documentation in a more or less formal way.

4.3.9 Documentation

One key challenge of architecture exploration is to understand and document the reasons different designs are chosen over other. In engineering, design rationale captures the reasoning underlying the creation and use of artefacts [26]; it subsumes the design decision, alternatives and the design decision rationale. Such a decision may be captured at several different levels of detail:

- R0 – artefact level: on the lowest level, there is no rationale, just the artefact that the rationale is attached to. This means that disregarded design alternatives are not captured. One example of this approach may be using iterative versions of the architecture, as is done in traditional version control systems such as Subversion.
- R1 – decision level: provides the different alternatives that were part of the design decision, and the taken design decision. This means that the design decisions are included to the level that discarded options can be seen
- R2 – decision rationale level: provides the different alternatives in each design decision and the reason for them. The reason may be quantitative (e.g. analysis results) or qualitative (e.g. textual motivation).
- R3 – rationale rationale level: additionally provides justification for the chosen quality analysis, quality models, metrics and the trade-off analysis that were selected as criteria for the decision.

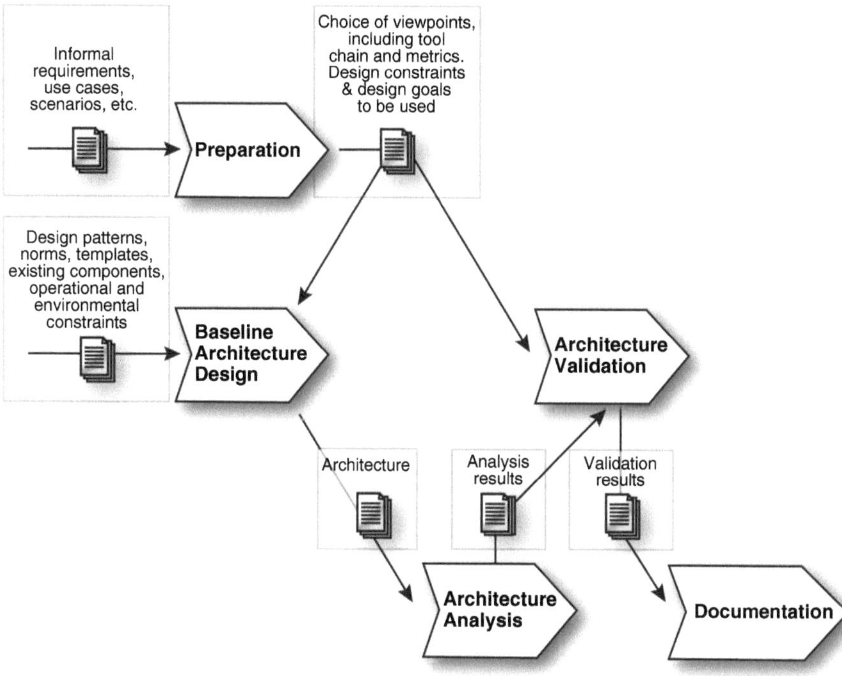

Fig. 4.4 An overview of process-instantiation in the special case of architecture validation

4.3.10 Discussion

4.3.10.1 Architecture Validation: A Special Case

Architecture validation – illustrated in Fig. 4.4 – can be seen as a special case of architecture exploration, where the design space has a cardinality of 1, i.e., there is only a single architecture to validate. Since there is only a single architecture to evaluate, the need to build a description of the design space disappears, and hence also to select a subset thereof as the search space. We can hence remove these work tasks and artefacts from the methodology. Since there is only a single architecture to evaluate, there is also little reason to do a complete trade-off analysis. Rather, the work step occurring after architecture analysis is rather an overall assessment of the single architecture option towards its design goals, with essentially only two possible outcomes: accepted or not accepted (i.e., meets or does not meet the given validation criteria).

Applicable modelling formalisms used:

- Baseline design: complete design to be validated
- Design variability/design space: N/A

4.3.10.2 Tool Integration Is a Key Success Factor

Efficient and accessible tool integration is a key factor in making architecture exploration a feasible approach in practice. The reason is that there is no single tool that is covering the entire design process. Although in theory it is possible to introduce architecture exploration without having integrated tool support, in real life the lack of tool integration will make architecture exploration impractical. The workload overhead implied by creating new architecture variants manually, possibly also ensuring semantic consistency between different modelling environments manually, is in practice not feasible. An integrated approach is needed in order to ensure consistent semantics in the explored design variants throughout the design space, implying that at least shared concepts have a consistent definition e.g., through the use of a standardised core set of concepts (e.g., through the use of a shared meta-model as discussed in Sect. 6.4). Hence, architecture exploration acts as one of the main applications showing the most benefit from systematic approaches to model and tool integration.

4.3.10.3 Inherent Challenges in Multi-view Architecture Exploration

One of the main challenges with any multi-view-based system engineering process is to ensure that all the views are consistent at least each time the architecture description has to be considered as a whole, single entity [43]. Moreover, in the case of architecture exploration, when change decisions are evaluated, it is also important to be able to capture their impact on the views of interest and the associated models. This can be useful to identify trade-off points, i.e., sensitivity points that impact more than one metric (and likely also several views). While consistency and impact analysis can be seen as two different concepts, they are in fact closely related, as they both rely on capturing what data is shared between parallel viewpoints and views. Hence, the same approaches may be used to handle both.

To support a multi-view architecture exploration process, consistency has to be guaranteed each time the architecture is analysed. Indeed, the point is to be able to compare and make decisions based on an overall view of the attributes of a complete baseline architecture instance. That means that some provisions have to be made so that:

- The initial baseline architecture description is consistent (or rather that all the views it is made of are consistent)
- Each subsequent baseline architecture instance after that is consistent when a trade-off analysis is performed. That means that any changes to the design derived from that analysis will have to be propagated across the whole architecture description as soon as the design team is committed to them.

However, there can be phases when ensuring consistency is not necessary, and could even be adversary to the overall process. This is the case if views are handled in parallel by different engineers, to explore possible solutions for a single criterion for instance.

There are several ways to handle the consistency problem (see e.g. [43] for a more in-depth discussion). How the views are chosen will influence the amount of relationships between the views. Using a more cleverly chosen decomposition into views may assist in reducing the consistency problem. Some of the approaches are discussed below.

- *Consistency by design* – Designers manually ensure from the start that all views are consistent, by agreeing on the overall structure and attributes of the system design. Each view is developed by consensus, so as no redundant data is included (e.g. properties describing the same attribute but having different names). When a view is updated, changes are propagated to others manually.
  While it is desirable in all cases to have a consistent design from the start, the fact that most operations are manual and/or involve manual information sharing and management can quickly lead to involuntary introduction of discrepancies in the design artefacts that may not be discovered before many design iterations have taken place, unless a tedious cross-checking process is set-up.
- *Multiview management by linking corresponding model elements* – Another approach is to concretise actual links between artefacts contained within models instead of trying to gather shared data. The links are fully semantic relationships (i.e. completely independent of syntax or even the names of the artefacts) between view model elements describing the same parts of the system design. For instance, one could directly link a component in an AltaRica dysfunctional model [9] used in a safety view to the corresponding hardware device in an AADL [5] model used for a physical view. Hence, when a change is made to any of those two models, it could be automatically propagated to the others. The consistency model can then be represented as a graph (not necessarily connected) where model components are nodes and links are edges. One issue is that the number of links can quickly become very large, as all nodes representing the same system design artefact could form a complete graph. However, some tools such as the link repository discussed in Chap. 6 could be used to implement such an approach.
- *Single source model* – Since the main issue of multi-view consistency is the management of the data shared across multiple views, an idea is to derive view instances from a single data source representing all the knowledge about a specific architecture instance. Typically, that involves having a "master model" of the design that contains either all the view information (the master model being the union of all the views) or only the information that is shared between the views, leaving to each view data that is not shared (the master model being the intersection of all the views). In any case, the master model has to be able to capture the concepts of its dependent views in a semantically consistent manner, meaning that some mapping has to be established between the view model concepts and the master model.

The approach retained within the CESAR project is the latter, where the master model contains only shared data as described in Sect. 6.4. The idea is to define a system engineering data model that captures the information shared between the different views so as to be able to import data from the views to the data model (i.e. updating the data model from the view models) and to export data from the data model to the views (i.e. updating the view models from the data model). This implies some ways exist to automate those transformations. Usually, the import from a master model is easier to automatically handle than the export, since the data model itself is generally much simpler that the modelling formalisms used for the viewpoints.

Two approaches can be taken as for the definition of such a data model. Either some effort is made to try to make it general and suitable to every possible architecture description or it is designed to be application-specific. While the former is more appealing from a theoretical standpoint, it would involve building an all-encompassing ontology for system engineering, which may not be achievable – a well-known example is the STEP standard [91] in the CAD area, which had insurmountable difficulties in achieving an overall agreement on shared concepts. Hence, the second, more feasible approach is preferred as the design team can precisely decide of what goes in or not. This does not preclude some genericness however, as domain-specific data model templates can still be defined to be reusable.

There are also multiple ways to approach the implementation of an exploration process. While the general objective and methodology is well defined, depending on the standard practices of a given organization trying to concretise it, the size of the project, the number of stakeholders and development teams involved, and the nature of the project (a modification of an existing system or a new 'from scratch' development for instance), the actual implementation may differ.

Setting aside preparation steps, which are deemed common to every approach, the main point is to identify the degree of parallel work that is desired or can be carried out. If the project is large enough to assign specific viewpoints to different engineers, a 'divide and conquer approach' is a possibility. In some other cases, a more holistic approach may be preferable. In the first case (Fig. 4.5a), each design engineer can work separately on the viewpoint he/she is responsible for, making local design change decisions according to its associated concerns, through viewpoint-specific sub-iterations. The trade-off analysis and the choice of architecture modifications is then taken by examining the change possibilities for each viewpoint, and propagating the validated changes to all the impacted views once to give out a new complete architecture instance. The main issue of this approach is that since different engineers work in parallel, they can make conflicting decisions. In that case, a prioritization of decisions could be made to decide which should be implemented and which should not during the trade-off analysis step. Another possibility is to only allow work on orthogonal viewpoints or local modifications on system features that do not impact other features worked on. In the second case (Fig. 4.5b), small incremental changes can be made to a specific viewpoint and the impact on other views immediately propagated, to have at all times a full overview of the architecture available for analysis. That approach is more in line with the usual software/hardware/allocation-based exploration process found in the literature, but can become rapidly overwhelming as the number of criteria, and thus viewpoints of interest, increases.

a

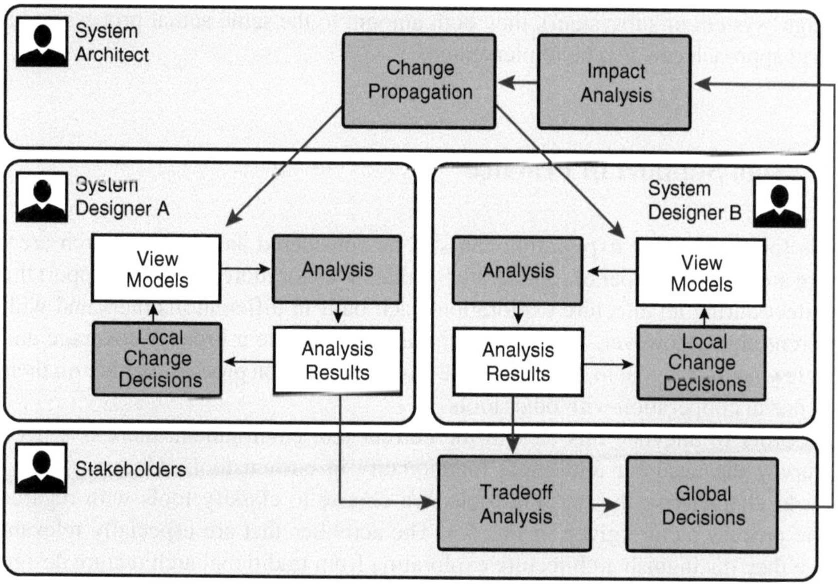

b

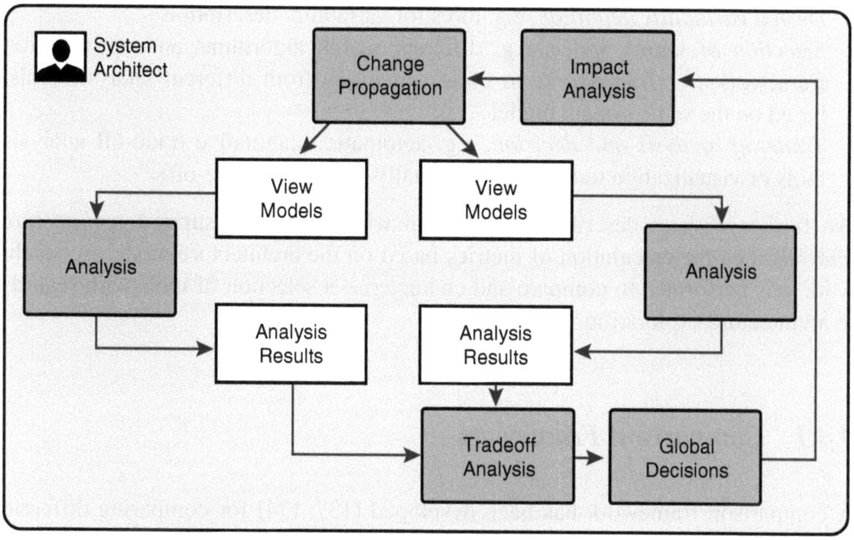

Fig. 4.5 Examples of possible search space exploration processes. (**a**) Divide-and-conquer approach. (**b**) Holistic approach

However, when only a single stakeholder is involved (in the case of a 'simple enough' system or subsystem), they both amount to the same actual process and a hybrid approach can also be implemented.

4.4 Tool Support in Practice

Tools for architecture exploration can still be considered an active research area. There are a large number of tools available that in one or more ways can support the architect during architecture exploration, each built in different manners and with different aims. However, very few of these tools provide a broader coverage and very few have as a goal to fully cover the entire exploration process, neither on their own nor in cooperation with other tools.

In order to alleviate this lack in the current tool environment, there is a need to survey the need for additional functionality in current tools. One way to do so is to characterise the existing tools. We choose to classify tools with regards to the process picture given in Fig. 4.3. The activities that are especially relevant (since they distinguish architecture exploration from traditional architecture design approaches) are mainly

- *Design variability definition*, e.g. tools for variability description
- *Selection of search space*, e.g. different search algorithms and optimization frameworks, or frameworks to collect analyses from different analysis tools, based on the same system model
- *Trade-off analysis and decision*, e.g. automatic quantitative trade-off analysis tools or visualization tools to help manually performed trade-offs.

We further wish to describe the extent in which the tools support *architecture analysis*, i.e., the calculation of metrics based on the architecture model. As such, work was performed to compare and characterise a selection of tools with regards to architecture exploration.

4.4.1 Comparison Framework

A comparison framework has been developed [137, 174] for comparing different tools with each other. The comparison framework provides a structured set of questions which help the user to evaluate the tool in the context of architecture exploration. It also provides assistance to the developers in the early stage of a project in the decision of which tool should be used for which task. As the result of the comparison framework one gets the degree of usefulness of the tool which is reflected in the amount of steps and sub-steps where the tool can assist the developer. This result can hand help to choose the right tool depending on which steps of the architecture exploration are the most important in a certain project and help form a tool chain with the strengths of the tools complementing one another. Evaluating a tool according to the framework can consequently also give guidance when adding new features to existing tools.

The criteria in the comparison framework were derived from different sources: the architecture exploration process (Sect. 4.3), other literature on architecture exploration, the early version of the comparison framework [137] and finally through a discussion with experts on the topic of desirable features of architecture exploration tools. The comparison framework was validated by applying it to a number of tools using the SDSCS experiment as a supportive argument.

4.4.1.1 Structure of Comparison Framework

The comparison framework consists of five high-level criteria. Four of them are directly derived from the architecture exploration activities which were described in (Sect. 4.3) whereby the activities *Design space Definition* and *Deduction of Alternative Architectures* were merged into a single high-level criterion for the reason of their strong dependency on each other. The fifth criterion is called *General* and captures information which is not directly related to the architecture exploration process but still has an influence on the quality of the tool.

These main criteria are broken down into more detailed ones. Apart from the general criteria (which are characterised by a free-form text), every low-level criterion consists of a boolean value (indicating the presence or absence of a certain type of quality of the tool) and a comment. The usage of a boolean is a partial simplification; after all, different tools may implement a feature more or less successfully. For that reason before one assigns the value true to a criterion one has not only to look whether the tool under evaluation possesses this functionality but also to take into account whether the usage of this functionality provides a benefit.

4.4.1.2 Criteria in Comparison Framework

The criteria in the framework are structured hierarchically as follows:

C0. General

Under the category general, the main focus of the tool is given a free-text description (C0.1 Tool focus), and its licensing model (C0.2) and capabilities for data in- and output (C0.3) to or from other tools are described.

C1. Architecture Design

A critical part of designing large software systems is to represent their overall structure and high-level organization. Tools may approach different parts of this work. They may handle requirements (C1.1), either simple management/storage

(C1.1.2), or possibly also allocation onto structural elements (C1.1.2). They may also represent the architectural structure (C1.2) of the system under design. They may do so at one or several different abstraction levels, e.g., feature architecture (C1.2.1), logical architecture (C1.2.2), or technical architecture (C1.2.3). Some architecture tools also provide modelling of behaviour (C1.3) in addition to structure. The modelling could be complete (C.1.3.1) in the sense that it fully describes the actions of the real system, or it could be partial (C1.3.2), in the sense that only parts of the behaviour is described, e.g., only timing aspects. The tool could further handle variability (C1.4), and could be based on different types of model syntax (C1.5), e.g., an industrial standard (C1.5.1), such as UML, or using some tool-specific format (C1.5.2).

C2. Architecture Analysis

Tools that provide architecture analysis may provide a possibility to get well-defined scalar metrics (C2.1), either pre-defined ones (C2.1.1) or customizable ones (C2.1.2). The feedback from the analysis may be either in graphical form (C2.2.1) or textual (C2.2.2).

C3. Design Space Exploration

The tool's approach to design space exploration, if present, is characterised mainly in how it defines the used search space (C3.1) – using a predefined, fixed definition (C3.1.1), or having a somehow customizable search space definition (C3.1.2). Several different exploration strategies (C3.2) may be used, e.g. exhaustive search (C3.2.1), genetic algorithms (C3.2.3) or some other strategy (C3.2.3). The tool may present a result that is giving an approximative identification of the Pareto front (C3.3) or using some other presentation.

C4. Trade-Off Analysis

The trade-off analysis is characterised by the availability of figures of merit (C4.1) – metrics on a comparable scale – that may be either pre-defined (C.4.1.1) or customizable (C4.1.2). If a scoring function (C4.2) is used to compound two or more metrics, it can also be pre-defined (C4.2.1) or customizable (C4.2.2). The feedback to the user from the trade-off (C4.3) may be graphical (C4.3.1) or textual (C4.3.2).

4.4.2 *Characterization of Existing Tools*

We have made an approximate characterization according to the comparison framework of a number of groups of existing tools depending on the intended role and purpose of the tool. Although most existing tools can be clearly classified into a single group, some fall into multiple categories. The profiles of the different tool groups are given in Tables 4.2 and 4.3, with the exception of the licensing (C0.2 which will vary individually from tool to tool) and the general focus, which is described in the introduction of the tool categories below.

The first category of tools are system modelling tools, that tend to focus on modelling/describing the structure of a system under design. Typically such tools operate on a version of an industry standard formalism, such as UML, SysML, AADL [5] or EAST-ADL [154]. Examples include e.g. AADL tools such as OSATE, UML tools such as Rhapsody, MagicDraw, Papyrus [124] etc. Most of these tools handle also related artefacts (e.g., requirements), but usually in a somewhat ad-hoc way.

The next category of tools are analysis tools, i.e., tools with the explicit purpose to analyse a very specific aspect of the architecture under design. These tools are usually built to cover only a very specific stakeholder concern. Examples include FaultTree+ [113] – to analyse reliability – and MAST [57] – to analyse timing.

The next category consists of generic tools that build on mathematical modelling. These are generally extremely general, with a large need for the user to build a suitable model abstraction from scratch as part of the work. The tools may use both symbolic representations of the system as well as rely on external models (e.g., provided from the previous category of tools). Example tools include Microsoft Excel, Matlab, and IBM ILOG Solver [80].

Variability modelling tools generally provide functionality to model variability during system development. To a large content, they may. Examples include e.g., PLUM [153], or tools implementing variability functionality, such as tools with support for the variability part of EAST-ADL.

The final category consists of tools that aim to support the overall architecture exploration process. The only commercial tool available in this category known to the authors is PREEvision [4]. It is likely that there is a need for such tools (or tool chains) to be further developed, as further discussed in Sect. 4.5.3.

4.5 Discussion

This chapter has treated architecture exploration for embedded systems and has provided

- Usage scenarios and concepts for architecture exploration
- A methodology for architecture exploration together with an experimentation where this methodology was used
- A survey of tools for architecture exploration

In this section, we discuss outstanding challenges in making further progress and the methodology, supporting tools and industrial adoption. We conclude by suggesting topics for further work drawing from the discussion.

Table 4.2 Comparison of tools for architecture exploration. A capital X means the tool of this category generally fulfils the criterion. A lower-case x means some tools fulfil the criterion. A dash means that the tools in the group typically do not fulfil the criterion at all

	System modelling	Analysis tool	Math & optimization	Variability modelling	Overall tools
C0 General					
C0.3 Data In- and Output	X	x	–	–	x
C1 Architecture					
C1.1 Requirements	X	–	–	–	X
C1.1.1 Administration	x	–	–	–	x
C1.1.2 Allocation	x	–	–	–	x
C1.2 Architecture structure	X	–	–	X	x
C1.2.1 Feature architecture	x	–	–	X	x
C1.2.2 Logical structure	X	–	–	x	x
C1.2.3 Technical structure	X	–	–	x	x
C1.3 Behaviour	X	x	–	–	x
C1.3.1 Overall	x	x	–	–	x
C1.3.2 Partial	X	x	–	–	x
C1.4 Variability	x	–	–	X	x
C1.5 Model syntax					
C1.5.1 Industrial standard	X	–	–	–	–
C1.5.2 Tool-specific	x	X	X	X	x
C2 Architecture analysis					
C2.1 Metrics	x	x	x	–	X
C2.1.1 Predefined	X	X	–	–	x
C2.1.2 Customizable	x	x	x	–	x
C2.2 Analysis feedback	x	x	x	–	X
C2.2.1 Graphical	–	x	x	–	x
C2.2.2 Textual	x	x	x	–	x

4.5.1 Industrial Needs for Architecture Exploration

Architecture exploration involves the development of system architectures in order to achieve the desired system properties. Architecture exploration is challenging since the architects have to reconcile the concerns and requirements coming from many stakeholders. These concerns, such as cost, performance, extensibility, reliability will tend to drive the architecture in different directions – i.e., they will be conflicting. Moreover, given more advanced embedded systems, the design space and system complexity is steadily increasing, driving needs for better tool support.

It has become apparent in the CESAR project that the process maturity varies among the industrial domains. While the aeronautics domain already has established use of validation methodologies and tool chains that are suitable for incorporation

Table 4.3 Comparison of tools for architecture exploration, continued

	System modelling	Analysis tool	Math & optimization	Variability modelling	Overall tools
C3 Design space exploration					
C3.1 Search space definition	–	x	x	X	x
C3.1.1 Predefined	–	X	x	x	x
C3.1.2 Customizable	–	–	x	X	x
C3.2 Exploration strategies	–	X	X	–	X
C3.2.1 Exhaustive search	–	x	x	–	x
C3.2.2 Genetic algorithms	–	x	x	–	x
C3.2.3 Other	–	x	x	–	x
C3.3 Pareto front identification	–	x	x	–	x
C4 Trade-off analysis					
C4.1 Figures of Merit	–	x	X	–	x
C4.1.1 Predefined	–	x	–	–	x
C4.1.2 Customizable	–	x	X	–	x
C4.2 Scoring Function	–	x	X	–	x
C4.2.1 Predefined	–	x	–	–	x
C4.2.2 Customizable	–	x	X	–	x
C4.3 Trade-off feedback	–	x	x	–	x
C4.3.1 Graphical	–	x	x	–	x
C4.3.2 Textual	–	x	x	–	x

into an architecture exploration process, other domains such as the automotive are not as advanced, making domain-specific demands difficult to assess and consolidate. Part of the reason for this is a perception that exploring a design space to find optimal or at least better architectural solutions can be detrimental in domains where development cost and time-to-market are important metrics of market success.

Architecture exploration as a structured approach to system design is still a rather new approach. This can explain some of the industrial reluctance to the approach. Moreover, a lot of architecture design decisions are not made based on technical merit, but are rather directed by intuition, common practice or even external management pressure (the decision may be "political" rather than technical).

There are a multitude of architecture analysis and optimization techniques available, and some tools, but these are according to the CESAR partners not used to a large extent in industrial settings. We do not really know what developers would do with capable tools since such tools do not exist today. Many tools only handle part of the problem or at best a specific aspect of architecting, such as hardware topology, and tool integration is still a challenge.

It can also be noted that the industrial concerns in the project have been largely targeted to other issues such as requirements engineering, design and to some extent verification and validation. This indicates that there other issues that currently have higher industrial priority.

However, with more and more features added to embedded systems, growing scales of the systems and with high expectations on their performance, more efforts will have to be spent on architecture exploration and the industrial priorities will change. From the processing platform level this will also be driven by multicore components.

4.5.2 Automating (Parts of) Architecture Exploration

The level of automation needed within a concrete tool chain is an important preparation step before choosing which tools to use. For full automation, a significant formalization of the architecture design is needed, along two axes:

1. The description of the architecture itself, including relevant variants, have to be sufficiently formalised such that it is possible to automate the creation of models of the design variants. One approach is through explicit enumeration of either full variants or alternative bindings of individual variation points, as applied in CVM and PLUM. Another possibility is to use a more generic variability modelling language that is, in the general case, only implicitly giving the variants. Such an approach is exemplified by CVL.
2. The description of qualities/metrics used for evaluation needs to be sufficiently described, either through internal formalization or through references to external tools and models, in order to describe which qualities and metrics are relevant for evaluation. This set of qualities and metrics further need to be sufficiently complete for the architecture exploration task at hand, such that no important concern for the architect is disregarded. Of course, some aspects may be hard or impossible to evaluate mechanically. In such cases, a manual or semi-manual approach should be chosen instead.

It is rarely the case that an architecture description encompassing variants is available. Typically, this means either that exploration will have to be done later on, or that more detailed models of the system will need to be deployed earlier on. Such models may need to be based completely on estimates; it is hard to calculate the resource usage of a component before it has actually been implemented. Additionally, it is uncommon that the intended variability between different design options/variants is sufficiently well-modelled.

If only the requirement on formalization is satisfied, (partly) manual evaluation has to be implemented. In the best case, however, the formalised models could be used later by evaluation tools. Alternatively, if only evaluation tools are provided, models have to be created.

Throughout the work in CESAR, several different viewpoints have been defined for description of systems and components. Some of these have received more formal descriptions and semantics, whereas others have more or less only been defined by their names. Since these viewpoints ultimately will be shared between several different tools, there is a challenge in that informal description is not sufficient to ensure semantic consistency. This could be achieved e.g. by giving the viewpoints precise, formal definitions, where several interpretations only are possible if the viewpoints are explicitly defined to be possible to interpret in several ways. Indirectly, formalization of the system descriptions will also be beneficial to architecture exploration, since it will lead to more clearly defined metrics and criteria for the architecture design.

A significant challenge with all architectural development is the issue of incremental and concurrent development. Cross-dependencies throughout the entire architecture may sometimes be triggered by only minor changes. Hence, component models able to handle such constraints would be very useful in the context of architecture exploration. In case any assumptions during architecture exploration cannot be met by the implemented components, this may for example retrigger the exploration, to see if the changed circumstances may make additional changes necessary.

4.5.3 Tool Availability and Limitations

Architectural design constraints and design goals should ideally be linked to the requirements specification process. As described in Chap. 3, requirements can relatively simply be formalised to different degrees. Ideally, these requirements should be directly applicable to the architecture models and could potentially even be evaluated automatically, if the architecture models are also formalised. The current tools for requirements handling and architecture development do not yet fulfil this requirement, mainly due to lacking tool integration between them.

The use of standardised or at least coordinated data models in different tools is an important enabler for architecture exploration to be practically usable. Even though it is possible to ensure consistent system views manually (either by manual checking or by explicitly choosing variability descriptions that are inherently consistent), this is a significant potential bottleneck for productivity.

One business issue for future architecture exploration tools has to do with whether tools will be generic or domain-specific. Are the embedded systems sufficiently similar enough such that "generic architecting tools" can be developed, providing a larger market for such tools or would tools have to be specialised to deal with characteristics specific to the application domain (however providing smaller markets and potentially less of an incentive for tool developers)?

4.5.3.1 Tool Limitations

Tool limitations form an important constraint for the process of architecture exploration. Tools having explicit support for architecture optimization and/or automated assessment of several different architectural options naturally are more easily deployable in an architecture exploration process.

Limitations come in different forms including limitations in

1. Analysis capability. A tool may be able to analyse a few properties. Another tool may be able to analyse some other properties. There is then the option of either extending one tool with more analysis functions if that is possible or to take benefit from both tools, however requiring an extra effort to make sure that the tools can be used as part of the same tool chain in order to avoid manual model transfer.
2. Modelling capabilities. Architecture exploration requires formalization of the system description as a basis for analysis. Given the many viewpoints, different formalisms are naturally used and not all will be supported by one tool. The need then arises to integrate or manage a number of system models (views) required for the different parts of architecture. Moreover, the design space also needs to be described. This could be done in different ways as exemplified by the variability description languages investigated (CVL, CVM and PLUM).
3. Gaps between modelling formalisms mainly used for describing architecture, and others that are more suitable for analysis. Unfortunately, these seldom meet, making the integration of them problematic in practice. Unifying different modelling paradigms still remains a challenge.
4. The detail and width of architectural alternatives (width referring to how many architecture alternatives are evaluated). For every exploration step, architecture analysis has to be performed, either manually or automatically. Potentially, partial design work may also be necessary. As the number of options grow larger, the effectiveness of these calculations grows increasingly important. Hence, an important issue is to build efficient and scalable algorithms.

There is a trend towards system level design tools as well as life-cycle management tools, but these are so far not yet directed to support architecture exploration, as already described in Sect. 4.4.

4.5.3.2 Improved Situation with the CESAR RTP and CMM

The developed technology, in form of the RTP (Chap. 6) and common meta-models (Sect. 6.4), together with the specification of interoperability (Chap. 8), and concrete integration software such as for example ModelBus, will contribute to reducing tool integration efforts. By building these integration frameworks, several different gaps mentioned above will be closed.

- By making it possible to automate model interchange between tools, the work needed to perform parallel evaluation of an architecture in several different tools, and in several different variants, is greatly reduced.
- By working on giving different models of the same system a consistent semantics through the proper use of semantically true model transformations, synchronization and model analysis is simplified. Since the work needed for translating models into the format needed by their analysis tools is decreased, this should reduce the effort to build model transformation by removing many of the technical representation problems [20], in contrast to the actual model content itself.
- Using standardised protocols and models and otherwise designing tools to be interoperable, they are more easily integrated into tool chains.
- To the extent that a common formalism (e.g., a formally defined meta-model) is used, this will help reduce the problem of ensuring consistency in different separate models, as will standards and specifications to ensure interoperability. By having built such formalisms modularly, different tools may choose to implement only a part of the specification. It is however no silver bullet – efforts are needed also in configuration management, process, etc.

4.5.4 How Should Industry Adopt Architecture Exploration?

A central tenet in this chapter is that there is a need for a better understanding of the architecture exploration process. For this purpose an emphasis was placed on methodology.

The main strength of architecture exploration and the suggested methodology is that it allows design teams to formalise and rationalise the evaluation of alternatives when architecting a system. Indeed, no system engineer will come directly to an ideal solution when designing a new system or updating an existing design: some alternatives will always be considered at each step of the design process, from the high-level functional design down to the detailed design. However, most often this is done in an ad-hoc, informal manner, where engineers select or discard solutions from experience, habit or familiarity considerations. Implementing an architecture exploration process can thus help design teams make better decisions about the architecture as a whole, enhance traceability of those decisions and rationalise them by expressing hard factual arguments, in the form of metrics, for each decision.

Industrial deployment of architecture exploration could target early stages, with little detailed information and formalised models, or later stages where more detailed architecture exploration and verification is possible. In the former case, rough assessment of feasibility and system properties would likely be sufficient as a basis for moving on from conceptual to further development; consequently, it may be sufficient with spreadsheets, Matlab or something similar to support the work. In the later stages, when more detailed analysis is desirable, there are stronger needs for formalised system models and more capable tool support.

In this discussion chapter, we have addressed key issues that we believe are important to consider before embarking on detailed architecture exploration.

- Consideration of usage scenarios, efforts and investments before acquiring tools and building models.
- Selection of tools and a tool integration approach that provide model and analysis reuse, adaptation to specific needs and ease of use for the actual engineering purposes.

4.6 Future Work

Systems architecting is an extremely important activity for embedded systems in laying the foundation of key structures that determine central system properties. Given the increasing sophistication of such systems there will be a need for improved theories, method and tools. As discussed in this chapter, this concerns a number of topics where we here highlight suggestions for future work encompassing both academic research as well as industrial evaluation of supporting methodologies.

- This Chapter has provided a methodology. This methodology should be exposed to further case studies at different abstraction level in order to refine it.
- Further work is required to investigate modelling techniques for architecture exploration that are scalable, support multiple viewpoints and expressing both design and product line variability.
- Related to modelling, there is also a need to adapt existing constraints and requirements languages such that they can be more easily integrated with architecture design and the design process as a whole.
- To support architecting there should be opportunities to refine existing and develop new heuristics and analysis tools that would be suitable for early stages of architectural design.
- Evaluating architectures from multiple viewpoints leads to challenging multi-goal optimization and multi-criteria decision problems for which multiple algorithms can be employed. Here there is a need to further these techniques such that they are applicable to embedded systems architecting.
- There is a need for applied research, using scenarios and prototypical tools, in order to develop a better understanding of suitable functionalities and abstractions of architecting tools tools at different design stages.
- Finally, to enable tool development and information reuse, there is a need for a systematic approach for information management and tool integration. With its many concerns, architecture exploration is a typical example of a tool functionality that is highly likely to benefit from integration (reusing information such as requirements, using other analysis tools, providing information to other tools).

Chapter 5
Component-Based Development

Kung-Kiu Lau, Marc Pantel, DeJiu Chen, Magnus Persson, Martin Törngren, and Cuong Tran

In this chapter, we focus on the use of component-based development (CDB) in CESAR. First, we introduce what we mean by components and how they fit into the product lifecycle. Then, we report on the two major technological innovations of the project: the X-MAN and HRC frameworks.

5.1 What Is Component-Based Development?

The aim of component-based development is to (i) build components and deposit them in a repository; and (ii) use or reuse these pre-existing components to build many different systems by assembling the components using pre-defined composition mechanisms. This is illustrated by Fig. 5.1.

For a given problem domain, components are identified and developed, using domain knowledge. Similarly, the composition mechanisms for the components are defined and fixed for the domain.

For hardware systems, component-based development is of course standard practice, and components are standard parts that can be assembled in pre-specified ways. For instance, chips can be assembled by wiring their pins together.

For software systems, the most widely accepted generic component is depicted in Fig. 5.2: it is a software unit with *provided services* (lollipops) and *required services* (sockets). These components are composed by connecting provided services with matching required ones, as shown in Fig. 5.3.

D. Chen · M. Pantel · M. Persson · M. Törngren
Kungliga Tekniska Högskolan, Stockholm, Sweden

K.-K. Lau (✉) · C. Tran
The University of Manchester, Manchester, United Kingdom
e-mail: kung-kiu@cs.man.ac.uk

A. Rajan and T. Wahl (eds.), *CESAR - Cost-efficient Methods and Processes for Safety-relevant Embedded Systems*, DOI 10.1007/978-3-7091-1387-5_5,

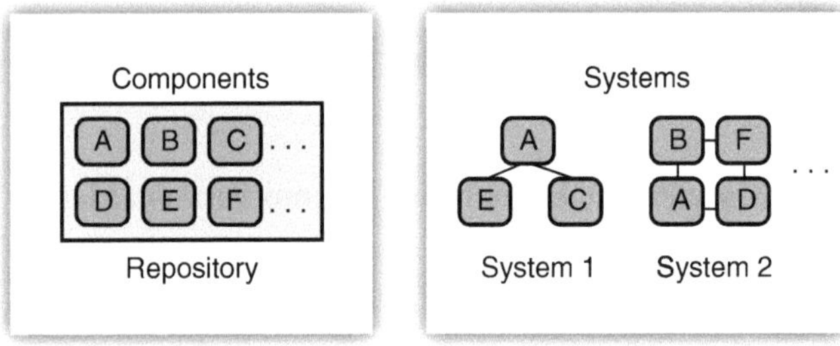

Fig. 5.1 Component-based development

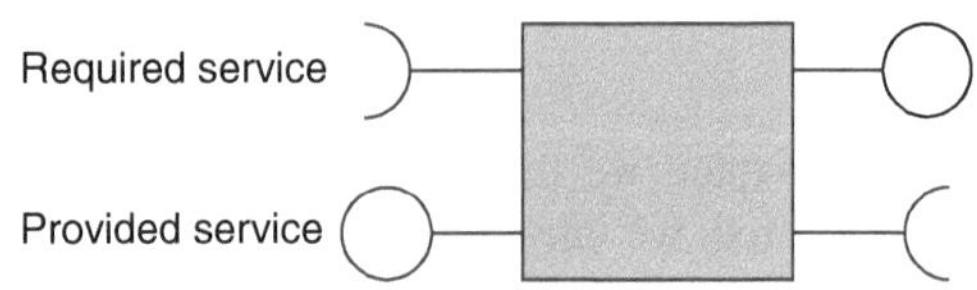

Fig. 5.2 A generic component

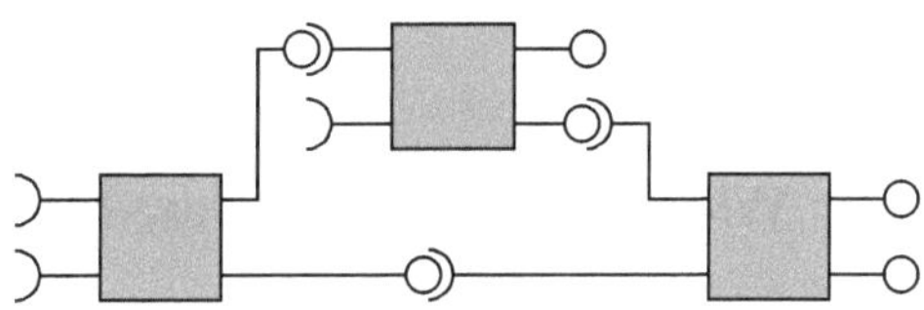

Fig. 5.3 Composing generic components

Different types of components are defined as different instances of the generic component, together with corresponding composition mechanisms. Such definitions are given by *component models*.

5.2 Component Models

A *component model* [77,105] defines: (i) what components are; and (ii) composition mechanisms for composing the defined components.

For software systems, there are three categories of component models [105] with (i) *objects*; (ii) *architectural units*; and (iii) *encapsulated components* as components, respectively. Table 5.1 shows these categories and examples of component models therein. The three kinds of components are illustrated in Fig. 5.4.

An object is similar to a generic component except that an object does not have or show any required services (hence sockets in dotted lines). An architectural unit is just the same as a generic component, except that an architectural unit uses ports instead of services. An encapsulated component has only provided services, but no required services; it therefore performs all its computation within itself, i.e., it does not call another (encapsulated) component.

Table 5.1 Categories of component models

Category	Components	Sample component models
(i)	Objects	EJB
(ii)	Architectural units	CMM, EAST-ADL, AADL
(iii)	Encapsulated components	X-MAN

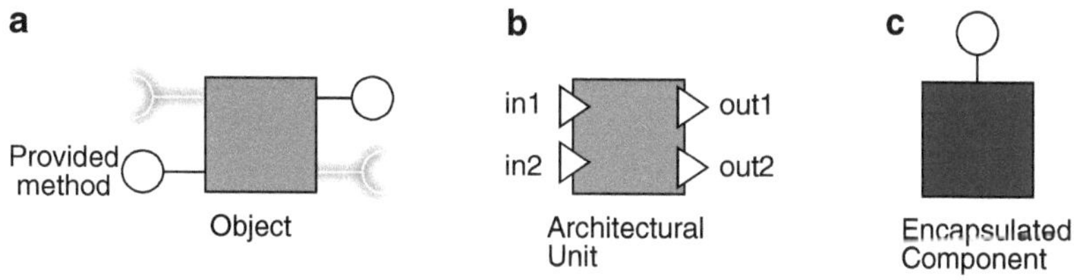

Fig. 5.4 Types of components

Table 5.2 Composition mechanisms

Category	Components	Composition mechanism
(i)	Objects	Method delegation
(ii)	Architectural units	Port connection
(iii)	Encapsulated components	Coordination

The composition mechanisms for these three kinds of components are shown in Table 5.2.

Objects "compose" by method delegation, i.e., direct method calls (or direct message passing). Architectural units compose by connecting their ports (or indirect message passing). Encapsulated components compose by coordination, i.e., exogenous connectors that coordinate control flow between components. These composition mechanisms are illustrated in Fig. 5.5.

For software systems, a number of development processes, or *life cycles*, for CBD have been proposed, e.g. [27, 29, 37, 100, 142], to name a few. (A recent survey can be found in [98].) Naturally these processes all reflect the desiderata of CBD [23], and converge on the general view depicted in Fig. 5.6.

The generic CBD process in Fig. 5.6 comprises two separate processes: one for *component development*, and one for component-based *system development*. Component development is also known as "development *for* reuse", since it is concerned with developing components that can be stored in a repository and (re)used to build different systems. Component-based system development is also known as "development *with* reuse", since it is concerned with developing systems by reusing pre-built components (the result of the component development process).

Each process beginning follows the same life cycle of "requirements analysis, design, implementation, testing and maintenance". For component development, implementation is a single activity, whereas for system development,

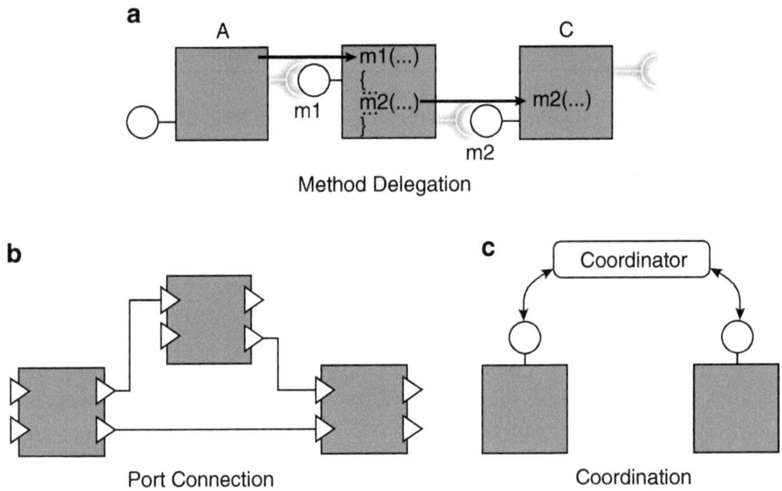

Fig. 5.5 Composition mechanisms

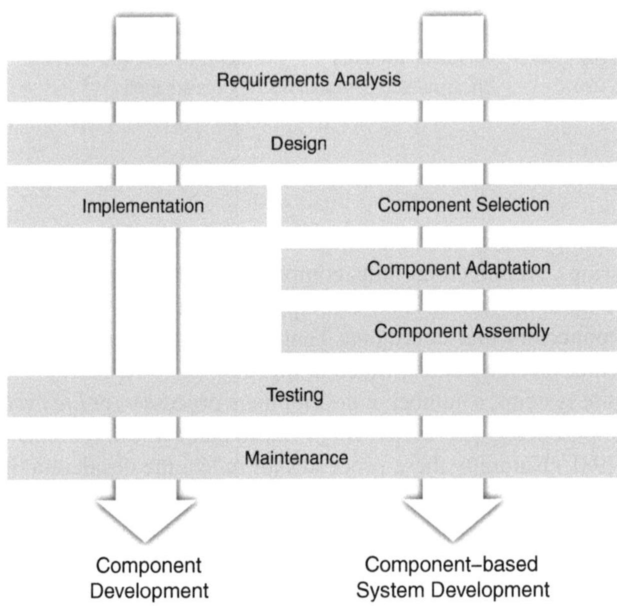

Fig. 5.6 Standard CBD processes

implementation is a sequence of activities based on pre-built components, namely component selection, adaptation and assembly.

Figure 5.6 does not differentiate between the context for the component life cycle and that of the system life cycle. In fact, the component life cycle is followed

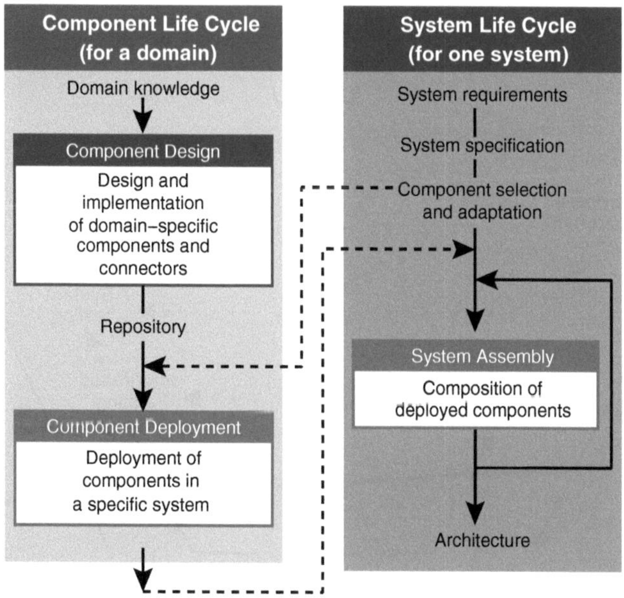

Fig. 5.7 Idealised CBD life cycle

by the component developer for the whole domain, whilst the system life cycle is followed by the system developer for each system to be constructed. This distinction is illustrated in Fig. 5.7.

Figure 5.7 differs from the standard CBD process (Fig. 5.6), in that its component life cycle is a more complete one, namely the idealised one [105]. The idealised component life cycle is called so because it meets all the desiderata of CBD that have been identified in the literature [23]. It consists of two phases: component *design* and component *deployment*, and is set in the context of a problem domain. In the design phase, components are (identified and) designed and constructed according to the domain requirements or knowledge [106], and deposited into a *repository*. Repository components are domain-specific but not system-specific. In the deployment phase, components are retrieved from the repository and instantiated into executable component instances which are then deployed into a specific system under construction.

The system life cycle also differs slightly from that in Fig. 5.6 in that system design is now replaced by a completely bottom-up process of component *selection* (from the repository) and *adaptation*, followed by (component *deployment* in the component life cycle followed by) *system assembly*, which is simply the composition of the deployed components. The bottom-up nature of this process is indicated by an iterative loop in Fig. 5.10. It is worth noting that within this loop, the component life cycle links up with the system life cycle, since deployed components (from the component life cycle) are iteratively assembled into the system under

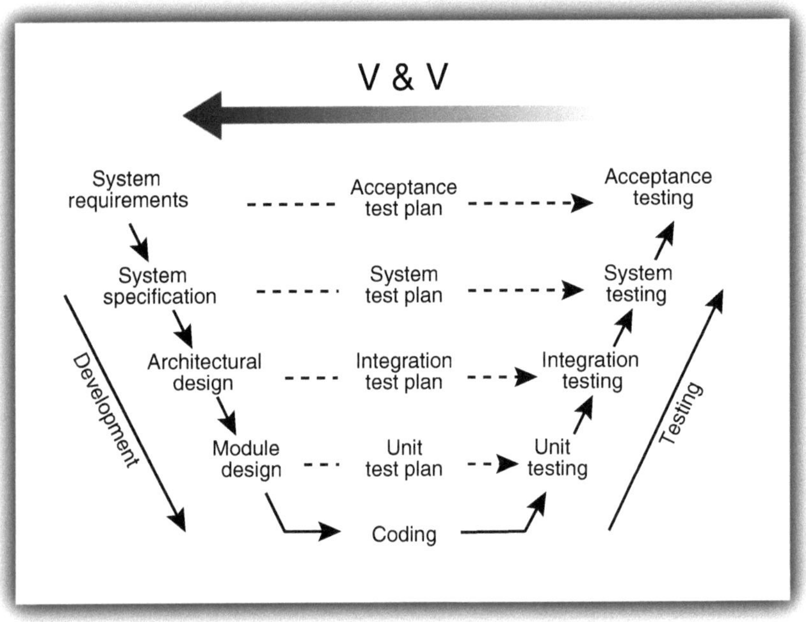

Fig. 5.8 The V-Model

construction (in the system life cycle). This link is denoted by the dotted arrows between the two life cycles in Fig. 5.7, via the step of component selection and adaptation, and the step of component deployment.

5.2.1 The W-Model

The CBD process in Fig. 5.6 does not explicitly address Verification and Validation (V&V). For general (modular) system development, the standard model for V&V is the V-Model [159]. The V-Model is an adaptation of the traditional waterfall model for *modular* system development. It defines a process consisting of phases for requirements, system specification, system or architectural design, module design, implementation and testing. Implementation consists of coding for the individual modules, and coding for integrating the modules into the entire system using the architectural design for the system. Testing follows coding [159] (Fig. 5.8). Thus, the coding phase divides the whole process into *development*, the left arm of the V, and *testing*, the right arm of the V.

The key property of the V-Model that is pertinent here is that it is a *top-down* approach to system design and development, as Fig. 5.8 shows. First, a top-level design is made of the architecture of the entire system; this identifies and specifies subsystems or modules, and their inter-relationships. Then the individual modules are designed according to their specifications in the top-level design. In general, this top-down approach may be applied successively, each time decomposing subsystems or modules in the current level of design into further sub-systems or modules. This decomposition is repeated as many times as is necessary, until a final design is arrived at in which the design of the system as well as all the individual modules is deemed complete, i.e., no further decomposition is necessary or desirable.

Compared to the standard CBD process in Fig. 5.6 and the idealised CBD life cycle in Fig. 5.7, both of which contain two life cycles,[1] one for component development and one for system development, the V-Model contains only one life cycle, for system development.

Furthermore, the standard CBD process in Fig. 5.6 and the idealised CBD life cycle in Fig. 5.7 also show CBD as an essentially *bottom-up* approach to system design, in the sense that components have to be developed first (in the component life cycle), and any particular system is constructed from these components (in the system life cycle). In contrast, the V-Model (Fig. 5.8) is essentially a *top-down* approach to system design: the system is designed first (thus identifying the required components), and then components are developed.

A straightforward adaptation of the V-Model for CBD would be to retain the top-down approach to system design but use a component as a module, as shown in Fig. 5.9.

For example, the V-Model adopted by the avionics industry as a CBD process (e.g., Airbus processes [64, 67]) is such an adaptation.

However, such a straightforward adaptation of the V-Model is at variance with the standard CBD process in Fig. 5.6, precisely because it does not include a component life cycle and consequently does not incorporate the bottom-up nature of CBD.

An adaptation of the V-Model for CBD that does incorporate the bottom-up nature of CBD is that of [36]. It does so by containing separate life cycles for component development and system development, like in Fig. 5.6. However, this adaptation really applies the V-Model only to its system life cycle; there is no evidence of the V-Model in its component life cycle (which is the same as the one in Fig. 5.6).

In our view, to adapt the V-Model properly for CBD, we need not only to incorporate both the component life cycle and the system life cycle, but also to apply the V-Model to *both* of these cycles. In addition we need to specify a component model that defines the components (and their composition) properly. (A definition and survey of component models can be found in [105].)

[1] We will use 'life cycle' interchangeably with 'development process' or simply 'process'.

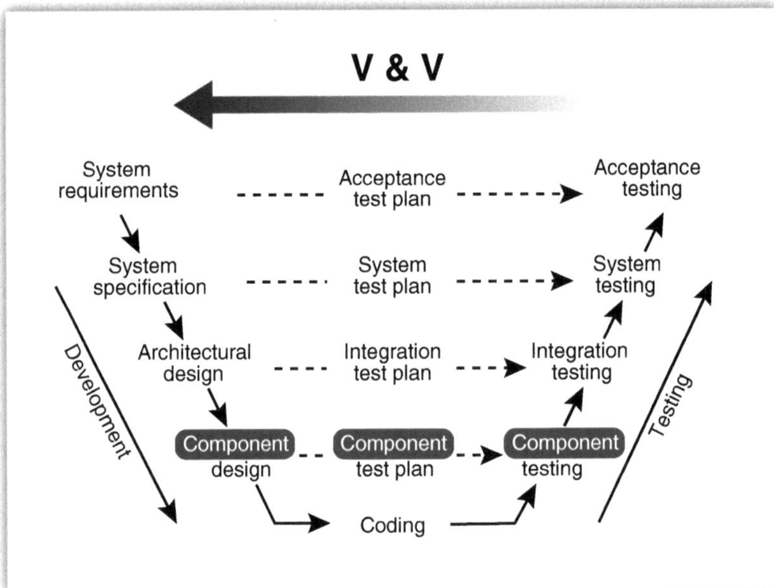

Fig. 5.9 Adapting the V-Model for CBD

We have defined such an adaptation, using a component model that we have defined ourselves. Now we describe our adaptation, which we call the W-Model,[2] for reasons that will become apparent later.

In CESAR, we use the X-MAN component model [107, 108, 169],[3] and we have defined a CBD process based on X-MAN. This process is the one shown in Fig. 5.7.

Applying the V-Model to both the component and system life cycles yields a CBD process with V&V as shown in Fig. 5.10.

Compared to the straightforward adaptation of the V-Model in Fig. 5.9, *component V&V* (which corresponds to component testing in Fig. 5.9) now occurs in the component life cycle, whilst *compositional V&V* (which corresponds to integration testing in Fig. 5.9) and *system V&V* (which corresponds to system testing in Fig. 5.9) occur in the system life cycle.

The X-MAN CBD process with V&V in Fig. 5.10 can be re-cast straightforwardly as a process with two conjoined V-Models, one for the component life cycle and one for the system lifecycle. These two V-Models are conjoined via the step of component selection, adaptation, and deployment. This "double V" process is

[2]The name W-Model has also been used in software testing [146] and product line engineering [112] in the context of traditional (i.e., non-CBD) software engineering.

[3]This name was not used in these papers, but has been created within the CESAR project.

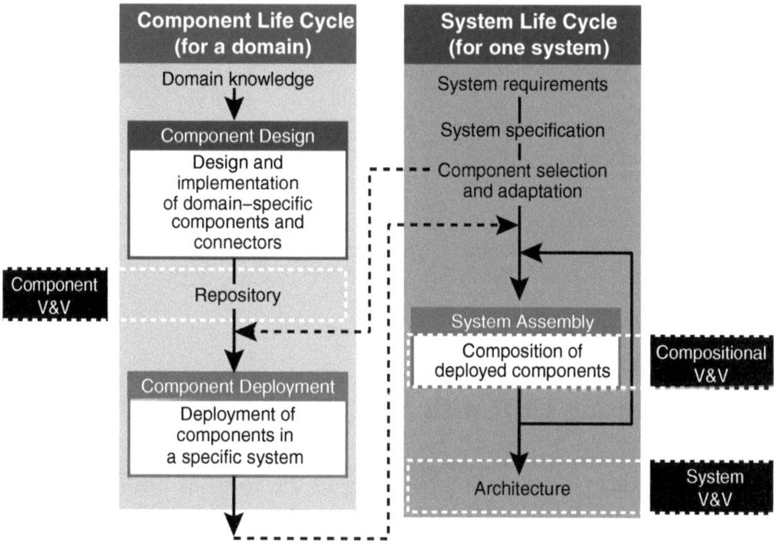

Fig. 5.10 X-MAN CBD process with V&V

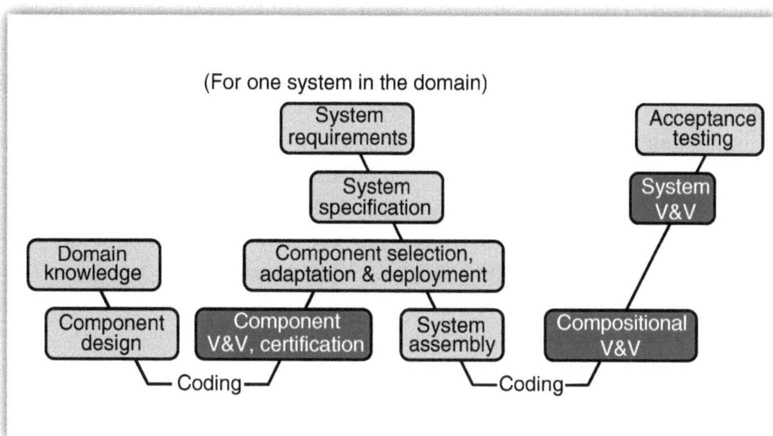

Fig. 5.11 The W-Model

shown in Fig. 5.11. We therefore called it the *W-Model* [109].[4] We have highlighted the V&V activities in the W-Model by boxes with black borders.

[4]In English, "W" is pronounced "double u"; there is no letter pronounced "double v".

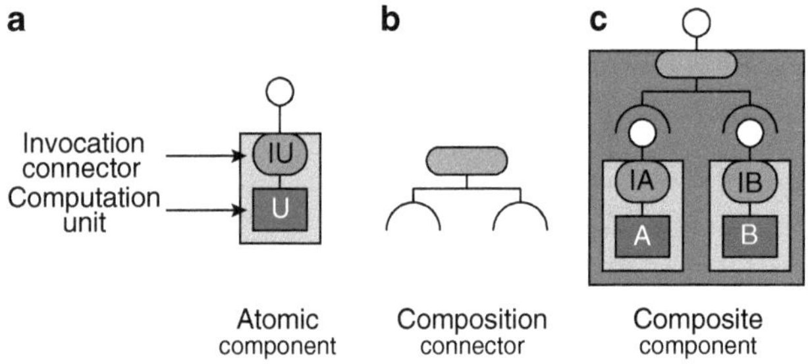

Fig. 5.12 X-MAN component model

5.3 Implementing CBD with X-MAN

In the CESAR project, CBD is implemented in the form of a tool for X-MAN. This tool implements the X-MAN component model as well as the W-Model. The X-MAN tool is implemented in a model-driven manner. It has been implemented using the GME toolkit [69]. In GME, meta-models that contain definitions of elements, structures and syntax first have to be defined using a class diagram notation similar to UML class diagrams. Models can then be created by instantiating the pre-defined meta-models. To provide behaviours for models, GME allows us to develop interpreters that can interact with models, i.e., execute or manipulate models.

To implement the W-Model, we had to implement: (i) the X-MAN component model, for defining and constructing components and their composition mechanisms; (ii) the component life cycle in the W-Model; (iii) the system life cycle in the W-Model; (iv) the link between the component and system life cycles; (v) component V&V; (vi) compositional V&V; and (vii) system V&V. For lack of space, we cannot describe all these fully; so we only highlight the key elements of the implementation, and briefly describe them.

5.3.1 *The X-MAN Component Model*

The X-MAN component model has been described in several papers, e.g., [107, 108, 169]. Here we give a brief summary.

In X-MAN there are two kinds of components: atomic and composite (Fig. 5.12). An atomic component contains a computation unit and an invocation connector. The computation unit provides methods or functions which can be invoked via the invocation connector. An atomic component is encapsulated in the sense that all its computation occurs within its computation unit, i.e., an atomic component does not invoke the methods of other components. Thus an atomic component has only provided services (denoted by lollipops) but no required services.

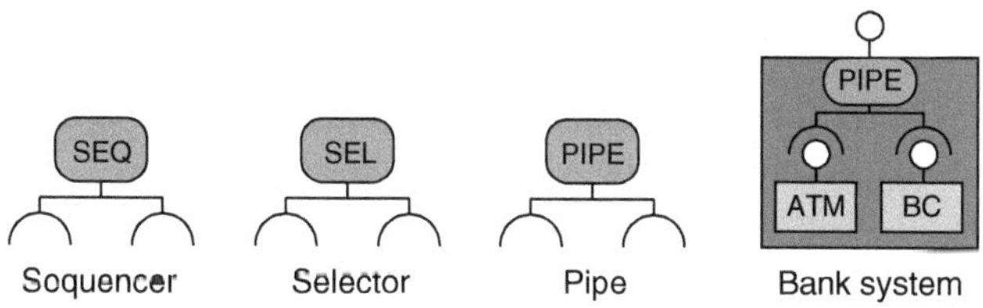

Fig. 5.13 X-MAN composition connectors

A composite component (Fig. 5.12c) is built from atomic components by composition connectors (Fig. 5.12b). A composition connector coordinates control (as well as data flow) between components, i.e., it coordinates the invocation of services between components. A composite component is also encapsulated, and it also has only provided services; this is a direct consequence of the encapsulation of an atomic component. In general, composite components are self-similar; this is important as it means that we have *compositionality*. In particular, this enables us to do *compositional V&V*.

The X-MAN connectors include the Turing-complete set of control structures: sequencing, branching and looping. Sequencing and branching are composition connectors for multiple components, whereas looping is an adaptor for a single component. For sequencing, we have the *sequencer* and *pipe* composition connectors; for branching, we have the *selector* composition connector (Fig. 5.13). A sequencer only passes control to the next component, whereas a pipe passes control and results from the first component to the next component, as shown in the bank system example in Fig. 5.13.

In the X-MAN Tool, the X-MAN component model is implemented as part of the component life cycle and the system life cycle (see below).

5.3.2 *The Component Life Cycle*

As mentioned in Sect. 5.2.1, the component life cycle in the W-Model is the idealised one, consisting of the component design phase and the component deployment phase. To implement the component life cycle in GME, we first need to define meta-models for these phases. However, as we also pointed out in Sect. 5.2.1, the component life cycle links up with the system life cycle during component deployment and system assembly. Therefore, it is more convenient to put the component deployment phase in the meta-model for the system life cycle. Consequently, the meta-model of the component life cycle consists of just the meta-model for component design phase.

To implement the component life cycle in GME, we define the meta-model for the component design phase and implement an interpreter for it. We also implemented a component repository including interpreters to support component deposit and retrieval.

5.3.2.1 The Meta-model

The meta-model for the component life cycle is presented in Fig. 5.14. It defines component design according to the X-MAN component model.

The top-level element is a *Design*, which contains the definition of a (atomic or composite) component according to X-MAN (Figs. 5.12 and 5.13). The property *ExecutableCode* of a *computation unit* provides computation implementation. Each *method* in a computation unit can be augmented with a contract that is defined by pre- and post-conditions. This contract is used for *component V&V*.

Each atomic or composite component must have one *interface* that consists of one or more *services* and *data elements*. A *service* contains references to methods provided by the component's computation unit. A *data element* denotes a datum that is required or provided by the component. As components in the design phase are templates, data elements are characterised with type, min, max and default values; actual values are provided when instantiating these templates during component deployment. Finally, a component can have an invariant, which is defined in its interface.

5.3.2.2 Component Designer

We have constructed a Component Designer that supports component design, V&V, and storage. Figure 5.15 shows a component, namely *Locker*, under construction in the Designer, with a design palette on the left, tree view of the design on the right, and the main design view in the middle. The main view shows that *Locker* is an atomic component built from connecting an invocation connector *Locker_INV* to a computation unit *Locker_CU*. The component exposes its interface containing the provided service *getLocking*.

When completed, a component can be stored into the component repository. This activity is triggered by selecting the *DEP* button on the toolbar (see close-up in Fig 5.15).

5.3.2.3 The Repository

For the component repository, we used the Firebird database [62] and for database access we used the SQLAPI driver [147].

5.3.3 *The System Life Cycle*

The system life cycle contains the iterative loop of component selection and adaptation, component deployment and system assembly. To implement this life cycle in GME, we defined a meta-model for component deployment and system

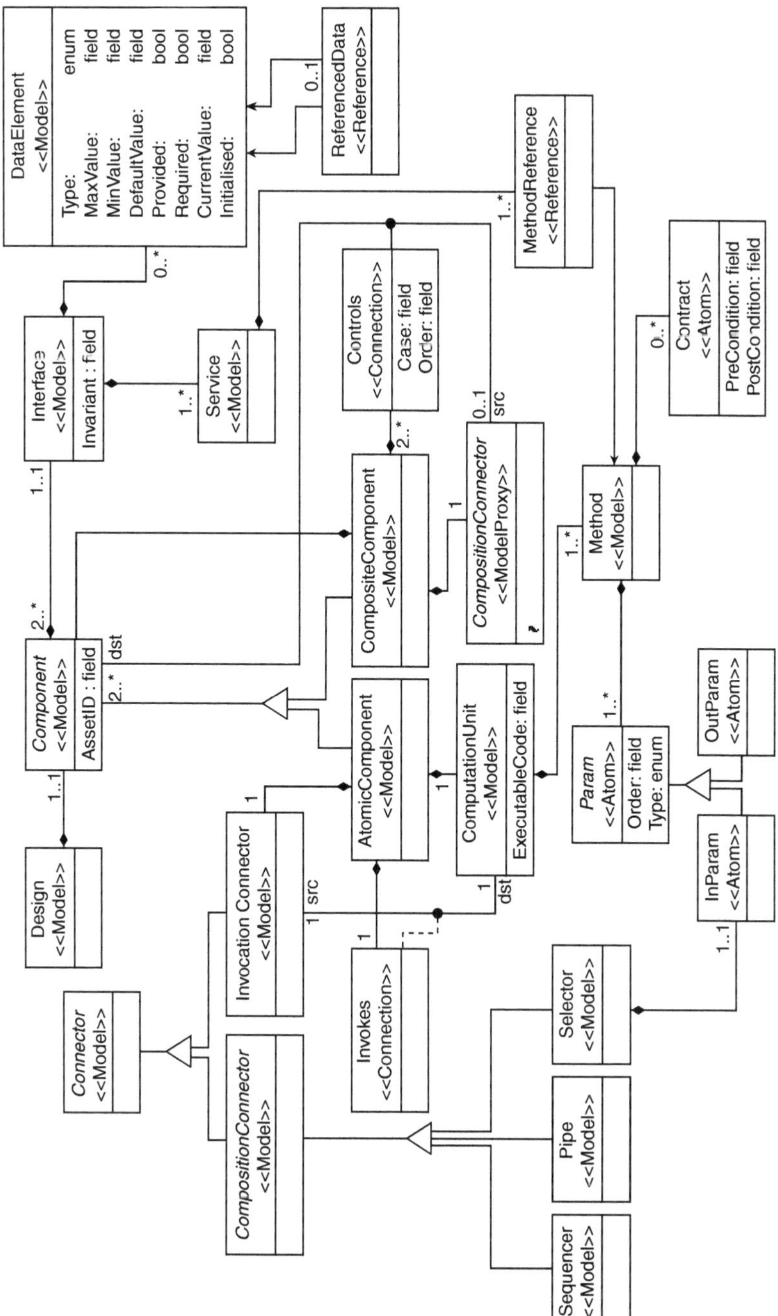

Fig. 5.14 Meta-model for the component life cycle

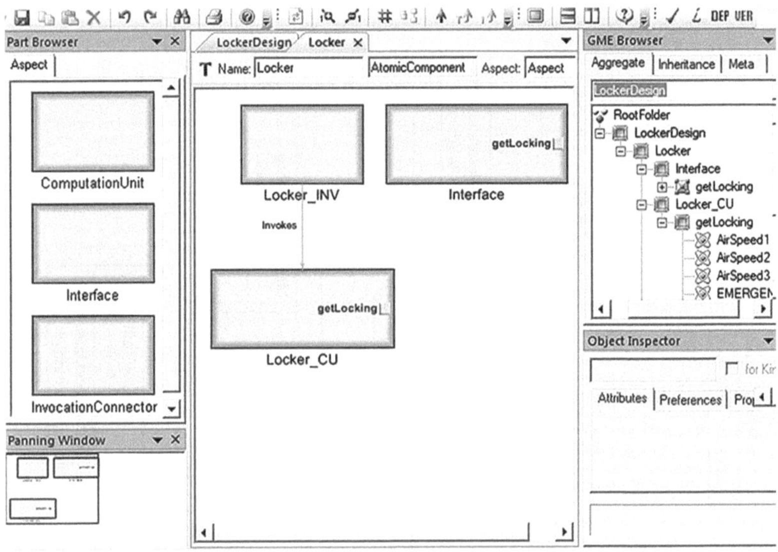

Toolbar close-up:

Fig. 5.15 Component Designer

assembly together, and provide an interpreter for it. This interpreter also provides for component adaptation. Component selection makes use of the retrieval function implemented for the repository in the component life cycle.

Systems constructed in the system life cycle are executable. To provide for system execution, we developed a system simulator in the form of another GME interpreter.

5.3.3.1 The Meta-model

The meta-model of the system life cycle is depicted in Fig. 5.16. It defines system assembly from (deployed) component instances according to the X-MAN component model.

The top element is a *System*, as opposed to (*Component*) *Design* in the component life cycle. Like a component, a *system* has one *system interface* that contains some services and data elements. A *system reference* denotes a reference to another system. Sub-systems, like component instances, can be composed using composition connectors. Therefore, systems can be constructed incrementally from (sub-)systems. Finally, like components in the component life cycle, component instances can have invariants and methods can have contracts. These are used for *compositional V&V*.

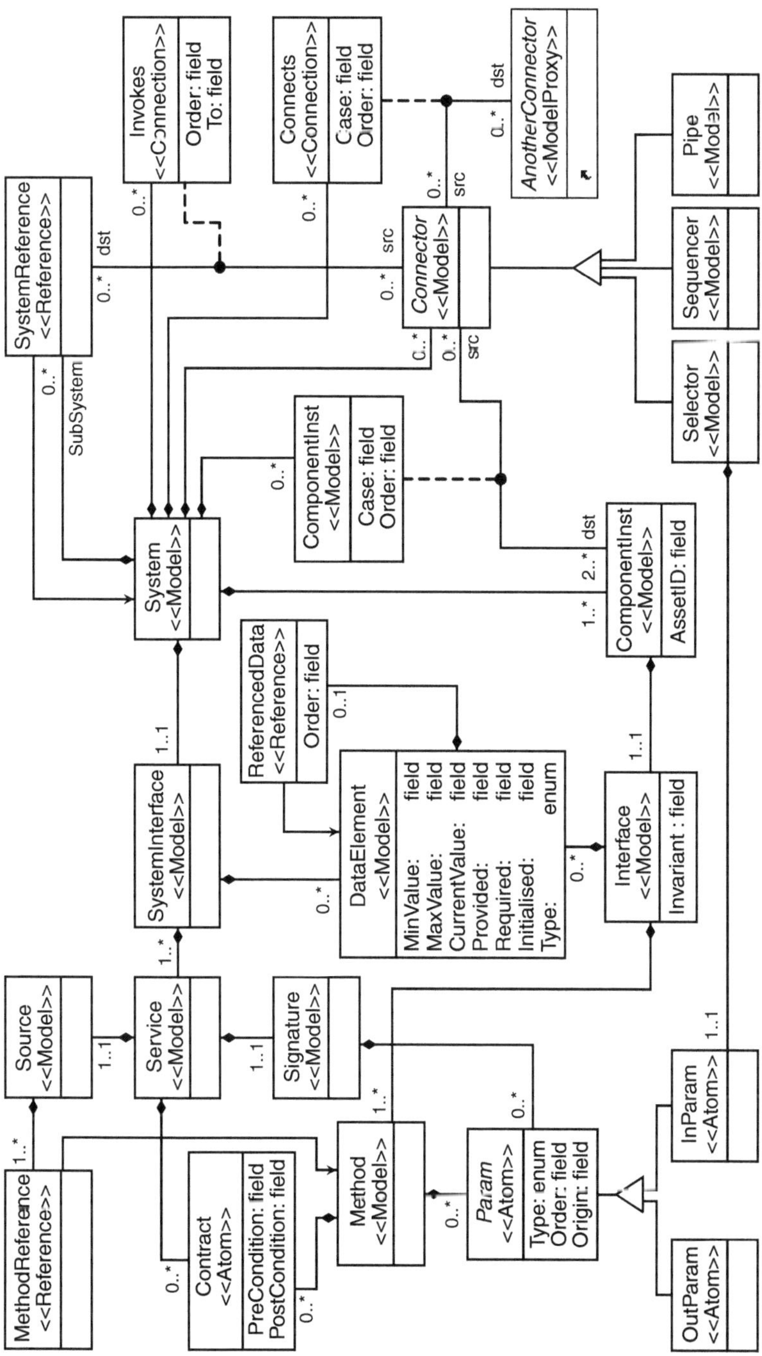

Fig. 5.16 Meta-model for the system life cycle

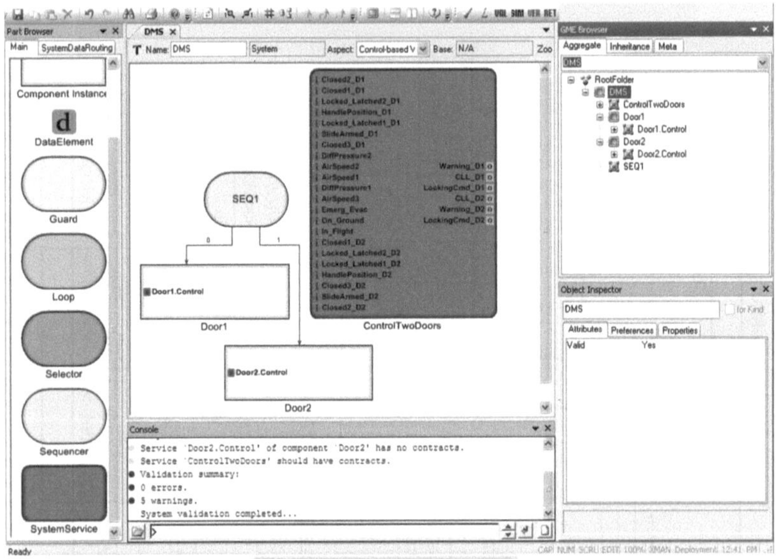

Toolbar close-up:

Fig. 5.17 System Assembler

5.3.3.2 System Assembler

We have constructed a System Assembler for composing selected deployed component instances. Components are selected from the repository and instantiated, and the instances are then deployed. Component instance composition, i.e., system assembly, to create the system follows after that. Similar to the Component Designer, the System Assembler offers a design palette, a tree view and a main view for system design. It is illustrated in Fig. 5.17 in which the developed system is built from component instances Door1 and Door2, using connector SEQ1.

5.3.3.3 The Simulator

The simulator asks a developer to select a provided service and supply the required inputs. It then traverses the system architecture and provides behaviour to our connectors (Pipe, Sequencer, Selector). Furthermore, when reaching a component instance, the simulator has to retrieve the component design (from the component life cycle) and traverses its structure with deployment data until reaching a computation unit to execute its computational source code to perform the desired computation. The simulator passes through the system and finally returns to the top-level connector with some outputs. Moreover, our simulator can take test cases that specify inputs and expected outputs, that are defined as simple assertions, e.g.

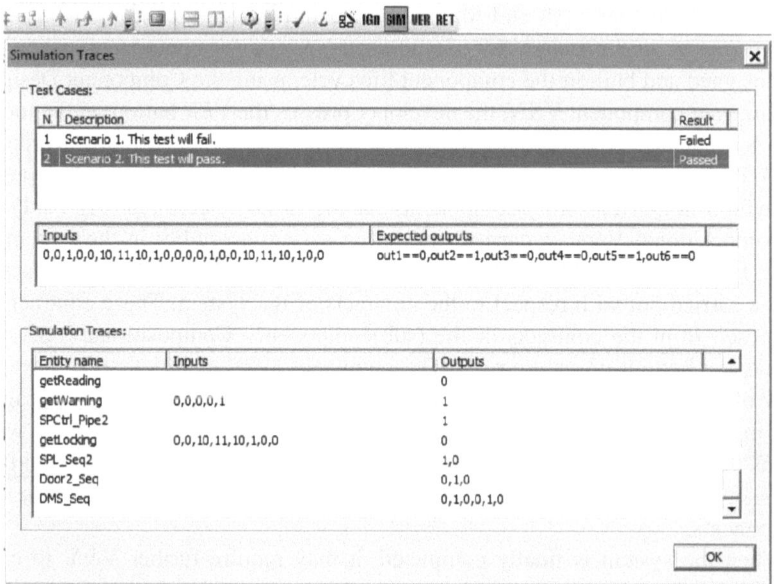

Fig. 5.18 Simulation results

output1 equals 1. The simulator then evaluates these assertions with system outputs and additionally provides simulation traces to support system execution analysis.

To implement the simulator, we assumed that computation unit implementation is given in the C programming language. We then used C++ to implement our connector behaviour and the interpreter itself. Moreover, to execute computation unit C code, we adopted the Ch library [158] to interpret without the need of compilation.

Figure 5.18 shows the screen after the simulation completes successfully. The screen depicts the two test cases description and the result. As we can see, the first test failed and the second test passed. When a test case is selected, the screen shows its specified input and output assertions. In addition, simulation traces are displayed to support further analysis.

5.3.4 *Validation & Verification*

All V&V activities rely on the CBMC model checker [33], and are carried out by a separate tool that is interfaced to the X-MAN tool. For activating this tool, the X-MAN Tool provides the *VER* button on the toolbars of the Component Designer and System Assembler (see close-up in Fig. 5.15). The X-MAN Tool exports the component design or system assembly into an exchangeable format, e.g., XML, and passes the design to the model checker.

As shown in Figs. 5.10 and 5.11, there are three V&V activities: *component, compositional, and system* V&V. Component V&V is for components when they are designed and built in the component life cycle, using the Component Designer. To carry out component V&V, the developer presses the *VER* button on the toolbar (Fig. 5.15). This V&V is carried out by the verification tool, and is with respect to the contracts of the component. After component V&V, components are deposited in the repository. Repository components can thereafter be certified (Fig. 5.11).

Compositional V&V is carried out during system assembly in the system life cycle: for any iteration of the system assembly process, V&V of the current system can be carried out with respect to the contracts of the system. These contracts are composed from the contracts of the (sub)components. Compositional V&V thus reflects the bottom-up nature of CBD: component V&V occurs first, followed by V&V of composites built from the components. It also reflects an important property of composition in the X-MAN component model, namely *compositionality*, which means that properties of a composite can be derived (via composition) from those of its sub-components. This provides a means of re-using the V&V of the sub-components.

When the system is finally completed, it may require further V&V to check for additional system properties that may not be compositional, such as emergent properties. The model checker can check such properties.

For compositional and system V&V, the developer presses the *VER* button on the toolbar of the System Assembler (Fig. 5.17).

5.3.5 An Industrial Example: The Door Management System

Now we present an industrial example from the aerospace domain that we have implemented using the X-MAN Tool. The example is a simple version of the Door Management System (DMS) on a civil aircraft[5]; it manages only 2 passenger doors (the most complicated system can consist of up to 14 doors including passenger, cargo and emergency doors).

Structurally, each door is equipped with a number of sensors that produce status readings: "closed", "lock-latched", door handle position, and "slide armed". In addition, there are inputs coming from other systems on the aircraft. The inputs are "on ground", "in flight", differential pressure, emergency and air speed signal readings.

Functionally, the system monitors the door status, manages the evacuation slide, and controls the flight lock actuator for each door. The door status is calculated from "closed", "locked", and "latched" signals read from two "closed" and three "lock-latch" sensors on every door. Also, a warning is activated when the inner door handle is moved and the evacuation slide is in the "armed" position. The evacuation

[5]This is one of the pilot applications in the CESAR project.

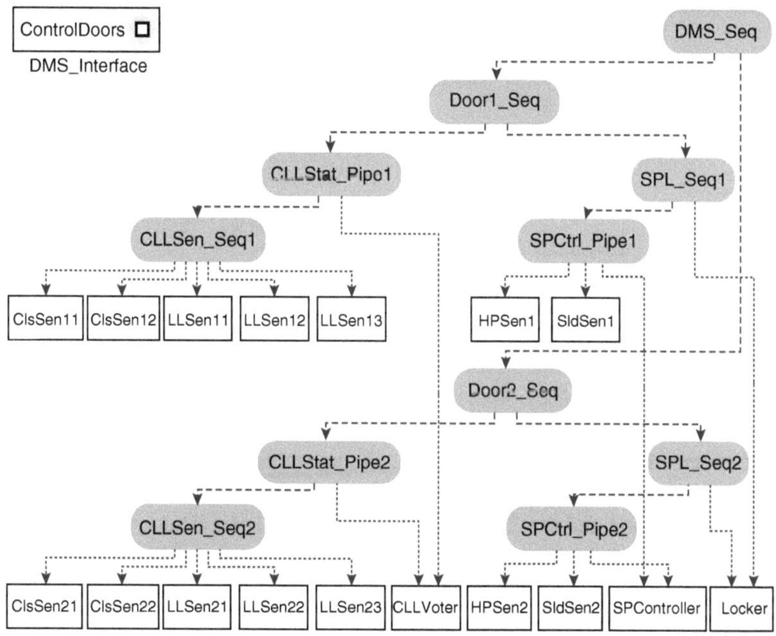

Fig. 5.19 The aircraft doors management system as implemented in X-MAN

slide management furthermore generates a warning if the aircraft is "on ground" and the cabin is still pressurised, but the slide is disarmed. Finally, the flight lock control only permits the door to be unlocked if the aircraft is "on ground".

From our analysis of the DMS system, we identified four components: Sensor, CLLVoter, SlidePresWarnController, and Locker. These components are common and can be used many times in possible variants of DMS, e.g. a DMS variant consisting of four passenger doors. Hence, in the component life cycle, we designed the four components and deposited them in the component repository. The Sensor component offers the *getReading* service to get sensor reading; CLLVoter provides *vote* service that calculates door status; SlidePresWarnController offers *getWarning* service that manages the evacuation slide; and Locker has *getLocking* service to control the flight lock.

In the system life cycle, we implemented the DMS by using pre-defined components. The complete system is depicted in Fig. 5.19. In the architecture, rounded rectangles are composition connectors and normal rectangles are component instances. The rectangle on top is the system interface that specifies the provided service of DMS. Our implementation of DMS uses four pre-defined components from the component repository. The Sensor is instantiated 14 times with different port values to implement different sensors, i.e., ones that read "closed", "lock-latched", handle position, and "slide armed" values. The other three components, which are CLLVoter, SlidePresWarnController, and Locker, are instantiated once to make CLLVoter, SPController and Locker respectively.

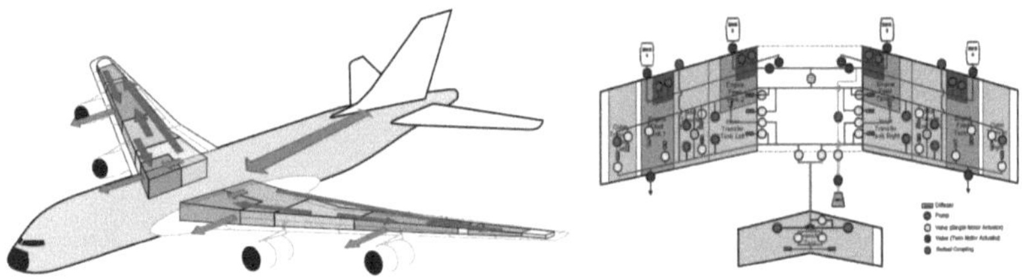

Fig. 5.20 The fuel management components in an aircraft

The architecture consists of two clusters. Each cluster represents one door. We executed the DMS using our simulator. We built a test case file with two test cases in which inputs and expected outputs are specified. These simulation parameters can be seen in the middle table in Fig. 5.18. The simulation outputs returned by the top-level connector DMS_Seq are presented in the last table in Fig. 5.18. The outputs are 0, 1, 0, 0, 1, 0 which means the doors are not closed, lock-latched, evacuation slide warning is activated and the flight lock cannot be engaged. These values match the expected output specified in the test case, hence the test passed.

5.4 Case Study: The Ground Fuel Transfer Function

In [74] we presented a case study on a representative avionics application – the Ground Fuel Transfer function of a large transport aircraft. Ground fuel transfer is a specific function of the fuel management system of an aircraft. Figure 5.20 shows a schematic overview of the components involved in fuel management on modern systems, involving multiple tanks, and numerous pumps and valves connecting these. There are several functions to manage, including safety critical ones. These functions include refuel/defuel, wing bending relief, and communications to and from several other systems. Ground fuel transfer implements the specific behaviours of the fuel management system when the aircraft is physically on the ground, as opposed to behaviours while the aircraft is in flight.

5.4.1 The Component Structure

Recall that the behaviour of an X-MAN component is defined by means of a set of operations. The ground fuel transfer system consists of six selective top-level operations which are mutually exclusive (Fig. 5.21): *Automatic Refuel* (AR), *Manual Refuel* (MR), *Defuel* (DF), *Ground Transfer* (GT), *Shut-Off Test* (SOT) and *OFF*. Each of them is further composed of a set of sub-operations. The sub-modules are shown for the *Manual Refuel* operation only.

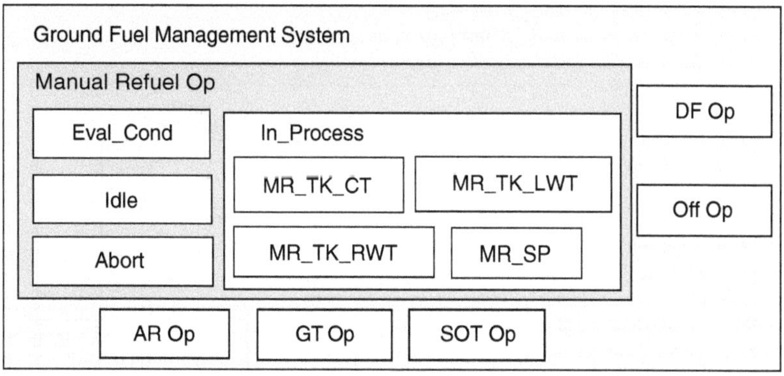

Fig. 5.21 A functional composition overview of a ground fuel management system

As given in the largest box of Fig. 5.21, the *MR* operation is composed of four sub-operations: *Eval_Cond*, *Idle*, *In-Process*, and *Abort*. The last three sub-operations are also selectable on a mutually exclusive basis. *Eval_Cond* shall be called first to decide which of the rest three sub-operations is chosen to be executed next at the runtime. Furthermore, the *In-Process* consists of another four sub-components in sequence. They determine the respective operations of the central tank (*CT*), the left-wing tank (*LWT*), the right-wing tank (*RWT*), and the surge tank (*SP*) under the manual refuel operation mode. All tanks provide both fuel output via pumps and fuel inlet via valves, each being independently switchable by an external controller. This controller monitors the fuel flow between tanks, calculates the required tank-to-tank fuel transfers and sends the appropriate control signals to the pumps and valves of all tanks.

The internal compositional structure of the MR component is typical for the ground fuel management system. The other operations have a similar compositional architecture; for brevity, Fig. 5.21 contains the details of MR only. Moreover, the component-based implementation of MR exercises most features available in the X-MAN framework, making it a suitable exemplar to illustrate the proposed component-based design and verification approach.

In order to clearly describe the properties under verification, we first clarify the differences between two terms used in the system requirement specifications of our case study. (1) *State*: describes the (durable) condition after the system performs one or a sequence of operations following the entry condition. For instance, the *Inlet Valve* state of each tank shall be either *"SHUT"* or *"OPEN"*. (2) *Status*: It describes the extant condition of some physical component. In this case study, the status of each component is evaluated every cycle and given the value *NORMAL* or *FAILED*.

For verification purposes, most properties that we currently check are *safety* properties, which state that *"something bad must never happen"*, such as:

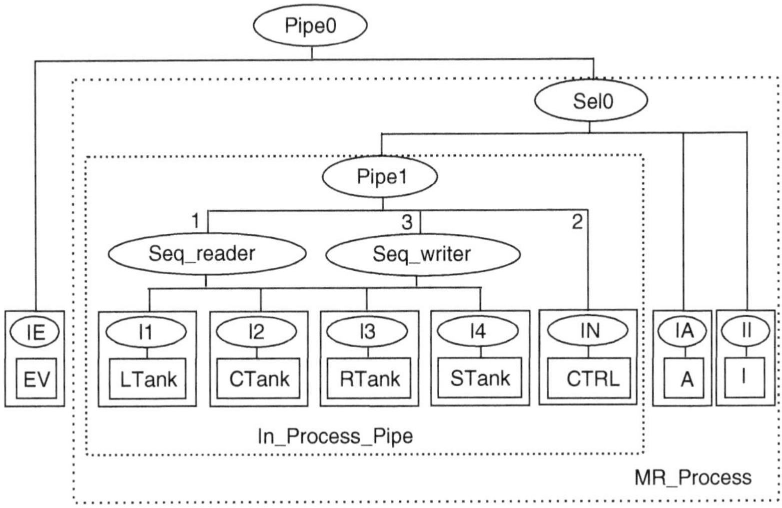

Fig. 5.22 A component-based manual refuel system

1. When the system *input overflow_condition* is *True*, *MR* must be in the *Abort* operation mode no matter what other inputs are.
2. When the *Inlet Valve* state of a tank is *SHUT*, but the status of this *Valve* is *FAILED*, the fuel mass of this tank must never exceed a certain constant value C.
3. When the status of *MR* is *NORMAL*, the execution of *MR* must never be engaged for longer than 55 s.

These example properties above are at the different abstraction levels. (i) and (iii) refer to the entire system-level properties of *MR*, while (ii) relates to the specific atomic operation of a tank which can be used directly to formulate the contract of an atomic component.

We implement a component-based version of the ground fuel management system by applying the X-MAN component framework, and verify the given properties of the system by using the compositional verification approach adapted to this framework. The details of the component implementation and compositional verification on the MR operation of this system are presented in the following paragraphs.

5.4.2 X-MAN Modeling of Manual Refuel

In X-MAN component modelling, systems are built by composing component instances which are instantiated from component designs in the repository. We will make use of five atomic component designs to model the component-based *MR* system in X-MAN (as shown in Fig. 5.22). These design components are *Tank*,

Controller, *Evaluator*, and *PrAbort*, and *PrIdling*. *PrAbort* models the computation when an operation mode of the ground fuel transfer system aborts. Similarly, *PrIdling* encapsulates the computation when an operation becomes idle. The final two components are relatively simple in our case study, basically returning the output of the component with some constant values, so, we omit their description here, and concentrate on the remaining three components.

- Tank: This component maintains some stored state variables which represent a mass of fuel in a particular tank. Fuel output and inlet are modelled by the setting of arguments corresponding to each of 3 ports. A positive mass represents fuel input to the tank, negative signifies outputs. The tank component contains an internally selected maximum-mass figure. Input of a fuel mass which causes this figure to be exceeded results in an overflow mass being outputted. The component also outputs the current mass, states and status of pumps and valves of each tank.
- Controller: calculates the fuel flow between four instances of the Tank component, modelling the functionality of the pumps, valves and pipes linking them. The outputs of each tank along with the command signals for each pump and valve are fed to this component.
- Evaluator: accepts the system inputs and evaluates the right operation mode based on the inputs. The evaluation logic is represented as the first-order logic formula in this case. It outputs the evaluation result.

X-MAN model system shown in Fig. 5.22 makes use of eight atomic component instances, and five composite components, involving three different kinds of composition connectors – *Pipe*, *Selector* and *Sequencer*. Firstly, we compose four instances of the design component *Tank*: *LTank*, *CTank*, *RTank*, *STank*, which model *LWT*, *CWT*, *RWT* and *SP* respectively, into a composite component named *TanksReader* using the *Sequencer* connector *Seq_reader*. In the same way, these four component instances are composed together into another composite component named as *TanksWriter* using the connector *Seq_writer*. Thus, every *tank* instance is accessed twice for each iteration of the model. Next, *TanksReader*, an instance of *Controller* named as *CTRL*, and *TanksWriter* are further composed together into a single larger composite component *In_Process_Pipe* by a *Pipe* connector *Pipe1*, which models the *In_Process* operation. Then, *In_Process_Pipe* is composed with the component *A*, which is the instantiation of the design *PrAbort*, and the component *I*, which is the intantiation of *PrIdle*, by the *Selector* connector *Sel0* to construct the more complex composite component *MR_Process*. Finally, by means of another *Pipe* connector *Pipe0*, *MR_Process* is composed with the component *EV*, which is an instantiation of the design component *Evaluator*.

The model behaves as follows at run-time: when a call (with inputs) reaches the top-level connector *Pipe0*, the connector first calls the atomic component *EV* with its required inputs and passes the outputs to the *Sel0* connector. Based on the values passed, *Sel0* evaluates the selection condition in order to choose among the composite components *In_Process_Pipe*, *A* or *I*. If *In_Process_Pipe* is chosen, *TanksReader* is first called and its monitoring outputs are passed to *CTRL*, then finally the computed command ouputs of *CTRL* are fed into *TanksWriter*. When either *TanksReader* or *TanksWriter* is executed, the *LTank*, *CTank*, *RTank*, *STank* are accessed in sequence, but the usage of the outputs from these *Tank* instances is different. For instance, *TanksReader* manipulates the outputs which represent the current mass, the state and status of the tank, while *TanksWriter* uses those outputs that correspond to the handlers of the pump and valve of each tank. If *A* is chosen, the model starts running as the *Abort* operation mode in *MR*. Otherwise, *I* is chosen to simulate the *Idle* behaviour of *MR*.

5.4.3 Component-Based Verification of Manual Refuel

Consider the atomic component *LTank* of the *MR* modelling system. It contains six contracts: three of these specify the integer ranges of the outputs parameters, the other two describe the conditions under which fuel may be added without overflow occurring, and the final contract relates to the minimum and maximum execution time of the component. We use one contract *cMassRange* for the component *LTank* as an example:

```
Contract(cMassRange) {
    Inputs:          iAddedMass : int
    Ouputs:          oMass : int, oMaxMass : int
    Pre-condition:   true
    Post-condition:  oMass ≥ 0 ∧
                     oMass ≤ oMaxMass
}
```

Two integer outputs are used in this contract. *oMass* reflects the current fuel mass of the tank, and *oMaxMass* represents the internally selected maximum mass figure. The *Pre-condition* of this contract is *true* which means no assumption of the component inputs is made. The corresponding *Post-condition* requires that *oMass* must always be a positive value and *oMass* must never exceed the value of *oMaxMass* after the execution of the *LTank* component. As a further example, the contract for specifying the extra-functional timing property of this component uses the special parameters *preTime* and *postTime*, which are globally accessible throughout the component, to describe the timing constraint as follows.

```
Contract(cTankTiming) {
    Inputs:
    Ouputs:
    Pre-condition:   true
    Post-condition:  postTime ≤ preTime + 5 ∧
                     postTime ≥ preTime + 1
}
```

According to this contract, the execution time (in seconds) of this contract must always fall within the range $[1, 5]$ under any combinations of inputs. In addition to the basic *Range-Contract* of output variables and *timing-Contract* as introduced above, some component contracts formulate the complex data constraints between model inputs and outputs that are derived from the system requirement. For instance, the contract for the *Controller* component instance *CRTL* is as follows:

Contract($cFuelConservationController$) {
Inputs: $iMass[4] : int,$
Ouputs. $oAddedMass[4] : int$
Pre-condition: $iMass[0] \geq 0 \wedge iMass[1] \geq 0 \wedge$
$iMass[2] \geq 0 \wedge iMass[3] \geq 0$
Post-condition: $oAddedMass[0] +$
$oAddedMass[1] +$
$oAddedMass[2] +$
$oAddedMass[3] = 0$
}

In this contract for *CTRL*, the input parameter *iMass* is an Integer array which is passed from the composite component *Seq_reader*. Each element corresponds to the output parameter *oMass* of *LTank*, *CTank*,*RTank*,*STank* respectively. The output parameter *oAddedMass* is also an Integer array which is passed to the composite component *Seq_writer*. And each element corresponds to the input parameter *iAddedMass* of four *Tank* component instances. This *Post-condition* of this contract requires that the fuel mass must be conserved in the *CTRL*.

5.4.3.1 Vertical Verification Phase

We apply the software Model Checker CBMC to formally verify that the implementation of every atomic component in *MR* (a total of eight components) satisfies its corresponding contracts. For instance, we check that the C implementation of the component *CTRL* satisfies the post-condition specified in the contract *cFuelConservationController* given above, under the assumption that the input parameters adhere to the *Pre-condition* constraint. Our verification tool first automatically instruments the following assumption statement at the beginning of the implementation code.

$$\textbf{assume}(cFuelConservation.Pre - condition);$$

and, the following assertion at the end of the code.

$$\textbf{assert}(cFuelConservation.Post - condition);$$

Then, CBMC is applied to verify the instrumented code.

Due to the exhaustive search within the unwinding depth and precise modeling of data variables, CBMC can detect non-trivial corner case bugs, which eluded discovery during testing by means of a conventional test-suite derived from the requirements. As an example, consider the following code fragment from the *CTRL* implementation where the variables have been renamed.

```
const int sink_num = 2; // * * * *
...
int val_a = a/sink_num; *
int val_a_remainder = a%b;
...
if(v_open){
  val_c = val_a;
}
else{
  val_c = 0;
}
...
if(val_c > 0){
  val_c = val_c + val_a_remainder;
}
...
```

The variable a models a total amount of flow that is distributed between a given number of sinks. The amount is given as an integer quantity, and there can therefore be a remainder (val_a_remainder), which is to be apportioned to the valve flow val_c. Consider the special execution trace when

$a = 1;\ v\_open = 1;$

Then,

$val\_a = 0;\ val\_a\_remainder = 1;\ val\_c = 0;$

i.e., the remainder is non-zero but is not apportioned to the valve flow val_c. The buggy value of val_c propagates to the outputs and causes the violation of the *Post-condition* of *cFuelConservation*. The variables a and v_open are both internal variables. It is difficult for a conventional testing technique to find the primary inputs stimuli that could observe this corner case scenario where a is set to 1 and v_open is set to be *True* at the same time. Consequently, the error was missed by the conventional test suite, while CBMC quickly identifies an erroneous execution trace as above. Analyses revealed that the bug occurred because the code was adopted from a similar algorithm using floating-point arithmetic, and did not handle integer division truncation properly.

5.4.3.2 Horizontal Verification Phase

After the contracts of all components have been verified, we can start horizontal verification, which checks the properties of the modeling system based on the verified contracts of the components and their composite relationships. For instance, given a system C composed of two verified components A and B connected by a Selector, we combine the post-conditions in the contracts of A and B as follows:

$$f = (selC - cond \wedge post_A) \vee (\neg selC - cond \wedge post_B)$$

where $post_A$ and $post_B$ denote the post-conditions of components A and B, respectively. *selC-cond* is the condition that the sub-component is chosen in the case of using Selector connector. In this example, if *selC-cond* is true, component A is selected for execution; otherwise, component B is selected. For the composite systems using other connectors, the X-MAN-Verifier can also automatically derive the formula f from the proved contracts of sub components, according to the composition behaviours of the corresponding connectors with respect to the properties under verification.

Then, the X-MAN-Verifier checks whether f implies $post_C$. If so, we have proven that the component C satisfies its contract; otherwise the contract may or may not hold. The X-MAN-Verifier can produce a trace that demonstrates how the component C fails to respect its contract. For a given timing property

$$postTime < preTime + 55$$

for *MR*, the X-MAN-Verifier returns *hold*. On the other hand, the timing property

$$postTime < preTime + 50$$

may be violated and the tool provides a counterexample. We can conclude that the *MR* system is guaranteed to finish execution within 55 time steps, but may exceed 50 time steps. The counterexample extracted is as follows:

Inputs :
  $in0 = 0, \ldots, in3 = 0;$
  $preTime = 0;$

CallingEC
Inputs :
  $preTime = 0,$
Outputs :
  $value = 1;$
  $postTime = 5;$

CallingLA
....

Outputs :
  $out0 = 1, \ldots, out4 = 1;$
  $postTime = 53;$

This trace shows the system inputs and outputs, *preTime* and *postTime* of *MR*, and the components on the execution trace with their inputs, outputs, *preTime* and *postTime*.

5.5 CMM and the Doors Management System

5.5.1 CMM

The concepts of the CMM are related to the Heterogeneous Rich Components (HRC) which were developed in the SPEEDS project [21, 38, 52, 96]. In CESAR, the contract-based specification methodology and data model for component based development has been adopted and refined to support the CESAR development process. Contracts refer not to the internal states or the internal behaviour of a component but to the observable interactions via its interfaces. The interfaces of a component are specified by a set of (observable) variables. Those variables, called flows, can be characterized as input, output or bidirectional from the components point of view. In addition also a data type can be assigned for each flow. Contracts argue about traces of this interface variables.

In the first place a component is a black box with a sound specified interface and without any dynamic aspects like behaviour or interaction protocol (no traces). Since the interaction is the observable part of the behaviour of the component it is specified in the same manner as the (non-) functional behaviour of the component itself. Behaviour is specified by mechanisms which decide if a certain trace ("run") is part of the behaviour or not. A run is a possibly infinite trace of variables assigned to the variables. Finite state machines are one possible mechanism to determine if such a trace is part of the behaviour or not. This generic approach allows to specify the behaviour of a very wide range of components. Such components could be software or hardware elements because the specification takes only the observable variables (software variables or hardware pins) at the interface of the component into account. The concept of a contract for the assumed behaviour of the environment of the component and the promised behaviour of the component allows to verify if a implementation fulfils the specification. Note that this implies that the execution framework of the component is in most cases essential to provide the promised behaviour. For example, a software function does not automatically calculates $a + b$ but does this only if it is called (e.g. "$c = f(a, b)$;").

A component is a black box with a generic and formal specification of behaviour based on traces of observable variables. Control events, like triggers, are modelled explicitly and also part of the contracts of a component. The assumption-promise pairs allow to verify the implementation of the component independent from a certain environment. Furthermore, this enables reuse of component implementations. The formal nature of the specification of the components allows a so called Virtual Integration Test (VIT) without the need to implement any of the involved components, see [40]. To support the creation of this formal specification a pattern-based specification language has been developed in SPEEDS as well as in CESAR. This language supports the user to formalize requirements, see requirement formalization in Sect. 3.3.

Composition of CMM components is the parallel execution of the subcomponents. What does that mean? First of all this is not true because the parallel operator is not the only composition operator defined in the semantic definition but it is the most important one. Second, not the components are executed but the sets of allowed traces are merged to result in the composed set. Since the formal specification is driven by the idea to have state machines which define these sets of traces the parallel composition of this state machines results in one state machine which defines the set of traces of the composition. Due to this approach, the connections between the subcomponents define which variables match between two or more component interfaces. This is necessary because the alphabets of the traces have only local scope. Note that this implies identity of connected variables. If data or event exchange has a certain behaviour this has to be represented by a additional component, e.g. a bus or queue.

To summarize, the specification is contract-based and supports reuse of a wide range of (hardware and software) components. The formalization of requirements enables early assessments of requirements as presented in Sect. 3.4.

The result is the definition of the CMM as common views to data through a heterogeneous tool and language environment, the RTP (see Sect. 6). Based on this common view to different development entities (software components, requirements, test cases, ...), a process library was developed to enable the users to instantiate a company specific development process for safety-critical embedded systems. A generic development process pattern containing four generic development phases was identified to be applicable in several domains and companies.

The following subsection relates the application of the CMM methodology and data model to the Doors Management System.[6] This example covers two of these phases to illustrate the CESAR methodology.

5.5.2 Application to the DMS

The Doors Management System example is used to show how this process could be instantiated. This system is highly safety-critical because significant aircraft accidents and incidents occurred due to uncontrolled decompression. In some of them doors not fully closed, latched and locked opened in flight, leading to an explosive decompression. The instantiated example development process contains four analysis and design phases, where all referenced development entities are described in term of the CMM.

The top level requirements of the DMS are listed below.

1. The DMS shall be able to provide to pilots the status of Aircraft Door Slides. (Functional-Perspective)

[6] Odile Laurent and François Pouzolz; Airbus France; Airbus generic pilot application Aircraft doors management system; CESAR Use Case; May 2009

2. The DMS shall inhibit pressurization of Aircraft Doors when Doors are not fully closed and locked. (Operational Perspective)
3. The DMS shall enable pressurization of Aircraft Doors when Doors are fully closed and locked. (Operational Perspective)
4. When in Flight, the DMS shall inhibit intentional or unintentional opening of aircraft doors. (Operational Perspective)
5. When on ground, the DMS shall allow opening of Aircraft Doors. (Operational Perspective)

The whole set of top level requirements addresses certain concerns of the DMS system. We focus on a subset only to simplify the example. The top level requirements are transformed to CMM *SystemRequirements* by specifying

- What are the assumptions to the environment?
- What is the promised behaviour which the component can guarantee?
- To which Perspective is the requirement associated?
- To which Aspect is the requirement associated?
- Which component(s) shall satisfy the requirement?

In this early phase, having only a very abstract representation of the DMS, all requirements are promises since the environment is not clearly defined. Also, the DMS itself is the components which shall satisfy all the requirements. The definition of the system boundaries, the refinement of the SystemRequirements and the decomposition of the system follow during the development process. The association of the requirements to Perspectives and Aspects is indicated in brackets after the textual representation of the requirements.

5.5.2.1 Operational Analysis

In this first phase of the development process the top level requirements are represented as SystemRequirements and are connected via *Satisfy* links to one *OperationalActor* the DMS representation in the operational Perspective.

The Operational Capability Specification is performed with the goal of deriving operational scenarios and functional requirements that cover the original requirements. Since in this step the DMS is sketched in its boundaries, realistically many functional and non-functional SystemRequirements will have to be reconsidered in later steps.

First, the OperationalActors are assessed from the SystemRequirements. In addition to the DMS the other OperationalActors Passenger, Pilot, Cabin Doors, Crew and Landing Gear System are identified.

As a second step, operational capabilities were derived from the SystemRequirements by a system engineer such as "Enable Pressurization when Doors locked" or "Keep door locked when in flight" (see Fig. 5.23). For completeness and traceability these SystemRequirements are related explicitly to the operational capabilities by Satisfy links.

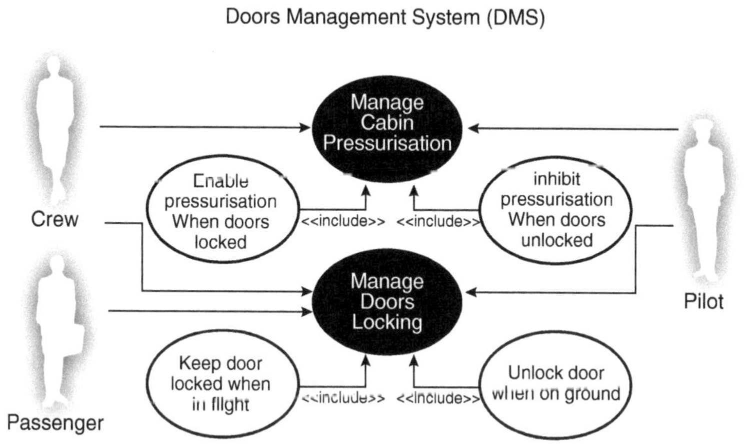

Fig. 5.23 Use case diagram of the operational capabilities of the DMS

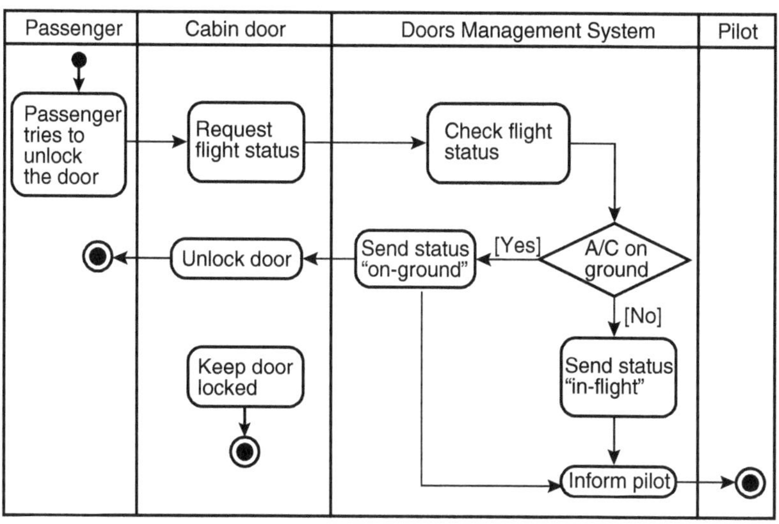

Fig. 5.24 Operational scenario of the DMS

The use-case-driven scenario description is refined by several scenarios which has to be supported. One of this scenarios is presented in Fig. 5.24.

Based on the identified activities, operational processes are specified which orchestrate these activities. For the operational process, the interaction between the operational activities is defined. The description of such uses cases allows sketching the DMS in its system boundaries. That means that the interaction with other actors such as passengers, pilot, crew and cabin door is identified. New SystemRequirements are derived form the decomposition of the operational specification into several activities. These new SystemRequirements (usually) refer to the functional Perspective.

In this design step also a preliminary hazard analysis is performed to identify the safety-relevant operational activities. You may find further information about this in Sect. 2.5.

5.5.2.2 Functional/Non-functional Analysis

Input of this design step is the result of the operational analysis. In particular this consists of operational scenarios, user requirements which could not yet be refined, hazards as well as their classifications and functional SystemRequirement identified during the operational analysis.

The goal of the functional/non-functional need specification is to identify and specify functions and where appropriate refine them by defining sub-functions. The DMS is decomposed into the functions "Manage Cabin Pressurization" and "Manage Doors Locking". For traceability all the defined functions have a relation (e.g. CMM *Derive* link) to the design entities defined in preceding phases.

One intermediate step in before mapping the functions to physical elements is to map them to logical elements. This logical components represent a preliminary architecture of the whole system before determining certain implementation solutions. One the one side the components are not characterized as hardware or software units but have interface discretions and allow to refine the expected behaviour in a abstract manner.

We illustrate this taking the "Manage Doors Locking" function as example. This function shall be represented by a logical component named "FlightLockController" which controls the flight lock on each door of the DMS. The purpose of this component is to signal to the flight lock actuators if the lock shall be enabled or disabled. Two of the top level requirements apply to this component and therefore the behaviour of the component shall be to enable the lock during take-off and landing/approaching and to unlock it during cruse and if on ground. This status is provided by the Landing Gears System, the Engine Status and the Air Speed Sensors. In addition three safety-related requirements enforce to disable the lock:

1. If the signals "landing gear status" (lgs), "engine running status" (ers) and "air speed status" (ass) are not available or
2. If the "emergency evacuation" signal (ees) is send or
3. If the differential pressure is less than 2.5 mbar.

The differential pressure is measured by two independent sensors and shall enable emergency evacuation on ground and doors can be opened without risk (low differential pressure) even if the other systems fail. Also some timing constraints are added to these components. The overall requirement for the "FlightLockController" is on the right side of Fig. 5.25.

The "FlightLockController" is decomposed into the "FlightLockLogic" acquiring the flight status and emitting the "flight lock status" (fls) which could be

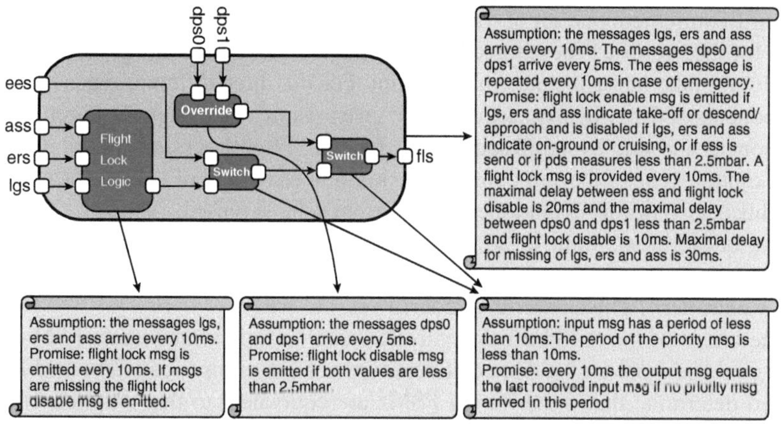

Fig. 5.25 Logical view to the "FlightLockController" and its requirement

"enabled" or "disabled", the "Override" component which sends the fls signal with content "disabled" if the differential pressure is low and two switches which pass through the message available at the "normal" port if no message is available on the priority port in a certain time frame. The priority message is passed if available and the other message is ignored. Note that no message queues are assumed and therefore only one message per port is recognized. For implementation, these messages and ports could be realized by variables with certain values as well as pins with high or low current. This abstract view of the "FlightLockController" decomposition is presented and also the requirements of the parts of the components are shown. All the requirements are formulated as SystemRequirement (with Assumption and Promise) to differentiate between what the components are responsible for and what they assume/need from the environment. The interfaces of the components describe what kind of exchange is possible, e.g., if boolean values (boolean variable, high/low current) or numerical values (integer, 8-bit coded data).

Note that this is specification and not implementation. The components mentioned can not be compiled into code or hardware but only refer to a implementation. Therefore not the components itself but the referenced implementation has to fulfil the requirements of the component. The SystemRequirements of these components argue about traces for the interface elements. If an implementation produces the same output trace as specified it fulfills the SystemRequirement. This could only be achieved if the input trace is also conform to the assumed one, since the implementation reacts to the inputs.

Note that in Sect. 3.4 a more detailed description of how a natural language requirement can be formalized into a formal (text-)pattern-based SystemRequirement. Such SystemRequirements describe sets of traces for Assumption and Promise pairs in a formal but human readable way.

In this example, we assume that such a formalization has been made and for each component the traces are defined e.g. using finite state machines (FSM). For each

SystemRequirement one set of traces has been defined for the Assumption and one other for the Promise. These traces argue about the values of the interface variables and do not take internal states into account. For the three different types of (sub-) components of the "FlightLockController", we assume one SystemRequirement per component and one SystemRequirement for the "FlightLockController" itself. In this example none of the sub-components is connected to all of the interfaces of the "FlightLockController" component and the "switch" is reused.

The question we want to address is: Does this composition of the four sub-components (three different types) fulfill the SystemRequirement of the "FlightLockController"? The answer is of course yes, but why?

The traces of the "FlightLockController" can be represented using FSMs. The same could be done for the three different types of sub-components. The problem is how to determine that the different FSMs are equivalent to the "FlightLockController" ones. Each FSM has its own alphabet since this relates to the interface of the component only. Connections between instances of this component types create a mapping relation between the different alphabets. In addition for each instance of the same type, an individual FSM is used. Build up on this the parallel composition of the FSMs can be created and this composed FSM defines the set of traces of the composition. At this point the FSM of the "FlightLockController" and of the composition can be compared if the composition contains at least the same traces as the "FlightLockController" FSM. We call this relation "entailment" and via the CMM the SystemRequirements and components are linked with each other to enable traceability of this relation. Note that because of merging the traces the connections between the components can not have certain properties.

The purpose of the entailment relation is not only to check if the composition is valid but to reduce the integration testing effort. If entailment is given, the individual sub-components have to be checked and not the composition itself. This also reduces the effort for verification of reusable elements since their implementation must only be checked once for each type and not for each instance in a certain composition as well. In addition this allows e.g. an OEM to contract several suppliers and to ensure that the integration is valid without exhaustive testing against each of the implementations.

This relation does not only support this design step but is applicable in all design steps of the CESAR design process. In Sect. 3.4, as mentioned before, guidelines as well as tools support this kind of (de-)composition also on requirements wrt. only a very draft architecture. The overall goal of this methodology is not only to reduce verification effort but also to enable traceability from top level requirements down to the implementations of individual components as well as the overall composed system. One last important aspect is the analysis of the impact of a change of requirements in all design steps (refer to Sect. 3.4 for details).

Chapter 6
The Reference Technology Platform

Alain Rossignol

6.1 RTP General Presentation

The CESAR *Reference Technology Platform* is a generic model-based tool integration platform composed of a set of interoperable tools, methods and processes designed to improve the development of safety critical embedded systems.

The objective of the CESAR RTP is to create an environment similar to Eclipse for software domain but for critical embedded systems development.

RTP shall provide a complete systems engineering environment with a set of management or engineering methods and processes, as well as engineering tools. The integration and interconnection of tools is needed in order to support collaboration between all relevant business processes of an enterprise for example interactions with customers, suppliers and partners. This generic integration platform will host tools from different vendors (more than 80 tools are in the scope) and must be able to host additional tools from end users. Furthermore the interoperability aspect between tools integrated in such a platform is getting more and more crucial for successful and efficient product engineering. It should be based on stable and non proprietary standards.

The main technical challenge in addressing this problem is the lack of open and common interoperability technologies supported by the different tools that generate and provide access to the information, and of the underlying IT infrastructure, whether these are based on COTS products or custom-built solutions. An Interoperability Specification (IOS) will guarantee these needs for interoperability and collaboration between tools across the entire engineering lifecycle. Furthermore the connection and integration of different RTPs based on the same IOS, allow composing a multi domain engineering environment.

A. Rossignol (✉)
ASTRIUM SAS, Toulouse, France
e-mail: alain.rossignol@astrium.eads.net

A. Rajan and T. Wahl (eds.), *CESAR - Cost-efficient Methods and Processes for Safety-relevant Embedded Systems*, DOI 10.1007/978-3-7091-1387-5_6,

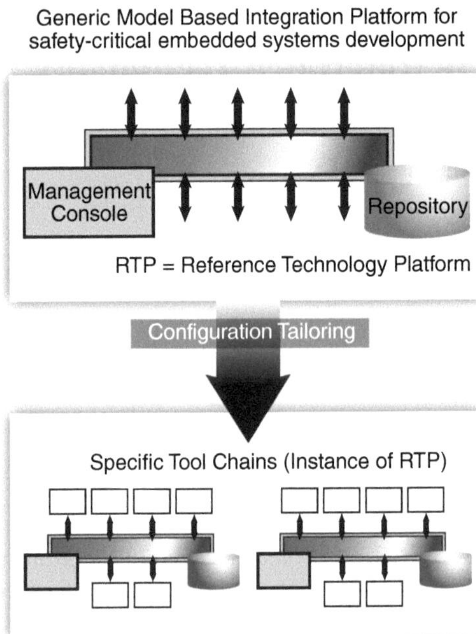

Fig. 6.1 The RTP concept

RTP as defined above is a generic platform which has to be tailored for a specific domain or end-user in order to create its own tool chain. The RTP user will have to enter the RTP tailoring process in which the RTP assets will be put together to create what we call an instance of the RTP with a specific customization. The Fig. 6.1 presents this RTP tailoring process.

Once a specific toolchain is created, the users of this integrated tool platform can use a generic interface for accessing it. This interface (called RTP Desktop, see Fig. 6.2) allows browsing artefacts, models and links. It allows the user of the toolchain to access its working project, open models, access to the different versions of the models, launch the tools on the models, create and navigate links. It can be used as the main entry point to access the RTP.

This chapter will present all major RTP concepts like tools, services, adapters and models with the following sections. Section 6.2 addresses the major requirements and basic concepts in system engineering which are applicable to RTP. Section 6.3 will present global architecture of the RTP and the main concepts. Section 6.4 will present CESAR Meta-Model definition and rational. Section 6.5 will explains main processes used in the definition and implementation of the CESAR RTP. Section 6.6 will introduce how the generic RTP is instantiated in the different Domain RTPs.

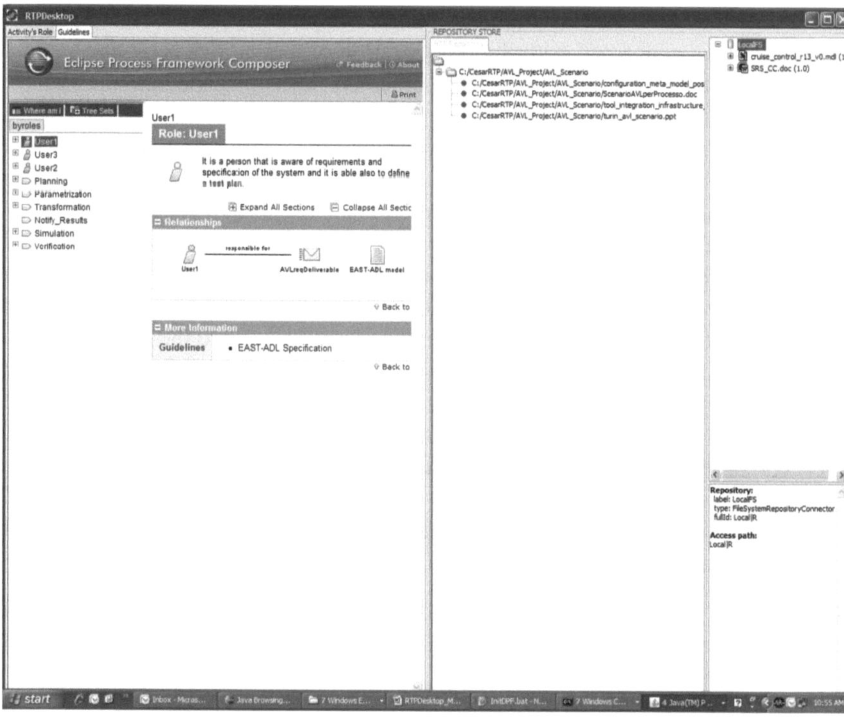

Fig. 6.2 RTP Desktop

6.2 General Principles for RTP Tool Chain Implementation

Since the RTP is a generic framework for integration of engineering tools, several aspects need to be addressed by an integration architect. Already in the late 1980s were described the five dimensions for tool integration: platform integration, data integration, process integration, control integration and presentation integration.

- Platform Integration: Is about giving the tools a common framework and a common layer of abstraction in order to provide network or operating system transparency. So tools can be placed on a common platform as a set of services they can rely on and to make them in that way interoperable.
- Data Integration: When integrating tools it is important to facilitate the integration of the data they work on. This mean to allow the exchange of data between tools but also to allow the definition of relationships among data objects produced by different tools.
- Process integration: Typically, in a development process there are several processes running either explicitly and formally defined or they are run implicitly and probably in an ad-hoc style. These processes are executed with the help of tools. Tool integration needs to take care of these different processes and shall

facilitate the common execution and integration of these processes. So process integration is concerned with the orchestration of the various activities and the coordination of the interaction with the humans executing the activities within the processes.
- Control integration: This is about coordinating the control flow between tools. So there is the need to notify tools about certain events in order to coordinate them. A special way of control integration is the use of tool functionality. Each tool offers a set of functionalities. Basically, this comprises the manipulation and analysis of data in various different ways. Control integration is concerned with offering functionality of tools in the control flow of another tool. So it is about the usage of tool functions without changing from one tool to the other.
- Presentation Integration: The interaction of human with the tools is a particular issue. Having adequate user interfaces for a tool is vital for the software ergonomics of the tool and therefore for the acceptance and productivity. Presentation integration of tools means to integrate their look and feel and the way humans interact with the tools. So users can work with various different tool via the same user interface with the same presentational elements regardless of the concrete tool functionality.

The CESAR tool integration is based on these general tool integration dimensions. In more detailed terms the main principles of tool integration from CESAR point of view are the following:

6.2.1 Model-Driven Development

The engineering of safety-relevant embedded systems isn't an easy task. An adequate development process is needed which satisfy specific requirements such as cost-efficient methods and processes should be part of the engineering task. Furthermore, the complexity of system engineering increased over the years in particular in the embedded domain. The implementation of a model-driven development process promised answers for dealing with this increasing complexity. Therefore, the CESAR RTP implementation task is focused on model-driven engineering aspects. Besides the benefits of a model-driven approach several issues have to be considered while implementing the RTP:

- Support of large and complex models
  One of the major problems when dealing with complex models is to keep track of all incoming and outgoing references (e.g. profiled UML models or diagram models). As models grow larger and larger it becomes more and more difficult to coordinate the workflow in a way that permits concurrent work on different parts of a model.
- Transparent model management
  Developers working with models in a development process often complain about a lack of integration and missing or inappropriate handling of modelling artefacts. "I simply want to use models, and I do not want to care about where they

come from and where they go to." The CESAR RTP should implement also a straightforward mechanism for model identification, which allows easy access to the right model from each of the connected modelling tools.

- Versioning
  Furthermore, a consistent versioning management in particular for arbitrary models is needed in a CESAR RTP implementation. In traditional development process (based on code instead of models), a lot of version control systems are established. However, in the context of model-driven engineering-specific methods are relevant in order to provide a consistent versioning of all artefacts as well as to provide support for typical synchronization tasks such as comparing models, computing differences and merging different models into a new one. The CESAR RTP implementation has to realized this mechanism across tool boundaries, due to the fact that some tools have not identified any specific version of their tool internal data or some tools do not come with an identity management of models and their elements which is a strong prerequisite for realizing model versioning in an integrated tool chain based on CESAR RTP principles.

6.2.2 *Distributed Development*

In large and possibly distributed development teams maintaining a conglomerate of different analysis, development, and documentation tools quickly becomes difficult. Especially tools performing operations on models (e.g. model transformations) have to be always kept up-to-date for all members of an engineering team to ensure consistent work. Using an extensible IDE does only partially do the job. The RTP implementation should allow the integration of heterogeneous tools covering the whole development lifecycle and comprising custom-made, proprietary tools as well as COTS tools. Therefore, the RTP implementation is based on Service-oriented Architecture (SOA) principles which enable a well-formed structure of the RTP assets with respect to their communication and specific reuse.

A further aspect of distributed development is the collaboration of several teams within development process. A team could consist of a huge number of engineers, which are not necessary belonging to the same organizational unit (Company, Department). The produced artefacts may not be merged automatically, due to the presence of developed conflicts. Therefore, the RTP implementation should provide a multi-user support which means the management of access rights and authentication mechanism. In addition to that the RTP needs to support the synchronization of development artefacts which are produced by different engineers.

6.2.3 *Information Management*

In a CESAR RTP different kind of information types are exchanged and shared across tools and services. Generally, the information is represented by artefacts which are created during the development process. Typically, models, meta-models or data are artefacts in the context of the RTP. Models and meta-models are well-structured and can be understood also on a semantic level. A data artefact contains closed information which could not be understood by everyone in the first instance such as binaries, generated reports or image files.

- Addressing
  An RTP instance consists of several components which represents the needed data types. The intention of the RTP Instance is to share artefacts between different tools in a distributed and heterogeneous environment. In the context of the RTP implementation an artefact is an abstract object. Every artefact that is shared between different tools should be uniquely identifiable by a corresponding namespace. The format of this namespace depends on the underlying RTP integration infrastructure. For instances, a namespace, used in a ModelBus environment, is represented by URL. The namespace could also be used to structure artefacts by grouping them into domains and sub-domains.
- Traceability
  A powerful traceability approach is needed as well in order to deal with the complexity of a distributed and heterogeneous development process. As one of the main and important concepts within CESAR RTP the traceability should allow to create links between different artefacts across tool boundaries. Furthermore, the traceability is mandatory for additional services such as impact analysis or corresponding generation of reports which summarize a specific state of the development process or a specific state of the integrated tool chain.
- Information Storage
  Generally, tools and services within the CESAR RTP can provide information. However, the RTP need a mechanism in order to access specific information within a tool chain. In traditional tool chains this is often realized by data bases, source code repositories or in worst case by common file systems. Nevertheless, the RTP should provide a generic approach for information storage. This doesn't mean that a central data repository is needed, but a common description of how to access information is mandatory.
- Information Mapping
  In CESAR RTP tools and services provide different information, but there might be the case that the information which a certain tool is providing cannot be understood by another one, due to different context or format of the information. The tools require a mechanism that helps to translate the provided data. As a basic principle two main approaches exist for information mapping. On one hand the mapping can be realized by definition of point-to-point mappings for different information types. On the other hand the mapping is part of a common semantic of the information. This means that all information provider (tools and services) agree on the structure and semantic of the shared information. Therefore, the RTP should support both information mapping approaches.

6.2.4 Process Automation

System engineering is often organized in an overall development process that takes into account different stakeholders of the engineering task, certain workflows or necessary development activities. If no user interaction is needed in order to perform an operation on the engineered artefacts a lot of tasks of this overall development process can be automated. Typical operations are code generation or static analysis in which no interaction with an engineer or administrator is needed. A high degree of automation has a lot of benefits. Besides the cost reduction of the whole development process also the fault-proneness of the process is less compared to a manual execution of a specific process step. Therefore, the CESAR RTP should be able to automate, as much as possible, process steps and workflows. In order to support the ability of process automation the RTP has to take into account the aspects of Notification and the Orchestration of process and workflows.

- Notification
  A notification message is a piece of information that can be used to transport information in an asynchronous way. Notification messages can be published by every stakeholder of a CESAR RTP. The messages are typed. The concept of having notification in a RTP based tool chain is mandatory in order to create choreography of different services, so that no corresponding user interaction is needed in order to start an orchestration that automated a specific development workflow such as a simple model transformation, verification of design artefacts or simulation of E/E models.
- Orchestration
  Orchestration is a specific sequence of invocations of services in that way that the output of one service could be the input of another service (service chain). The main goal is the grouping and automation of many simple steps to achieve a particular result. The CESAR RTP as a whole requires the ability to be able to orchestrate many services and events acting across RTP tool chain. While Orchestration itself is outside the scope of RTP implementation, it needs to provide the core capabilities to support such a thread of execution across many tools, many locations, many users, and potentially a long duration.

6.2.5 Extensible and Customizable

The CESAR RTP is not a monolithic approach for building tool chains. Rather the CESAR RTP is a generic concept which involves customization principles in order to build a satisfactory integrated tool chain. In addition to that, new tools, new methods and new processes will be used in future RTP tool chains. This requires an approach for extending the RTP by new concepts. The usage of (open) standards and the usage of existing technologies, which are already established and accepted in an industrial context, are general indicators for achieving extensibility and the possibility to customize a specific technology. Furthermore, well defined APIs eases the use of the underlying integration infrastructure.

- Open Standards
  The usage of open standards is crucial for a successful implementation of the CESAR RTP which should be built on accepted industry standards such as REST, SOAP, Web Services, or OMG standards like MOF, UML or OCL. The RTP implementation should not focusing on realizing of proprietary protocols for communication or data exchange, but rather it consider the potentials and advantages of widely spread technical approaches, which are already approved by industrial case studies.
- Adoption of Existing Technologies
  The used integration infrastructure is mainly responsible for the transportation of information from one stakeholder to another. It is obviously that this could be realized in different ways with different technologies. However, in order to ease the extensibility and in particular the acceptance the RTP implementation should be based on widely-used and well-known core technologies for the corresponding tasks of an integration framework such as notification, transportation or distribution.

6.3 RTP Architecture

Agreement on RTP architectural principles is a technical and industrial challenge. From an industrial perspective, a long term agreement must be obtained between tool providers, tool integrators, and end users on some simple architectural principles, leaving room for evolution and adaptation. Hence, even though the technical solutions to interoperability might be diverse and relying on innovative software engineering concepts, they have to be translated into simple, sustainable industrial schemes. From a technical point of view, the experience of CESAR end users and technology providers shows that no simple, unique integration framework is able to cover the whole scope of the RTP, in terms of variety of capabilities and variety of tools.

The Fig. 6.3 identifies classes of major RTP components and functional interfaces related to CESAR consortium background.

The different types of RTP components are:

1. Tools: in this class we found different types of tools covering the whole engineering life-cycle. Among them we can identify: Requirements management tools (requirement capture, organization, analysis), encompassing some de facto standard commercial tools as well as CESAR innovations from SP2, Design tools (architectural solution definition at different levels of detail, component based design, analysis with respect to different viewpoints) encompassing some commercial tools, as well as CESAR innovations from SP3, Software design tools, for which Eclipse tends to be a de-facto standard framework, Development, test

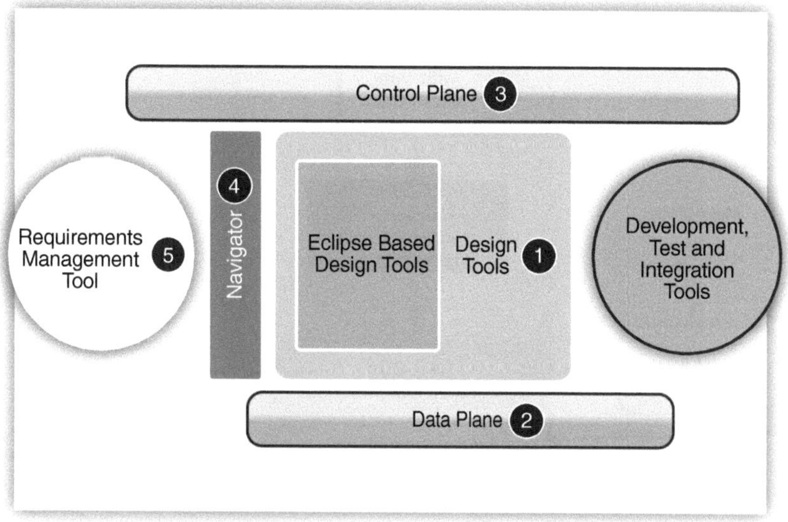

Fig. 6.3 Global architecture principle of RTP

and integration tools (including implementation level analysis, test management and test generation, integration, monitoring and debugging tools).

2. Data Plane: technical solutions to store and exchange data.
3. Control Plane: process orchestrator supporting the engineering process by activating different tools according to particular design process and rules.
4. Navigator: allows the navigation between tools, models and the traceability while ensuring the interface to requirement management.

If we go more in detail into the architecture, we will define several components presented in Fig. 6.4, which will be implemented and connected to create an RTP instance.

The different kind of components could be classified in following concepts:

- Tool: requirement management tool, design tool, development tool, test and integration tool
- Adapter: interface to connect a tool to the RTP Bus
- Meta-Model: language that supports a model
- Model: artefact manipulated by the tools
- Data: artefact manipulated by the tools
- Service: transformation or verification
- Repository: allowing to store models and development artefacts
- Process Orchestrator: supporting the engineering process by activating different tools according to particular design process and rules
- RTP Bus: allowing the data exchange between tools

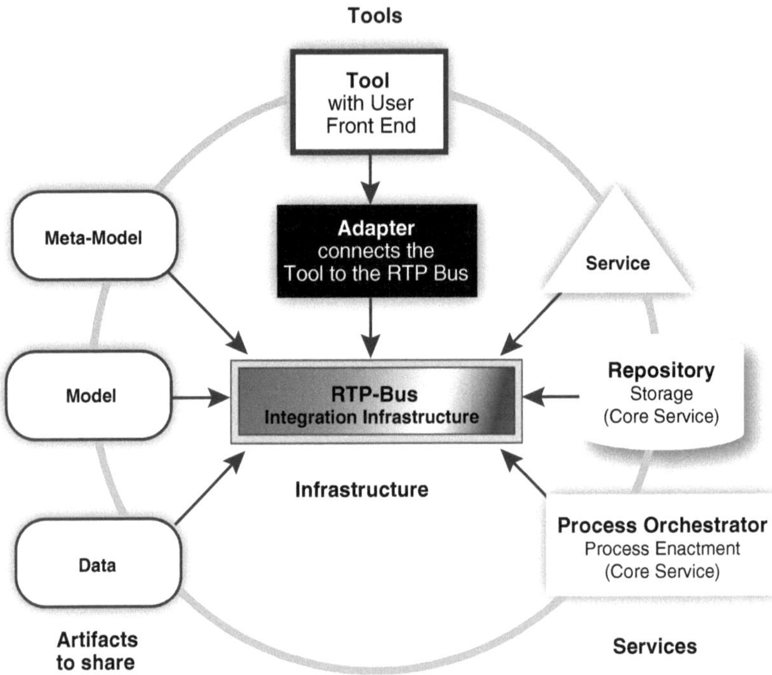

Fig. 6.4 RTP components

All of these components will be connected together in order to constitute an RTP platform instance. For exchanging data between tools, several links and integration implementation were defined and experimented in the different RTP versions and RTP instances for experimentations. These different integration schemes are not exclusive: they can co-exist within a RTP instance.

- Point to point data conversion: interfaces between tools (through a common repository or as point to point interactions) can be implemented without the need for a common meta-model, to the extent needed for scenarios of use of interest. High-level, between concepts may be sufficient to export or make visible the information that tool A needs from tool B.
- Data integration by linking: This is a lightweight data integration mechanism. It allows the creation of links between elements of models and the persistence on these links in a dedicated link repository, without impacting the models themselves. These links can be navigated by using a programmatic API or a GUI viewer.
- Common meta-model integration, based on the mapping between tool native models and the CESAR common meta-model (see Sect. 6.4).

The Fig. 6.5 presents the different interactions and links between major RTP components.

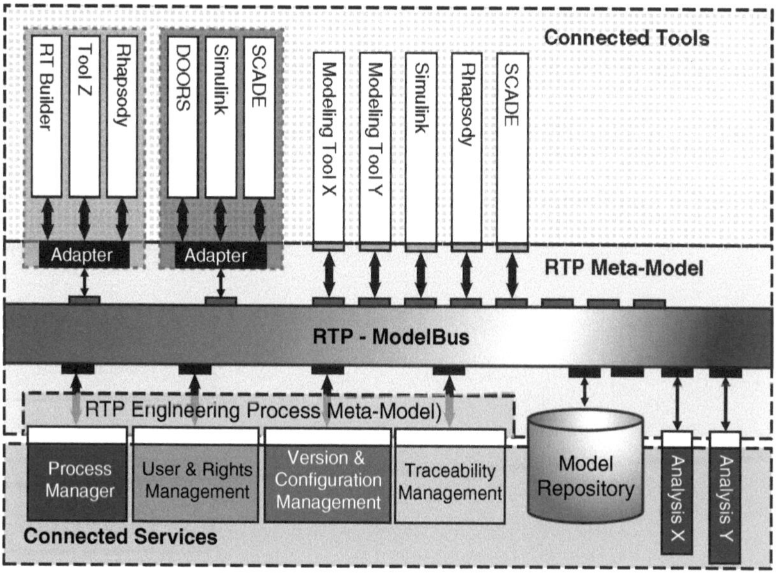

Fig. 6.5 RTP component interactions

All the details of the RTP Architecture, how the different concepts were implemented and integrated in the CESAR RTP versions are presented in Chap. 7.

6.4 CESAR Meta-Model

CESAR reflects several meta-model aspects.

The CESAR Meta-Model is a semantically founded conceptual meta-model that is independent from a concrete implementation. It defines a common terminology with conceptual elements and relationships allowing a common understanding of information that is exchanged among system engineers of different domains. It contains modeling concepts that are of common interest for the different industrial domains aerospace, automotive, automation and railway. It does not depend on any concept that is specific to an industrial domain (Fig. 6.6).

The CESAR Meta-Model is composed by a Core Meta-model supporting basic modelling concepts (allowing embedded system modelling with rich component models in conjunction with requirement descriptions, traceability links and verification/validation along the development process), and other common domain independent modelling concepts.

These core meta-model concepts are then realized in a Meta-Model Specification that will allow the implementation of this meta-model into the RTP.

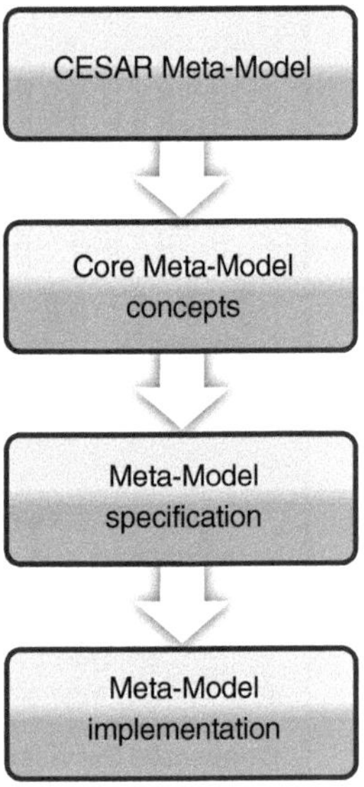

Fig. 6.6 CESAR meta-model aspects

Several implementations of the CESAR Meta-Model are possible and all are mappable to these concepts. An implementation of the CESAR Meta-Model can be used as a pivot meta-model to transfer information from one modelling language into another without defining a direct mapping.

Furthermore, an extension mechanism is defined in order for the end-users to define additional concepts and to be able to implement a domain-specific meta-model. The RTP can then be customized to this domain-specific system engineering process for which the CESAR Meta-Model enables common understanding of engineering information and therefore allows reuse of tool integration and analysis techniques (Fig. 6.7).

The purpose of CMM is to provide a technique to access common concepts, artefacts and relationships in models of domain-specific meta-models which are integrated into the RTP.

Such meta-models are integrated into the RTP by performing a tailoring process in which domain-specific meta-model artefacts and relationships are mapped to the concepts of the CMM of the RTP. The RTP meta-model provides core structure and semantics. Models can therefore be completely stored in the RTP for full model exchange and have furthermore a generic interface as defined by the RTP CMM.

The CMM will enable the user to have a common view on different tools regardless of the underlying specific format. This representation abstracts from some details but has the benefit that interconnections between different tools can be easily defined. CMM will provide a mean to connect and select different artefacts and items implementing and supporting all engineering views needed by several abstractions levels (From operational and functional levels to logical and physical levels), several perspectives and several points of view for properties verification (Safety, Real-Time, Variability, Behaviour) (Fig. 6.8).

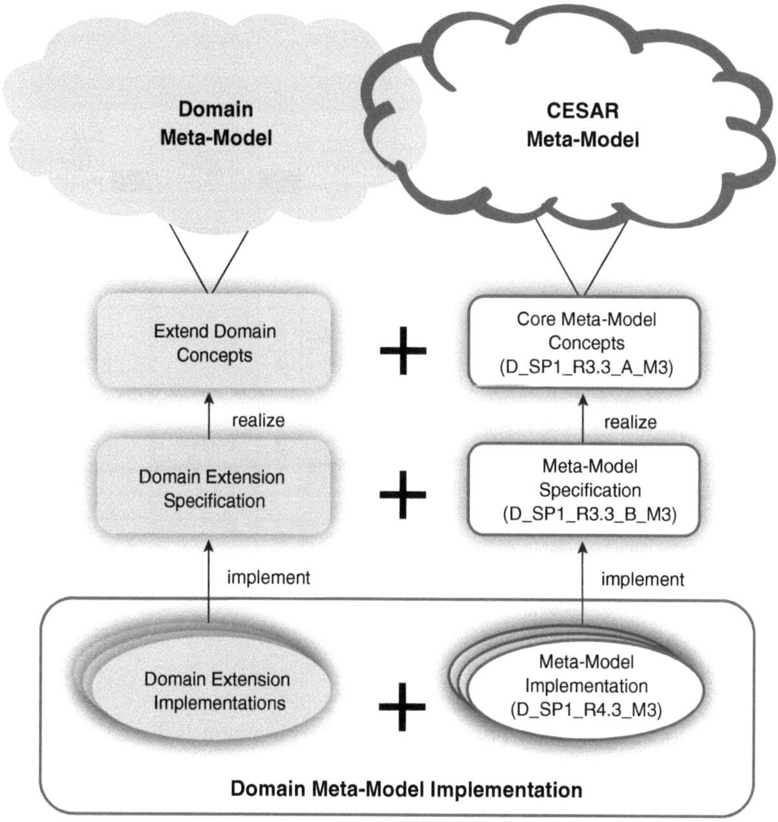

Fig. 6.7 Implementation of a domain-specific meta-model

6.5 RTP Processes

RTP is a very specific concept compared to classical engineering or software development environment targeting following objectives:

- RTP is an integration framework of many different components coming from various domain partners and heterogeneous activities including commercial and academic tools, open-source or proprietary solutions, which could be at different level of technology maturity and quality, but aiming at supporting industrial use-cases;
- RTP aims at be enhanced and modified all along its life-cycle, being able to add, change, remove some components, but also to be adapted to some new core technologies.
- RTP will never be used as a whole and as a "black box" but has to be instantiated to a domain, an industrial company, or even a project. This will be developed in Sect. 9.4.

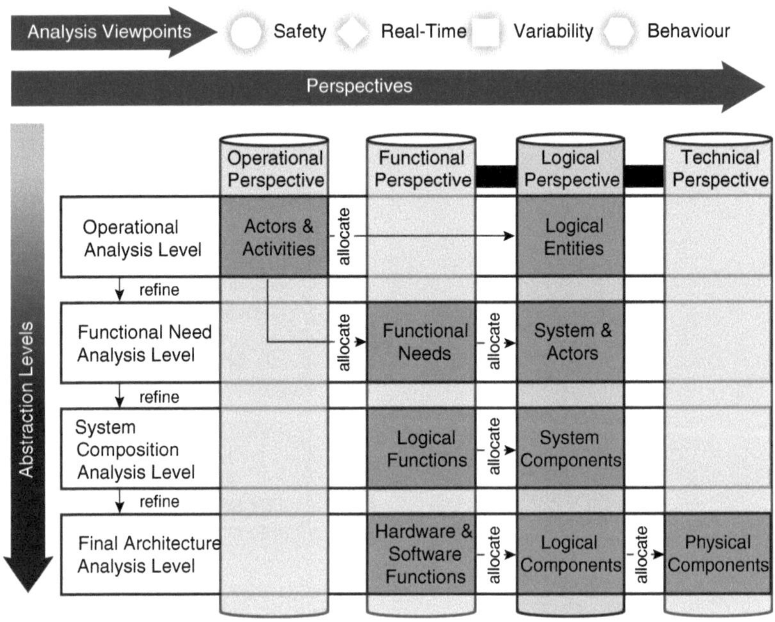

Fig. 6.8 Engineering multi-views needs

In order to achieve such objectives, RTP development is based on following defined processes:

- Development process including RTP definition, specification, integration, and delivery activities is based on an iterative and incremental approach.
- Quality process addresses quality assurance and control, configuration management.
- Assessment process covers both RTP component TRL evaluation (Technology Readiness Level) and RTP industrial assessment based on some criteria in different improvement areas like Cost, Schedule & Risk Reduction, and System Quality.

The following process describes main activities in the CESAR RTP definition and evaluation process (Fig. 6.9).

6.5.1 *Development Process*

6.5.1.1 RTP Definition Process

The RTP is defined in an iterative "Top-down" and "Bottom-up" combined process.

The "Top-down" approach is managed and supported by the TPC (Technical Project Committee) and RTP Roadmap Working Group together with the RMT

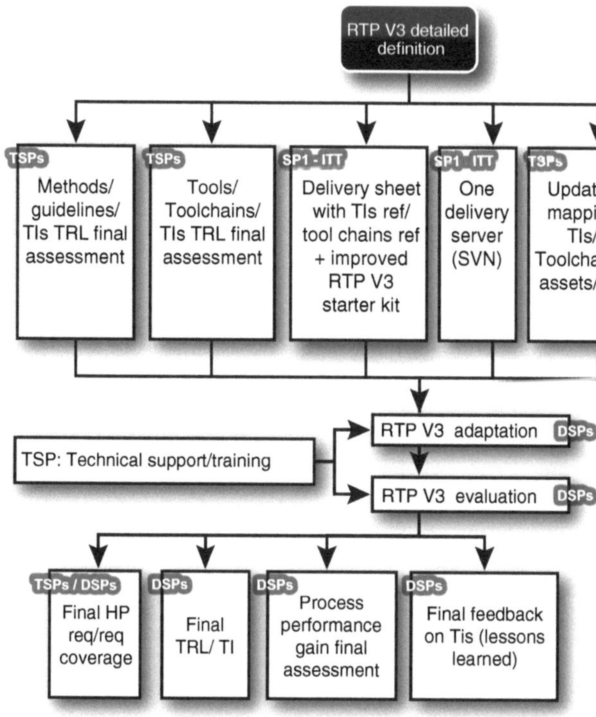

Fig. 6.9 RTP definition and evaluation process

(Requirements Management Team), from user needs and expectations in term of tools, gateways and features, towards the specification of the RTP.

The "Bottom-up" approach is managed by TPC and TI (Technical Innovation) leaders, from existing assets (Tools and existing connection between tools) and Technical Innovation planned within the project towards the requirement in order to identify what requirements are not covered by assets or TI.

6.5.1.2 RTP Assets or Components

The assets, which contribute to the definition and implementation of the RTP, can be classified as follows:

- Tools, which range
  - From tools used in the context of research labs, not yet widely used in the industry, but highly relevant to the objectives of safety relevant embedded systems,

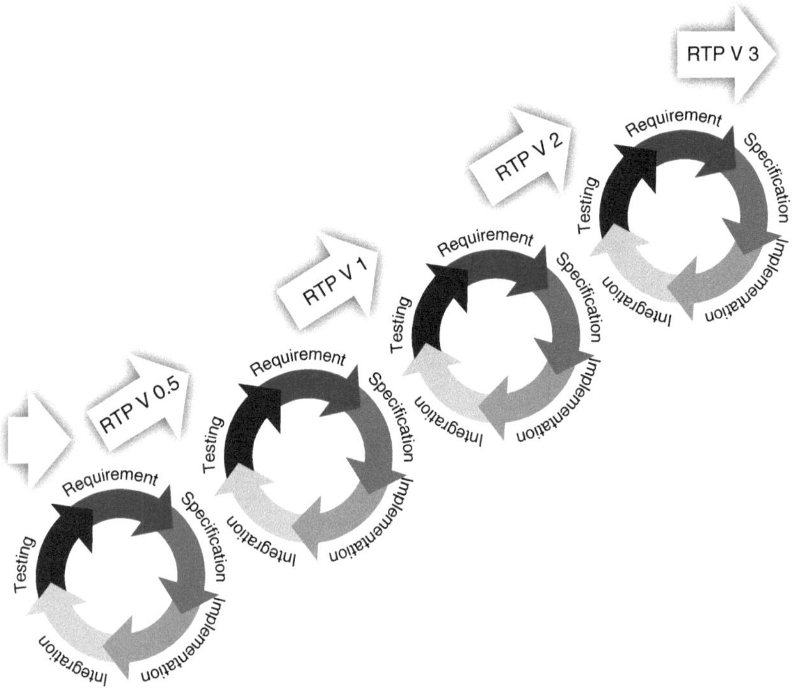

Fig. 6.10 The incremental and iterative development process of the CESAR RTP

  - To commercially available tools, or open source tools with industrial acceptance or industrial support, that are candidates to be "integrated", at some level, in the RTP.

- Integration support assets, such as: toolset architecture principles, interoperability interfaces, interoperability technologies, which define the possible schemes for "integrating" the tools in the RTP
- Results of previous research projects, which are not yet translated to specific tools or integration schemes, but which go beyond the state of the art and should therefore be considered as candidates with respect to the objectives of the RTP – e.g., emerging technologies which support the "tailoring" process of the RTP.

6.5.2 Incremental and Iterative Process

This process iterates for each version of RTP: during the CESAR project, four successive versions (V0.5, V1&V1.1, V2, V3) were defined, developed and delivered following an incremental and iterative approach (Fig. 6.10).

6.5.3 Delivery Process

The delivery of each version is done via the CESAR Communication Platform File Explorer mainly for all documents and a special SVN server provided by Fraunhofer FOKUS for tools source or executables and some model examples. In addition a bug tracking (TRAC) tool is available for all RTP users. Each CESAR member has a personal account on both platforms. Within these platforms a separate RTP Vx file structure is created. Each RTP asset is bundled with an installation instruction and user manual.

6.5.4 Quality Process

6.5.4.1 Quality Policy

CESAR is a project that groups different types of partners (industries, academics, SMEs) in an attempt to define a platform and associated components (tools, methods, etc.) that will support and enhance the design, development and V&V activities of safety relevant embedded systems. As it will be used for Aerospace, Automotive, Automation and Railway, the components will have to be compliant with the standards applied in those domains. Moreover, CESAR RTP is a platform integrating different kinds of components and tools: commercial, R&D, open-source. For all those reasons, the quality policy aims to define two levels: the component level and the CESAR RTP level.

CESAR RTP level is under the responsibility of the CESAR TPC and SP1 WP1. These organisation bodies write the plans, define and ensure proper application of the agreed development process, ensure the communication of the project (web-site, international meetings, …).

The component level is under the responsibility of the components' managers in SP1, SP2, and SP3. The managers shall ensure the quality of their products including all the quality controls. The SP1 WP1 verifies that the requirements are respected for each major release; it is the evaluation of the component. According to the results of the evaluation, the CESAR TPC accepts or not the component release in the CESAR RTP version.

The quality assurance is ensured by:

- Definition and respect of CESAR RTP development process
- Mentoring of new contributors.
- Definition and use of methodology and tools to realize the technical activities
- Continuous improvement of the processes ensured by analysis of the main problems encountered on the project and analysis of the discrepancies in the application of the Quality Plan.

6.5.5 Configuration Management

Each asset is identified with the following information:

- Its component,
- Its path in the repository (including the file name),
- Its version and revision.

The source code, when applicable, is maintained in configuration.

The documentation shall also be managed in configuration management, for both CESAR RTP level documents (plans, features, design, and test plan) and each component document (User guide, Test plan and tests results, specifications, release note).

When an RTP Version is ready to be published, there are different kinds of information to publish which are summarised in a RTP delivery sheet managed by ITT:

- The RTP version identification,
- The list of packaged components available in the release and that can be directly downloaded,
- The main contacts for each component
- The links to available documents for each component

6.5.6 RTP Assessment Process

6.5.6.1 Assessment Process

The RTP assessment is based on:

- Quality assessment at component level when applicable & RTP level
- Components Technology Readiness Level (TRL) evaluation for components integrated in the version
- Industrial evaluation both at component level & RTP level

6.5.7 RTP Component TRL Evaluation

A frequently used mechanism to measure the maturity of a technology is the Technology Readiness Level (TRL). Figure 6.11 presents its scale with nine levels.

A TRL scale is a means to quantify the maturity position of technologies, starting from basic scientific knowledge up to their use by customers embedded into product applications. It is used in particular to manage R&D and technology development (TD) processes with decision gates at key steps.

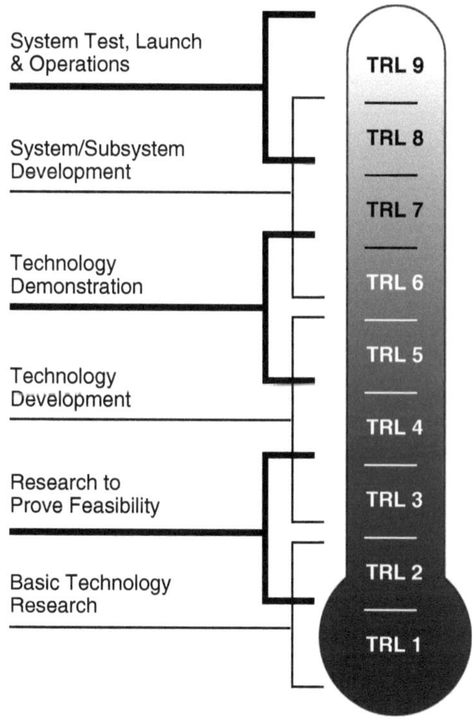

Fig. 6.11 TRL definition

Most national and international (e.g. European) institutions and agencies also use such a scale as a common reference in their relationship with their technology suppliers.

This scale is defined and used to evaluate each component of RTP.

6.6 Instantiating the RTP into Domain RTPs

6.6.1 Principle

The purpose of RTP Tailoring is the systematic planning of a tool chain (RTP Instance) from RTP assets according to a development process description. Not only a set of tools and additional modules that facilitate tool integration are part of an RTP instance but also a repository structure model that integrates the meta-models involved in the RTP instance. RTP Tailoring is partly tool supported by a configurator (Fig. 6.12).

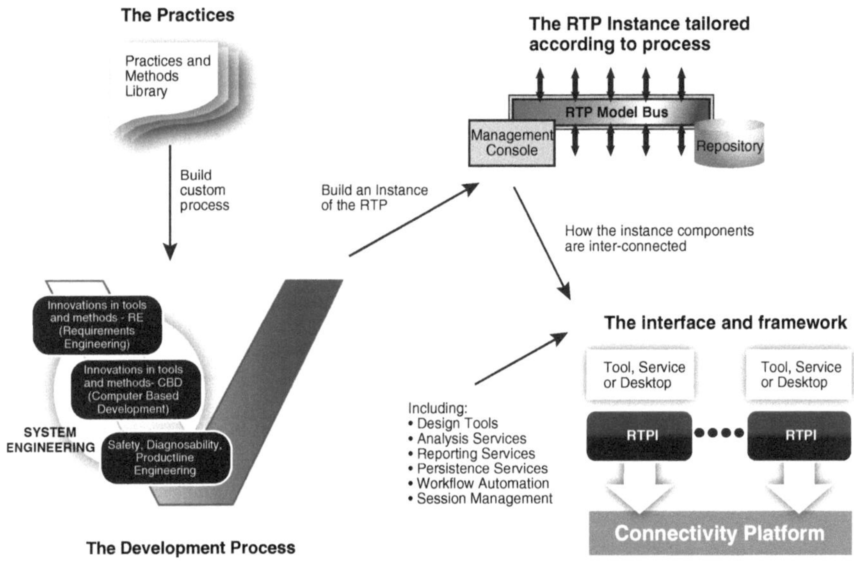

Fig. 6.12 RTP instantiation methodology

6.6.2 RTP Integration of Other Languages

When tailoring the RTP for a domain-specific development process, meta-models and respective access methods have to be adapted in a way that these meta-models can be used in the RTP. In order to integrate a foreign tool or domain-specific meta-model while tailoring the RTP, two steps have to be performed: First, to enable meta-model integration, a mapping between the foreign meta-model and the CESAR Meta-Model has to be defined. This step is crucial for the second step when an implementation is chosen to make a mapped meta-model usable in the tailored RTP.

6.6.2.1 Meta-model Mapping

Let's illustrate this with the mapping for EAST-ADL. Rich components are a general concept of the CESAR meta-model and Function Types from EAST-ADL are related and can therefore be mapped. Other elements can have a more complex mapping such as EAST-ADL Function Flow Ports that are related to a CMM port being typed by a Port Specification with a Flow (Table 6.1).

Based on such concept identification, an implementation for the integration of the respective meta-model can be performed.

Table 6.1 Example of mapping between EAST-ADL 2.1 and CMM

EAST-ADL				CMM			
Element	Reference name	Num. of ref. elts.	Referenced element	Element	Reference name	Num. of ref. elts.	Referenced element
Analysis function type	Part	0..*	Analysis function prototype	Rich component	Part	0..*	Rich component property
Hardware component type	Part	0..*	Hardware component prototype	Rich component	Part	0..*	Rich component property
Function type	Connector	0..*	Function connector	Rich Component	Interconnection	0..*	Interconection
Function FlowPort	Type	1	EADatatype	Flow in port Typed by port specification	Type	1	DataType
Function ClientServer port	Type	1	FunctionClient ServerInterface	Port	Type	1	Port Specification.
Requirement	Text	1	String	System Requirement	Weak assumption/Strong assumption/guarantee	1	Assertion Specification
Satisfy	Satisfied req'ment	0..*	Requirement	Satisfy	Satisfies	1..*	System requirement

6.6.2.2 Meta-model Implementations in the RTP

In order to make a mapped meta-model usable in the RTP, that is tailored to a domain-specific development process, an implementation of the meta-model integration is needed. A concrete implementation of the integration depends on the respective needs for tool interoperability and the used data-format. In CESAR several integration needs were identified:

- The tool internal meta-model can be user defined (e.g. UML tools with user defined profile like Papyrus, Topcased).
- A tool only uses a specific meta-model and does not provide any extension mechanisms (e.g., analysis tool without API and with a fixed but standard data format).
- The tool meta-model can be linked on a basic MOF level (e.g. CVL Tool).
- Data must be extracted from a tool without knowledge about tool internal data structure (e.g. DOORS).

For the respective integration needs, several implementation approaches can be considered like particular or full model-transformations, adaption mechanisms, element linking, encoding or extensions of the CMM by specialized meta-classes and associations of the mapped meta-model. Furthermore the CMM can be used in several ways to exchange model-information such as using a CMM implementation as a common exchange format, using the CMM for identification of common concepts when transforming between foreign meta-models or providing a common interface to instances of foreign meta-models. Possible implementations for tool and domain-specific extensions are discussed in the following subsections.

6.6.2.3 Specialization

Specializing the CMM is one possibility to implement an integration of a mapped meta-model into the RTP. The advantage of such an implementation is the direct relationship to the CMM. A CMM view on a specialized meta-model is provided via inheritance. Domain-specific information can be stored in elements or attributes inherited from CMM or stored in dedicated domain-specific model elements and attributes. The disadvantage of this kind of implementation is the need for a new implementation of the domain-specific meta-model. If the domain-specific meta-model changes, the CMM implementation is to be changed, too.

Such an implementation can be useful for tools where the implementation of a meta-model can be modified, such as UML modelling frameworks. UML modelling frameworks provide profile mechanisms. The CMM is implemented as a profile with stereotypes to give CMM semantics to UML elements. Domain-specific extensions are also provided in a profile with domain-specific stereotypes. Mapped elements are implemented in a way that the respective stereotypes specialize stereotypes from the CMM profile.

6.6.2.4 CMM Model Transformation

Model transformations allow the extraction of information that is mappable to the CMM into the RTP respectively back to a DSL. The advantages of this approach are scalability regarding the set of relevant information. Furthermore no new implementation of the domain-specific language is needed. The DSL may even evolve. Only the transformation has to be modified if it is affected by changes in the DSL. The result of a transformation into CMM is always a valid CMM model and therefore a CMM view. The disadvantage of transformations is the need for synchronizations. Transformations always produce new models which need to be related to the original models. If the original model changes the transformation has to be triggered again and the target model must by updated with the transformed model. Tools using the transformed model must explicitly synchronize their data with this model.

Such an implementation can be useful when domain or tool specific information shall be extracted to the RTP and the tool does neither provide an own data format nor an own repository. The extracted information is available as a CMM model and therefore a CMM view is automatically provided.

6.6.2.5 CMM Model Adaptation

CMM Model adaptation provides a means of accessing an adapted model in a way as if it was a CMM model. This enables a CMM view on a DSL model. When accessing specific model elements of a DSL the CMM view is generated using the mapping between the DSL and the CMM. Changes to the CMM view do directly lead to respective changes in the DSL model. If a domain-specific model element itself is needed, the adapter implementation can additionally deliver it in order to allow direct access to the domain-specific information.

6.6.2.6 CMM Model Linking

Linking on the one hand external model elements from CMM and on the other hand linking CMM elements from other data formats is another possibility to implement integration with CMM.

Some CMM elements allow storing references to external models in specification attributes. The interpretation of this specification is given by another attribute specification type. The specification type can state that the specification is a reference to an external file, model or model element. CMM elements with such a specification and specification type pair are this VVProcedure, BehaviorImplementation and Resolve. For example, in order to provide test vectors given in TDX format to an analysis task, the respective VVProcedure contains a specification with the reference to a TDX file and a specification type “TDX File”.

For other tools the CMM provides a data structure that has linkable elements. Being a meta-model any instantiation of the CMM is also an instantiation of a MOF meta-model. There are frameworks like the Eclipse Modelling Framework (EMF) that provide simple means for link implementations to refer from one MOF model to another. In fact it boils down to simple references. By using MOF reflection a tool can apply knowledge of the more specific MOF model (like CMM) without the tool itself being CMM savvy.

6.6.2.7 CMM Model Encoding

A very simple but particularly unclean possibility to integrate domain-specific data in CMM is to choose a fitting encoding. This approach requires that the semantics of the used CMM elements is still correct in order to preserve interoperability with other tools and data formats. The encoded domain-specific information can only be interpreted by a tool that exactly knows this encoding. This solution can be very dangerous when the encoding is equal for different domain-specific concepts with incompatible semantics.

Thus, a domain-specific language element which can be mapped to a rich component contains internal attributes which are not known in CMM. These are mapped to rich component variables with the respective attribute names. A domain-specific analysis tool would look for rich components having variables with the domain-specific attribute names. Other tools that are connected to CMM only see a rich component with a variable.

Chapter 7
Implementation of the RTP

Christian Hein, Tom Ritter, and Michael Wagner

7.1 Introduction

As a basic principle the CESAR RTP is a generic integration platform which focuses on a model driven development process for safety-critical embedded systems. The process of realizing and implementing a specific tool chain (RTP Instance) is defined as RTP Tailoring. The RTP overview is shown the Fig. 7.1. Furthermore, the general RTP concepts like Tool, Service, Adapter or Model were described in the previous chapter.

Based on the previous explained concepts, this chapter describes principles which are needed in order to implement the generic concepts and assets of the RTP, the RTP Tailoring process and as well as the realization of an instance of the Reference Technology Platform (RTP) by taken into account the underlying Interoperability Specification (IOS) which is described in Sect. 8. However, the IOS does not cover all aspects which are relevant for an RTP implementation. In fact, the IOS identifies the problem space and in particular, the IOS provides an approach how information can be shared between different tools and services in principle by defining structure and syntax of common information. The IOS does not define characteristics like semantic interpretation and processing of information in the different tools and platforms. Nevertheless, the semantic interpretation is mandatory for satisfactory RTP tool chain. Therefore, this chapter introduces general principles for RTP-based tool chains in several directions like data integration, process integration or support of multi user collaboration with focus on structural and semantic aspects.

C. Hein (✉) · T. Ritter · M. Wagner
Fraunhofer Gesellschaft zur Förderung der Angewandten Forschung e.V., Munich, Germany
e-mail: christian.hein@fokus.fraunhofer.de

A. Rajan and T. Wahl (eds.), *CESAR - Cost-efficient Methods and Processes for Safety-relevant Embedded Systems*, DOI 10.1007/978-3-7091-1387-5_7,

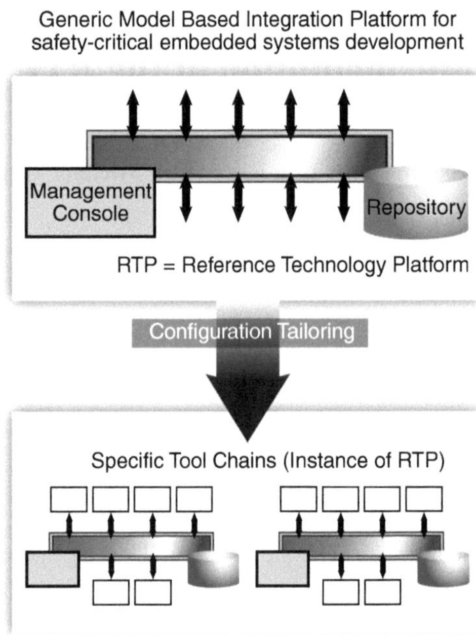

Fig. 7.1 RTP concepts

7.2 Generic Reference Technology Platform

One aim of CESAR is to provide a generic RTP. This RTP consists of different types of assets or items which are not integrated initially, due to the fact, that several kinds of stakeholders can provide content without focusing on integration aspects. This generic RTP and its assets are the basis for each derived RTP instance. However, how to implement such a generic RTP and what are the basic requirements for the assets?

Figure 7.2 depicts the main categories of assets and the mandatory parts of a generic RTP. First of all different stakeholders are involved in a development process. We distinguished between two different roles of engineers. The first role of engineers contains users which produces information with tools. The second group is dealing with the maintenance of the RTP itself. Therefore the second role contains administrators and users who control the configuration and the process flow of RTP Instances.

Since the RTP architecture is based on SOA principles particular assets are Services. In terms of RTP Services provide functionality on data that can be multiple reused in different tools. Services are running in background and can be orchestrated. Typical RTP services are model transformation, consistency checks, code generation, simulation or verification and validation. Beside services, Tools are also relevant RTP assets. Generally, a tool produces information. In the sense of RTP implementation produced information could be binary data, model or meta-model.

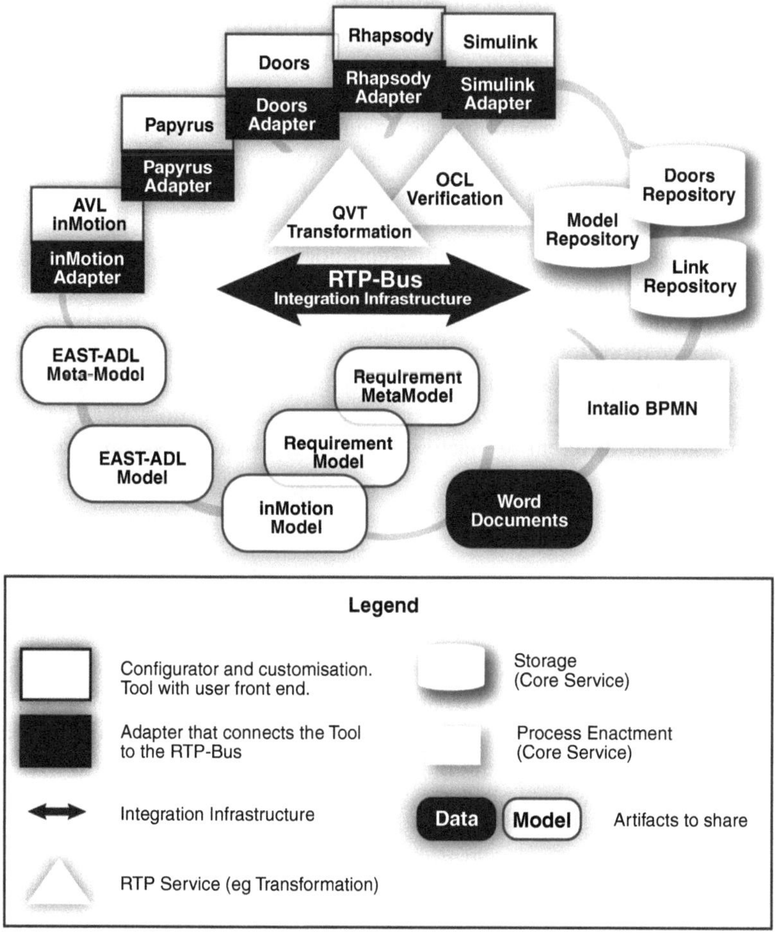

Fig. 7.2 RTP assets

However, a significant characteristic is that a user interacts with a tool. The tool presents information to the user. It allows interaction, for instance the execution of user operations on data and in particular it allows the editing and the creation of information. Tools are manifold and they can be categorized in varies ways, of course. With respect to CESAR the mostly considered ones are Matlab Simulink, Microsoft Office (Word, Excel), Eclipse based tools (Topcased, Papyrus) or IBM tools like Doors or Rhapsody.

Tools are integrated into the RTP with the help of Adapters. An adapter taking care of the integration aspects handled in the RTP. It coverts tool internal data into a format which can be processed within other elements of the RTP (data integration). An adapter also facilitates the control integration as it relays control flows to and from the tool. To certain extend the adapter covers the platform integration as it in general realizes the communication of the tool with the other elements of the RTP.

The communication of the tools via their adapters and with the other elements of the RTP is coordinated with the help of the RTP-Bus. It provides a technical framework for exchanging information. Examples are ModelBus, Enovia or Jazz.

Another part of the RTP is the Process Orchestrator. It can be used for the definition and execution of certain development processes steps. In that way it facilitates the automation of the development process.

An important part of the RTP is the Repository. It is used to store artefacts that are relevant in the development process. This comprises structured and unstructured data and in particular models conformant to meta models. There is a specific interface for a repository participating to the RTP and therefore the repository technology used is transparent to the RTP. So it can be relational databases as well as advanced object or model storage systems. But the repository must be capable of storing all the different artefacts which are relevant to systems engineering processes.

7.2.1 RTP-Bus in CESAR

As outlined before, one of the categories of RTP assets is the RTP-Bus. It plays a central role in the RTP and in particular in the RTP instances as it basically links all the other elements. The concrete properties of the RTP-Bus chosen for a RTP instance have an important impact on the complete RTP instance. The idea of the Reference Technology platform is to have also for the RTP-Bus a selection of comparable assets – multiple RTP-Buses – which can be used in the RTP instances. For this reason a specification of the architecture and interfaces of the RTP-Bus and its related assets (e.g. Tool Adapters) has been defined and has been used as input for the CESAR Interoperability specification as well. So in consequence, many implementations of the RTP-Bus assets are potentially available for creating RTP instances.

As the implementation of an RTP-Bus is a complex issue CESAR has used the ModelBus as available technology for most of its RTP instances created by the domain specific sub projects. ModelBus is the technology which provides the demanded functionality and adheres to the API definition for RTP-Bus. It also comes with extensive documentation and tutorials and fulfils CESAR requirements with respect to stability and performance. The following section gives a short introduction to ModelBus as an implementation of the CESAR RTP-Bus.

7.2.1.1 Overview on ModelBus

ModelBus is a model-driven tool integration framework which allows you to build a seamlessly integrated tool environment for your system engineering processes and is built to address the challenges listed beforehand. ModelBus is based on SOA principles, COTS components, and existing standards to the largest possible extend. ModelBus provides transparent and homogeneous access to heterogeneous and possibly distributed data sources and supports working on these data source with the formalism linked to model-driven engineering approach. The ModelBus core components are provided as Open Source and free software.

Concerning the dimensions of integration, ModelBus addresses Platform, Data, Control and Process Integration. It does not address Presentation integration as this is a problem space which can only be appropriately addressed by the tool vendors themselves.

7.2.1.2 ModelBus Architecture

Figure 7.3 shows the ModelBus as integration and communication platform connecting different services offered by tools connected to the ModelBus. Based on SOA principles it offers a model repository, a service registry and notification service as core services. Besides the core services, ModelBus offers also a set of specific and generic services. For example a model verification service, which can be used for verifying intermediate modeling results with respect to some modelling guidelines. Or model transformation services can be used to transform the results created with one tool to be usable in the context of another tool.

Keeping track of the relationships between the artefacts within a model and across model boundaries can be supported by a traceability service. Regarding the Process Enactment, Workflows – executed automatically – can be defined using orchestration tools. Specific or customized control and management tools help to manage the processes running on the ModelBus and to manage participating resources, like developers or computing nodes. Finally the tools which are used during the lifecycle of a product development are also attached to the ModelBus. These are typically tools, which have a user interface, so humans are interacting with these tools and do creative work with them.

7.2.2 RTP Repository in CESAR

As outlined in the previous figure, the RTP Infrastructure is connected via an RTP Bus (like ModelBus) with a set of core service. In general, services like discovery, registration or notification are provided by the bus technology, which is based on a SOA. However, CESAR RTP requests a further core service for storing all artefacts which are produced within the RTP. This service is provided by a Repository. This repository service can be applied in different use cases with different aspects. A lot of repository tools and services are available on the market. Repositories can be used to store source code, models or only traces and links. Several tools already

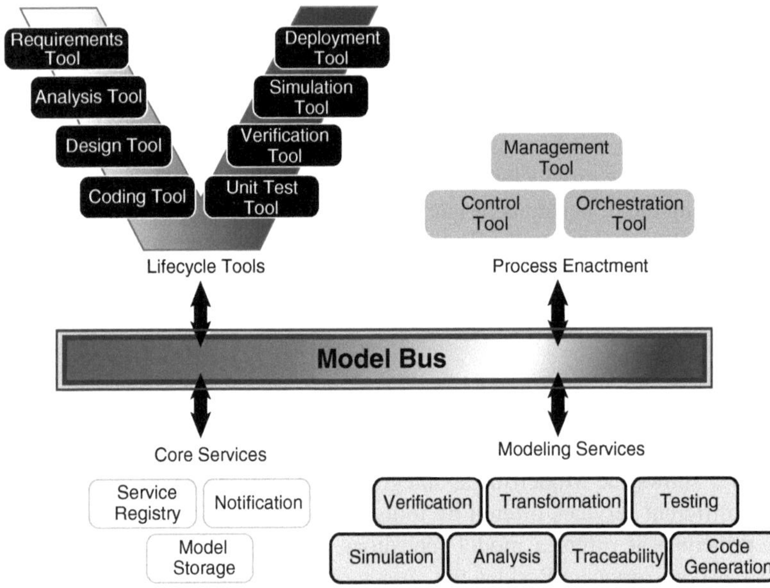

Fig. 7.3 ModelBus general structure

provide own repositories in order to store tool internal data such as DOORS and the corresponding DOORS repository.

The CESAR RTP is focusing on model-driven development process. Therefore, a model repository is mandatory for model management. The idea of a specific model repository services is that repository can be browsed through a tree-view like interface. The service should follow the check-in and check-out paradigm which is an existing approach in source code management systems like CVS or SVN. In addition to that paradigm a lockout mechanism should work at the level of the elementary modelling element. This feature is flawless in large teams and a high-concurrency environment. Model management deals also with configuration management of all kinds of artefacts. The configuration can be done on the level of specific modelling elements like an attribute of an UML Class or on a global level capturing the heterogeneous modelling objects that may have cross relationships with other modelling entities like EAST-ADL and CMM relations.

Besides the data storage and model management a RTP repository should also address the feature of element locking. Assume the situation that more than one user want to work on the same files or models at the same time. In this situation they can use the RTP repository, store their files and models there and synchronize their work using the lock mechanism offered. The lock will be indicated to the user who set it and is still able modify it. While the second user only have read access on the locked element until the first user unlocks the elements.

7.2.3 RTP Process Orchestrator

In CESAR, RTP services are specified as web services. There interfaces are described by a WSDL and can be invoked using web services mechanisms. This offers the chance to use all well-known and proven methods and tools for orchestration in the web services area. In addition the RTP repository emits notifications whenever its contents e.g. a model has been created, updated or deleted. Both features are very useful to automate workflows in the model-based development environment based on CESAR RTP. In the web services area at least two approaches for orchestration exist:

- BPMN (Business Process Modeling Notation), a graphical modelling approach for business workflows, defined in an OMG specification (http://www.bpmn.org/)
- BPEL (Business process Execution Language), an XML-based language specified by OASIS (http://docs.oasis-open.org/wsbpel/2.0/wsbpel-specification-draft.html)

BPMN is a graphical notation which is not directly executable, meanwhile BPEL workflows can be executed but there is no standardized graphical representation for them. Some BPEL tools have invented their own graphical representation. On the other hand BPMN includes the specification of a mapping from BPMN to WS-BPEL.

There exist not only tools that allow the specification of workflows in BPMN or BPEL separately but also tools that allow the graphical modelling of workflows using BPMN than generate BPEL from it, deploy it and then execute it using a BPEL engine. Due to the fact that a huge number of orchestration tools exist on the market (open source, free or commercial), RTP implementation is not focused on providing just another orchestration engine. Using existing technologies seemed to be an adequate approach. Therefore, RTP implementation assumes that each RTP service is based on web services and thus it is ready for orchestration. The principle approach of defining a workflow starting with BPMN is mostly the same in different tools supporting it:

- Specify the workflow using BPMN
- Add information about the service operations, notifications, data used based on the WSDL
- Deploy the executable workflow to a server
- Execute it

A small example should illustrate the use of a very simple workflow in the context of RTP-based services. The workflow should illustrate the combination of a model editor that changes a model, model repository to store it and a transformation tool (QVTService) to execute a transformation if necessary. Assume the following situation:

- A modelling service (e.g. some editor) is doing some changes within a specific model. It stores the updated model to the repository.

- The Repository will emit an update Model notification.
- Anybody interested in this event might receive it and react on it.
- A workflow in an orchestration tool will react on the update notification and depending on some condition (e.g. if it is a specific type of model) invoke a transformation service (QVT Service).
- If the transformation is invoked, it will transform the model (from the repository to the repository.

7.2.4 *Overview on CMM-API*

The CESAR RTP is intended to perform steps like assessment, analysis, and tracing of system specifications and design based on a common understanding of data artefacts in the RTP as shown in Sect. 3.5. The CMM-API is a realization of the CESAR reference technology platform (RTP) and eases integration of tools and methods with their data-structures avoiding expensive point-to-point connections as described by CESAR partners [16]. Instead, this approach supports interoperability of tools and methods by defining common interfaces for data-access, -linking and -modification. It consists of an API that establishes a semantically unified view on system artefacts based upon the common meta-model (CMM) as defined in Sect. 6.4. The CMM-API provides such a CMM-view is on the whole system developed with different tools during the development process. Traceability between data-objects from different tools can be achieved independently from the native data-representation. Interoperability between data-assets (e.g. tools, methods, data-repositories and data-formats) can be established by defining CMM-compliant data-services. This compliance provides a homogeneous data-view on all connected data-assets. Tracelinks that are specified by the CMM such as satisfy, refinement or allocation can be established between elements in the CMM-view. This concept enables the usage of common traceability means for different engineering-tools, analyses, data-repositories and data-formats. Figure 7.4 illustrates a CMM-view on the tools Simulink [114], DOORS [81] and Rhapsody [82] as well as System Desk [45]. A homogeneous view of requirements and components with their interfaces and connections, becomes visible based on the native data structures of the different tools. In order to establish interoperability between tools, methods and data-formats the concepts of the respective data-models have to be mapped to those of the CMM.

7.2.4.1 Architecture for Distributed Heterogeneous Data

The CMM-API supports different data-models which are tied to native tools. All artefact defined by these data-models can potentially be distributed across multiple and kinds of repositories. Figure 7.5 depicts a general overview of the basic RTP concepts for retrieving data-artefacts from their storage locations in repositories and how a unified CMM view is established.

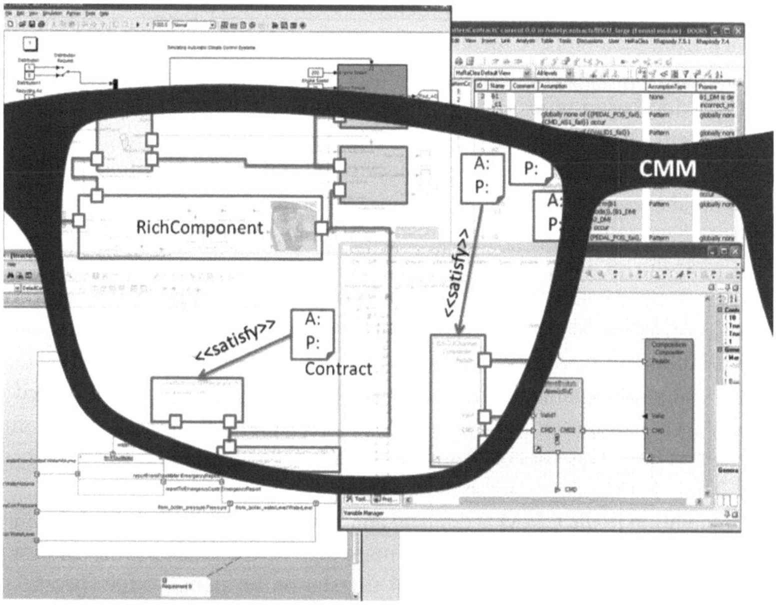

Fig. 7.4 CMM view on different tools with heterogeneous data-models

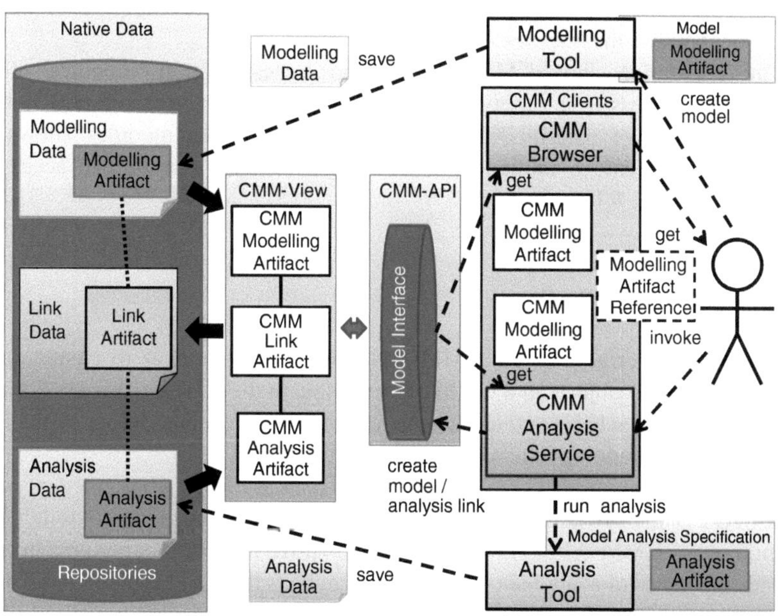

Fig. 7.5 CMM data manipulation

In order to work with data stored in the RTP, all addressed data-artefacts need to be fetched from their storage location first. Within the RTP, we distinguish between file-based and application-based repositories. The local file-system is the simplest version of a file-based repository. In most cases files are managed by a distributed storage system (like ModelBus or Subversion). In case of an application-based repository, native data is only accessible through the API of an application (e.g., DOORS databases). All types of repositories are unified within the RTP by so called access-support modules. These modules provide the functionality to access specific types of repositories and to make data locally available.

Data-artefacts throughout the RTP are objects identified by unique references. These references encode the location and type of the repository storing the objects, along with the identification of the individual data-artefact. Given such a reference the access-support modules can determine if a particular object is located in a repository of the supported type and provides the necessary procedures to retrieve the corresponding data. Access-support modules are not bound to specific data-models. They do not adapt data to the CMM.

The adaptation to the CMM is done by adapter-components called CMM-support modules. These modules implement the mapping between the specific data-models and the CMM. They perform the model translation on demand for specific data-artefacts into a CMM-representation. Having this representation a CMM-view for accessed data-artefacts is established and CMM-based data-navigation is possible. While navigating the CMM-view all addressed data-artefacts are accessed and their CMM-representation is integrated into the CMM-view. The technique used for this behaviour follows the *Adapter Pattern* [170].

The separation into CMM-support and access-support modules allows CMM-based access to a variety of combinations of repositories and data defined by different native data-models. The integration of these modules allows the CMM-API to create a unified view on supported data available in accessible repositories of the RTP.

7.2.4.2 Common Interfaces

The common interfaces of the CMM-API define basic services to create, query, modify and delete objects in a CMM-view on data in the RTP. Further services can be build upon these basic services to support engineering methodology in the RTP. For CESAR methodology this will be shown in Sect. 7.4.2.

Since unique references are used to identify data-artefacts, a base method of the CMM-API is the provision of a persistent reference for an object. Another base method of the CMM-API is browsing through native repository structures in order to find references of available objects (these can be folders and projects in DOORS or directories, files and models in file-based repositories). An update-method allows to write changes to the repositories and a delete-method allows to remove objects from a repository supported by the RTP.

We consider specific methods for creation and retrieval of data-artefacts in a CMM-view on native data. The creation-methods allow the creation of an artefact based on a native data-model of the referenced repository data-structure. After the creation a CMM-representation of the data-artefact is returned. Each created data-artefact has a unique reference which is crucial for interoperability to enable linkage of objects. Data-retrieval methods allow to get CMM-representations of data-artefacts and of other objects linked to these artefacts in the CMM-view.

The CMM addresses different topics of system engineering and several common modelling concepts. Therefore, we grouped the CMM-related methods in multiple interfaces. These groups are e.g. requirements, components, verification & validation, declaration, data types and extensions. They define get- and create-methods for e.g. components, different kinds of requirements and links like satisfy as specified for the CMM.

7.2.4.3 Support for Heterogeneous Data-Models

As discussed in Sect. 7.2.4.1, we distinguish between models in file-based repositories and application-based repositories. In this section we explain how the corresponding CMM-support components can be created for each type.

Repositories that provide an API for direct access to their data can be integrated into the RTP by a direct implementation of the CMM-API methods. Such an implementation invokes the application's API to access the native tool data and generate corresponding CMM representations. In this case the access-support and CMM-support modules are combined in the implementation and cannot be separated. The CMM-view is generated on the fly while the data-artefacts of the application are navigated. A unique reference can be generated in an arbitrary way for each object that is accessible via the CMM-API, as long as the part that is used to identify the CMM-support module is unique. There are no constrains on the other parts of the reference, so that native references from the repositories can be reused.

As an alternative for the application-based CMM-adapter, the CMM-API can be implemented for file-based repositories. For this kind of repositories access-support modules allow to access files inside the file-repositories and CMM-support modules perform the mapping of data-artefacts in files to CMM-representations. Typically all data-artefacts in a file have a unique identifier. This identifier and the reference to the file inside the file-repository provide a reference for the CMM-API in the RTP. When an element is created in the CMM-view actually an data-artefact in the native data-structure is created and a CMM-representation is returned. CMM-support modules for different data-formats and access-support modules for different types of file-repositories can be integrated to access data of file-based repositories via one CMM-view.

Table 7.1 Roles and stakeholders

Role	Description
Tool/Service Provider	Is an external stakeholder which provides a tool or a service which is integrated to the RTP
Standardization Body	Is an external stakeholder which provides standards and specification relevant to the RTP and the development processes executed with RTP instances. Standards may cover implementation aspects of particular RTP assets (e.g., communication protocols, data encoding) or they might be in particular relevant for the creation of RTP instances (e.g. certain UML Profiles)
Adapter Provider	Is an internal stakeholder which provides Adapters for tools and services provides by external stakeholders (Tool/Service Provider) in order to adapt the respect tool or service to the RTP and make them usable for RTP Instances. In some cases, the Adapter Provider and the Tool/Service Provider can overlap
RTP Integration Engineer	Is an internal stakeholder, which tailors generic assets and building RTP Instances
RTP Administrator	Is an internal stakeholder, which supervises RTP Instances
Integration Platform Provider	Is an internal stakeholder which provides the integration platform – the RTP-Bus

7.3 RTP Implementation Roles and Phases

During the execution of the CESAR project several roles and stakeholder have been identified participating to the creation or usage of the RTP in one or the other way. Some of them are involved in the RTP implementation process while others are not. So we differentiated internal and external stakeholders. An internal stakeholder is directly involved in the RTP implementation process while an external stakeholder provides assets which are relevant for the RTP implementation. A typical example of an external stakeholder is the Object Management Group (OMG). The OMG, as a standardization body is responsible for a lot of widely-accepted specifications like MOF or UML. The results of such a standardization activity are of great importance for the RTP. However, the work is performed without any focus on the RTP. Furthermore, a provider of a tool or a service could also be an external stakeholder, because the development of the tool was realized without any focus on the RTP as well.

Based on this, Table 7.1 lists all relevant roles and stakeholder with a relation to the CESAR RTP implementation.

7.3.1 RTP Implementation Phases

The development of the RTP is structured in three different implementation phases, which reflects the creation of the generic assets, the tailoring and the instantiation aspects.

7.3.1.1 Creation of Generic RTP Assets (Tool/Service Provider, Standardization Bodies)

The RTP consists of a set of technical assets and concepts as described in Chap. 6 which are provided by internal and external stakeholders. The assets are created by the stakeholders without a specific RTP instance in mind or even without knowing the RTP concepts at all. So external stakeholders like Standardization Bodies have created new modelling standards or new tool version is provided by Tool/Service Providers. In parallel, also the internal stakeholders create RTP assets specific the RTP concepts, e.g. Adapter providers do create adapters for a given tool. For each assets which is relevant for the RTP and which shall become a part of it, an assessment of its maturity and its integration status needs to be done. With such an assessment it is easier to select appropriate assets for creating the RTP instances.

7.3.1.2 Customization/Tailoring Phase (RTP Integration Engineer)

Based on requirements for a specific RTP instance, the RTP customization phase is the process in which an integration engineer is tailoring generic assets in order to build a specific RTP instance. This phase consists of identification of the overall purpose of the tool chain, the selection of the underlying integration framework, the definition of semantic for information mapping, the management of information and the specific adaptation of tools and services. Unfortunately, no automatic approach exists that eases this tailoring phase, in fact this phase is an engineering task which should performed by a corresponding expert. However, several best practices are established in order to deal with tailoring complexity.

7.3.1.3 Deployment and Configuration Phase (RTP Maintainer)

The last phase of RTP implementation deals with the deployment and configuration of the RTP assets. A running RTP instance consists of different tools and services. Several stakeholders are involved. Therefore, the RTP users need adequate access rights in order to put and get corresponding information from RTP. In addition to that, the distributed RTP environment must be configured in terms of firewalls, development processes or definition of developer/user groups. In a typical RTP instance, this configuration management is realized by a model-driven approach. Thereby, a model contains all relevant information of the current RTP configuration. The model is updated by increment transformations which are triggered by corresponding notifications within the RTP. Most of the configuration work can be done by a tool. In context of CESAR, the RTP Desktop and the ModelBus Manager can be used for this purpose.

7.3.2 RTP Tailoring

The main engineering part of building a running an RTP is the tailoring process. The tailoring includes the customization and configuration of an RTP instance. Thereby, RTP integration engineers are responsible for the definition of specific information mapping, roles of tools, the selection of the used integration platform and basic services in order to build an RTP instance, which satisfies the overall purpose of the integration. Therefore, the first lesson learned from CESAR is that an integration engineer is mandatory to realize a appropriated RTP-based tool chain.

In terms of RTP implementation the phase of RTP tailoring can be divided in several activities. Generally, the activities can be seen as a stepwise approach. The different steps are summarized in the next section. However, this is more or less a best practice instead of an absolute methodology.

7.3.2.1 Step 1: Identification of the Purpose of the RTP Tool Chain

The first step of the RTP tailoring process is the identification of the overall purpose of the RTP tool chain. The tailoring itself is an engineering task. Several aspects should be considered like the overall process, used methods and specific workflows. Therefore, the RTP integration engineer has to make a plan first of all. Of course, that's nothing new. However in the context of an RTP implementation the plan should address some specific topics. In particular, the overall goal of the RTP instance must be clear before starting implementation. Usually, the user of an RTP instance is not starting from scratch. A lot of tools and services are already used in the user development process. In fact, problems are caused by existing gaps of tool integrations or missing tool interoperability. Thereby, not every tool needs to be integrated with every other tool. Therefore, the integration engineer has to analyses the missing pieces of the current development environment with respect to tool integration or interoperability aspects on the one hand. And on the other hand, the engineer needs to take into account the development process and specific workflows which can be automated.

7.3.2.2 Step 2: Selection of the Integration Platform

An integration platform is responsible in particular for connecting tools and services of a RTP tool chain. Different integration platforms have different goals and aimed different targets. Therefore, the selection of a certain platform implies a lot of implication for RTP tailoring process and hence also to the underlying RTP instances. In the context of CESAR several integration platform approaches of were used in order to realize corresponding tool chains. The main important ones are ModelBus, OSLC and Jazz as well Enovia. ModelBus is used in most of the realized RTP instances. Therefore, this integration platform and its implication are more described in detail in this chapter.

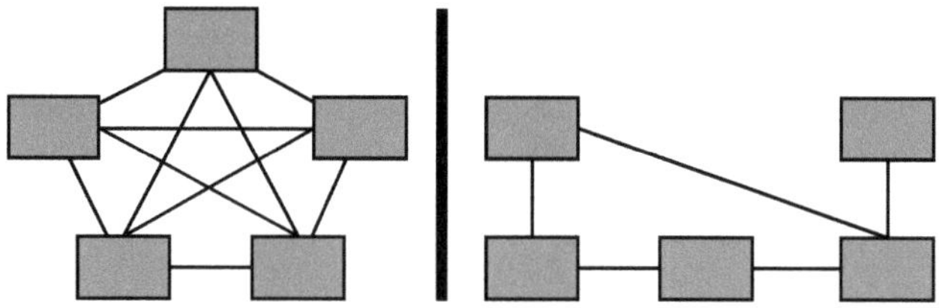

Fig. 7.6 Generic examples of point-to-point integrations

7.3.2.3 Step 3: Information Mapping and Agreement on Semantic

Once the used integration platform is fixed the integration engineer has to decide which level of information exchange is needed for a corresponding RTP instances. Either it is necessary to implement direct mapping between tools or it could be useful to rely on a common semantic approach in which every tool is required to exchange its data in an agreed format. Unfortunately, there are arguments for both approaches and no general accepted answer can be given. Therefore, the main approaches will be introduced in order to outline guidance for integration engineers.

Point to Point connection and RTP example Basically, the point-to-point/direct integrations, between tools and services, are almost easily and fast realizable. The information mapping is organized along the relevant aspects for the coupling of two tools. Figure 7.6 depicts two generic examples of point-to-point integrations of five different tools or applications.

As shown in Fig. 7.6, it might be the case that there is no need for a direct coupling of two tools, due to the fact, that this integration is not necessary useful from a certain point of view. In fact, integrations are based on a specific purpose. For instance, a generic tool chain could consist of the three tools DOORS, Rhapsody and Simulink. There might be the need to exchange information between DOORS and Rhapsody and between Rhapsody and Simulink, but not from DOORS to Simulink. Nevertheless, in worst-case the number of integration is equal $n(n-1)$ in which n is the number of the tools to be connected. Therefore, the complexity exhibits quadratic growth in a point to point integration when all tools should be connected (or $(n(n-1))/2$ in case of undirected connections).

In the context of tool integration a point-to-point approach is realized in different ways. A typically and widely applied method is the usage of import and export techniques of tools. Originally, a lot of tools provide already functions to import data from other tools or to export data to a corresponding format of another tool. Enterprise Architect or Magic Draw can export their internal data to an EMF-based UML representation. In addition to that, several tools provide also an API in order to offer specific functions for external tools. Integration engineers are requested to access tool data and functions via this API. However, not all tools are able (or not willing) to publish all needed information via an API. Furthermore, model transformations are another approach for a point-to-point connection. Model

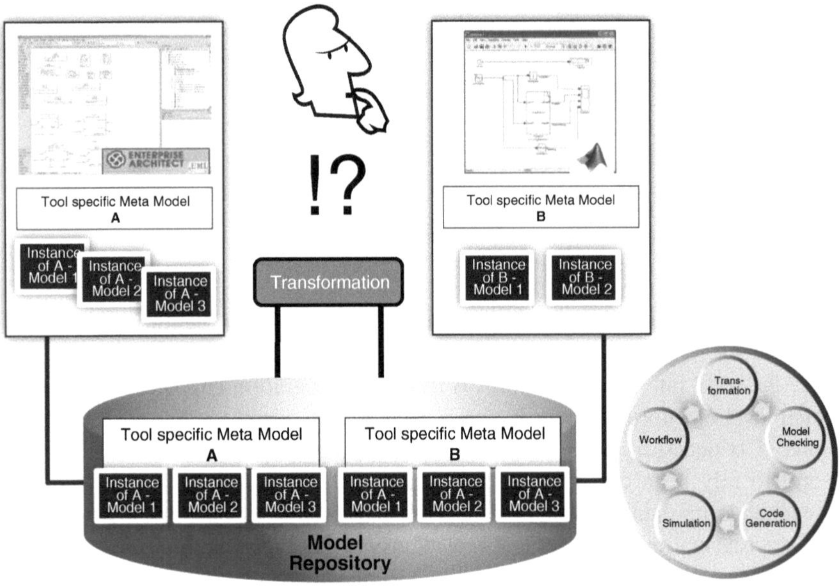

Fig. 7.7 Example model transformation between two tools

transformation is a key concept in model-driven engineering. The transformations are focusing on models which are specific representations of data. Basically, a model transformation converts a source model into a target model. In terms of tool integration a model transformation can be used as a mechanism to translate the data model of one tool into the data model of another tool by using specific transformation rules. Of course, model transformation techniques can also be applied in other integration aspects, such as code generation or parsers.

With respect to tailor generic RTP concepts and realization of tool-to-tool mapping, the RTP implementation follows the provided approach by IEEE in which tool internal data is represented by modelling concepts. And only these modelling representations will be exchanged and transformed. Once, the integration engineers decided to build the RTP instance on a corresponding integration platform, the integrator also implicit selects a corresponding modelling framework for expressing the tool internal data. In a ModelBus-based RTP instance the Eclipse Modelling Framework (EMF) is used for this purpose. EMF and its Ecore is an implementation of the OMG specification MOF. A lot of widely used modelling specifications are expressed with MOF concepts like UML, OCL, SysML or EAST-ADL. The basic principles of this tool-to-tool approach are shown in the following Fig. 7.7.

The Fig. 7.7 shows an example of the two tools Enterprise Architect and Matlab Simulink. The tools are integrated via a model transformation. The tool internal data is expressed by model that conforms to a corresponding tool specific meta-model. In a ModelBus-based RTP instance the models are usually exchanged via a model repository and the transformation functionality is provided as a modelling service.

The transformation service gets a notification from the repository whenever a corresponding tool model has been changed. The transformation rules are expressed on the tool meta-model level. The semantic of the rules differs from the purpose of the tool integration. The integration engineer has to decide what kind of information or data is needed for the exchange in a specific RTP instance.

An alternative integration pattern is based on an agreement on a common semantic across several tools. Theoretically, this reduces the integration effort to a linear complexity in the best case in comparison to a point-to-point connection. According to this idea, if a tool is integrated to this common semantic representation it is integrated to all other tools and services. This leads to a potential number of n tool connectors in order to fully integrate n tools. Such a perfect situation in which all stakeholders can agree on one realization is of course not realistic due to various reasons. In fact, the bottleneck of this solution is to represent all needed information in the common representation by focusing only on the purpose of specific tool integration scenario. In worst case all information of all tools and services needs to be represented and defined in the common specification. Otherwise, if the common semantic is to specific not all tool data can be exchanged, due to the fact that they cannot be expressed in the common representation. However, this approach already works for many closed domains, for example the requirements interchange format ReqIF or UML testing profile for testing purposes. Therefore, the agreement on a common semantic is helpful for specific domains with closed sets of tools.

As one of the lessons learned from RTP Implementation a promising strategy is to mix the different approaches of point-to-point and common semantics. In several domains already a lot of domain knowledge is included in standardized models such as requirement descriptions with ReqIF, testing concepts with UML Testing Profile (UTP) or business process descriptions with SPEM and BPMN. Therefore, the potential numbers of connections can be reduced by introducing domain specific languages which eases the finding of common agreements on semantics concepts. Moreover, the risk that specific elements of a tool cannot be represented in a common model is reduced as well. In addition to that, the information mapping between the different domain models is realized by transformations as well. This concept is depicted in the next Fig. 7.8.

7.3.2.4 Step 4: Information Management

The next step is dealing with the mapping of the several information which will produced by developers or services within the engineering process. This means, that an integration engineer has to decide how the information is stored and management within an RTP instance. In generally, two approaches exist that handle the problem of information management. The first approach is based on a central data storage in which all relevant information is available. In contrast to that, the linked data approach follows the idea of leaving the information at provider side. Basically, the information is only managed by linking the information. Every tool is responsible for the management of its own information that should be shared with others.

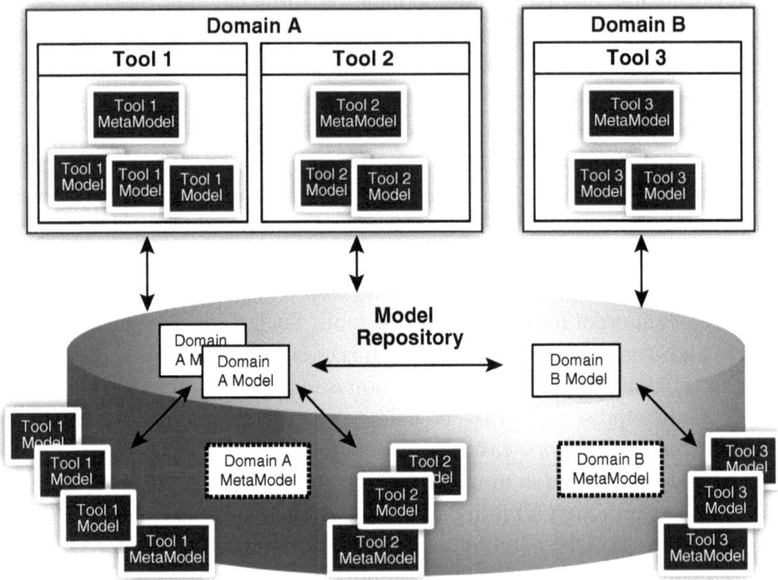

Fig. 7.8 Combined integration approach

However, information is not exchanged between the tools. Therefore, the following aspects should be taken into account by integration engineer in terms of information management.

Conceptual, there is a difference between information sharing and information exchange with respect to tool integration and tool interoperability. Information exchange can be defined as a transmission of information between two or more entities. The transmission can be bidirectional. It means practically, that the data from tool A is transferred to tool B. The data can be interpreted by tool B without any reflection to tool A. In contrast to that, information sharing is not focusing on exchanging data. Moreover, information sharing is an approach in which both tools are working on the same data. A change of information in tool A is also a change of information in tool B. The tools are sharing the same information.

The differentiation is important for the next tailoring step concerning adaptation of tools. Unfortunately, not all tools support an information sharing approach that implies the awareness of identities of specific development artefacts or a mechanism to synchronize with other tools. This needs to be considered while implementing a tool adapter. Furthermore, the design decision of central data storage versus linked data approach depends as well on the integration requirements for the RTP instance. In fact, a typical RTP instance is a distributed tool chain with specific characteristic concerning consistency, availability and partition tolerance. In that case, the CAP theorem indicates that only two of these three aspects (consistency, availability, partition tolerance) can be guaranteed concurrently in a distributed environment.

Thus, it is difficult to guarantee the consistency of the data by using the linked data approach. In contrast to that, it is difficult to guarantee the partition tolerance in case of a crash of the central data storage.

7.3.2.5 Step 5: Adaptation of Tools

Conceptual the tailoring phase ends with adaptation of tools and services. It is obviously that this is an iterative process which can be performed several times, due to the fact that in worst case a huge number of tools need to be connected to the RTP. Basically, a tool adapter was built in a generic way in order to ease to effort of customization for a specific purpose or in best case without any customization effort. This is an ideal point of view in an ideal world, of course. The real world is different form that ideal world. The adaptation of a tool is highly dependent on the purpose of the tool chain and the gap in a development process that should be closed. Therefore, a generic adaptation of the requirement tool DOORS is taken in the following section as an example in order to make these things clearer.

DOORS is a requirement engineering tool provided by Telelogic which is a subsidiary of IBM. DOORS is an acronym. It stands for Dynamic Object Oriented Requirements System. It is used, in particular in automotive and avionics domain, for structured definition of requirements. The requirements can be linked and based on this, a lot of analyse functionalities are provided. The DOORS tool is based on client-server architecture. DOORS data is stored in a database that can be accessed by a script language called DOORS eXtension Language (DXL).

In terms of adaptation, first of all, we analyses the purpose of a potential integration of Doors in a tool chain. DOORS could be integrated in different modes, such as a front-end, as a service or as an information repository. An integration as a front end means that a user/engineer is working with DOORS. The engineer is defining the corresponding requirements for a certain development project. The requirements can be shared with others within a RTP tool chain. For instance, a designer that uses an UML modelling tool can make use of the requirement specification in order to create a specific aspect of the later system. The designer can link a UML use case with a specific requirement in DOORS, for instance. This implies that the DOORS GUI needs also be adapted to enable the feature of linking directly a requirement to a specific UML use case. This must be reflected in the corresponding adapter for DOORS. Furthermore, the analyse functionalities can be offered as services within the RTP instance. This can be realized via the DXL script language. In that case, the developer of an adapter must be implement the corresponding DXL operations for accessing the analyse functions. The developer could realize the adaptation in two different technical ways.

No matter what kind of solution the engineer prefers, the different ways have different operational and technological implications. For instance, an extraction of all data within the database can be done either as a single short or as an incremental operation in which only differences of the model will be transferred. Both solutions have implications in terms of performance or consistency. In addition to that, the integration engineer has to decide what should be the information mapping of requirements within an RTP instance. There are two alternative ways of implementing the DOORS adapter with respect to semantic agreement. On the one hand, requirements can be understand only in the context of DOORS. This means that every other requirement tool need to migrate or transform its data into the DOORS requirement representation. On the other hand, a requirement centric approach could be used in which all requirement tools, such as DOORS, are transforming their tool internal representation into a common format.

In summary, this last step of tailoring is completely an engineering task. The adaptation of a tool highly depends on the integration needs for a certain tool. First of all, the adaptation of a tool is highly independent from the underlying integration framework. Therefore, the adapter implementation should be realized without focusing much on the used integration framework. While the integration frameworks are mature and stable, the tool adapters are very error-prone, due to the fact that tool providers often hide specific functions or operations which are useful for integration aspects. Furthermore, the access to the tool internal data is not sufficient in terms of information exchange with other tools and services within an RTP instance.

7.4 RTP Instance

7.4.1 ModelBus-Based RTP Instance

A typical ModelBus-based RTP instance consists of several core RTP components. The baseline for the integration framework is the ModelBus. In fact, ModelBus is based on a service-oriented architecture. Therefore, this integration solution provides a repository and notification as core services. In a ModelBus environment tools are integrated via adapters. The adapters must be able to connect to the core services. ModelBus focused on a model-driven development process. This means that the integration infrastructure has knowledge of the shared artefacts, in particular when models are shared between several stakeholders. In a ModelBus-based RTP operations that are useful for a number of tools are offered as services.

The key concept of model sharing in ModelBus is realized via a model repository. This repository interface is open and allows straight forward addressing of models via URLs. This addressing schema also results in simple service interfaces, because only model references instead of models are transmitted. Repository vendors can implement this interface in order to be ModelBus conform. ModelBus itself is delivered with a built-in model repository, which supports versioning, partial check-out of models and coordinates the merging of model versions and model fragments. The model repository implements the operations which are common for all repositories involved in CESAR RTP. The model repository is a web service and provides the following services and notifications.

Service	Description
checkInModel	The checkInModel operation stores or updates the model into the repository at the given modelURI and it creates a new revision. The operation returns a checksum in order to verify the successful transmission of the model. If the model does not exist within the repository the model will be created initial by the operation
checkOutModel	The checkOutModel operation gets a model from the repository according to the given modelURI. If a specific revision of the model is requested the URL needs to contain the corresponded revision number or peq revision number
checkInFile	The checkInFile operation stores the file into the repository at the given fileURI. The file will initial created if it does not exist already
checkOutFile	The checkOutFile operation gets a file from the repository according to the given fileURI. If a specific revision of the file is requested the URL needs to contain the corresponded revision number or peq revision number
delete	The delete operation removes the corresponding artifact according to the given URL
exists	The exists operation checks whether the corresponding artifact exists at the given URI with the given revision number
getDir	The getDir operation gets information of the RepositoryDirNodeKind according to the given URL and revision's version number
info	The info operation gets information of the artifact according to the given URL and revision number.
lock	The lock operation locks the artefact according to the given URL. The timeout parameter specifies the timing end of the lock. If the timeout value is 0 the lock is permanent and has no timeout
unlock	The unlock operation unlocks the locked artifact according to the given URL
getLocks	The getLocks operation gets information of all locked artifacts according to the given URL prefix

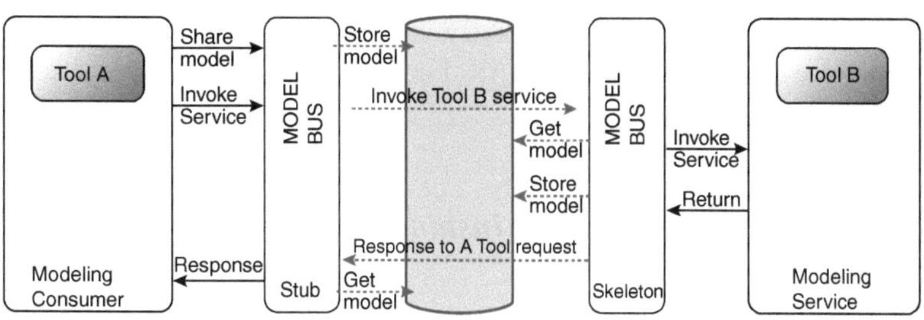

Fig. 7.9 ModelBus interaction pattern

ModelBus provides an interaction pattern in order to enable model sharing in a distributed and heterogeneous model-driven development process. Figure 7.9 depicts the general interaction pattern in a ModelBus integration scenario. This pattern is integrated in each adapter and it is completely transparent to the user.

The basic idea is that Tool A interacts as a consumer of the provided service of Tool B, which interacts as a service. If Tool A wants to consumer the service of Tool B it simply invokes the corresponding operation in the ModelBus Stub. The generated stub will offer by the ModelBus infrastructure automatically and it is based on distributed OSGI. The stub stores the model into the repository, for instance by calling checkInModel operation of the repository. After that the stub invokes the service skeleton which is also offered by the ModelBus infrastructure. During this invoking stub and skeleton only exchange the references of the models. This means that the consumer only specify of URL of the model, whereas the service skeleton only expects this URL. On the other hand the service skeleton takes this URL and gets the model from the repository by calling checkOutModel. The skeleton, itself, calls the corresponding implementation of Tool B.

The response is analog, due to the fact that the result data of the service operation is a model as well. The response model will also check into the repository by the skeleton. The skeleton returns to the stub only by transferring the URL of the response model. The stub checks out the response model again and delivers it to Tool A. The advantage of this is interaction pattern is that service interfaces become simple and only use standard WSDL data types. Thereby, external tools, like Intalio BPMS Designer can be used in order to orchestrate ModelBus services as well as exclusive or in combination with other non ModelBus services.

7.4.1.1 Typical ModelBus RTP Instance

A typical ModelBus-based RTP instance consists of several tools which are connected by adapters. Each adapter provides tool internal data as a model which is stored in the model repository. The data mapping is realized by transformation. Several adapters already exists which can be easily integrated in a ModelBus-based RTP. The following list gives in overview of existing adapters. However, not all adapters were developed during CESAR. Furthermore, not all adapters are free or come with an open source license.

7.4.2 CMM-API-Based RTP Instance

In this section we show how a an RTP integration of Papyrus EAST-ADL, DOORS and a requirement service is achieved using the CMM-API.

EAST-ADL is a language for multi-level system modelling in automotive domain. Papyrus is a modelling tool that stores UML files with applied EAST-ADL stereotypes in a file-based repository. EAST-ADL can be mapped to CMM. Therefore, a CMM-view on UML elements with applied EAST-ADL stereotypes can be provided compliant to the mapping of EAST-ADL and CMM. For instance, Analysis Function Types become Logical Components, Error Models become Rich Components, Quality Requirements become System Requirements and Satisfy-links keep the same. We implemented a CMM-adapter for EAST-ADL

Adapter	Description
Doors	Doors Service providing analysis services Doors get and put of Requirement/Link information
Rhapsody	Providing Rhapsody Models and storing them in model repository
Simulink	Check-In/Check-Out of Simulink Models including diagrams
Eclipse-based tools (Papyrus, Topcased,...)	ModelBus TeamProvider as Eclipse-PlugIn
Enterprise architect	Check-In/Check-Out of EA models including diagrams
Microsoft Office (Word, Excel)	Check-In/Check-Out of Office documents as models
AVL InMotion	Check-In/Check-Out of InMotion models

that is integrated into the CMM-API for file-based repositories. The CMM-view represents, e.g., UML packages with contained UML classes that have ports and applied *AnalysisFunctionType* stereotypes as CMM-packages with contained Logical Components and Ports. As described in Sect. 7.2.4.3 a link between a CMM-representation of an EAST-ADL element to another element in the CMM-view is established to the corresponding native EAST-ADL data-artefact in the file-based repository.

In contrast to a file-based CMM-adapter, as described above, the CMM-adapter for DOORS implements a combination of access-support and CMM-support. Modules in DOORS are configured in a CMM compliant way to allow a seamless connection into the CMM-API. The tables that are used to store the requirements in DOORS are very customizable. The implementation of the DOORS CMM-adapter provides set-up scripts to configure a folder with a special module structure that allows a one to one mapping between DOORS and CMM System Requirements. For each supported kind of requirement (Natural Language, Boilerplate, Pattern) a module is created that has a separate column for each relevant attribute of the System Requirement like name or comment. This allows the user to edit most of the attributes directly in the user interface of DOORS. Some attributes of the requirements that have a complex structure cannot be accessed by the user in DOORS directly. This applies especially to requirements that are instantiations of templates. In this case the instances of the parameters of the requirements are stored as XMI serialization in a hidden field in DOORS. The template itself is stored in a file-based repository and is referenced by the parameters that are stored in DOORS.

Figure 7.10 shows an example RTP containing a DOORS repository storing requirements and a ModelBus repository containing EAST-ADL models, which demonstrate two user requests on the CMM-API:

1. Receive a system requirement
2. Find all components satisfying this requirement

Initially, the user calls the `getPattern()` function of the CMM-API and provides the unique reference of the requirement storage location. The Model Interface Dispatcher (1.1) identifies DOORS as referenced repository and passes

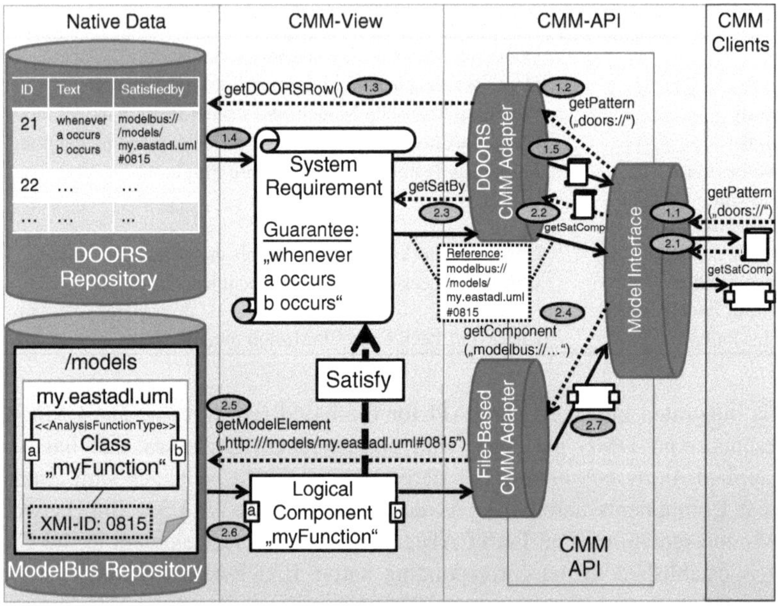

Fig. 7.10 CMM API scenario

the call to the corresponding CMM-adapter (1.2). Using this CMM-adapter, the request can be translated into native DOORS API calls (1.3). The requested requirement data-artifact is found in the DOORS database and adapted into a CMM *System Requirement* using the CMM-support for DOORS (1.4). Finally the System Requirement is returned to the user as a CMM-representation of the DOORS requirement data-artifact.

In the second request, the user asks the CMM-API for components satisfying the retrieved System Requirement (2.1). System Requirements and Components are connected using *Satisfy* links in the CMM-view. Typically, these links can be stored in arbitrary repositories. However, in this example all links are stored in DOORS to keep it simple. The access request is passed to the DOORS CMM-adapter (2.2 and 2.3) which returns the reference of the satisfying component. The Model Interface Dispatcher resolves this reference and identifies the file-based ModelBus repository as storage location of the referenced component (2.4). The corresponding access-support module performs a checkout of the file containing the referenced component (2.5). In this example the containing file is a UML file. The component reference points to a UML class with an applied EAST-ADL stereotype. The EAST-ADL CMM-support module adapts the referenced EAST-ADL component data-artefact as CMM *Logical Component* (2.6). Finally this CMM-representation is returned to the user (2.7).

7.5 Conclusion

At the beginning of the CESAR project, the term Reference Technology Platform was quite differently understood by CESAR partners. It was important to come to a common understanding about the various different terms. It was very helpful to classify the various assets of the RTP (e.g., RTP-Bus, Service, Tool, or Repository). And also it was important to clearly describe what the RTP is good for and what problem it solves. And whenever a RTP instance is created it is vital to be clear about what integration platform needs to be addressed before the RTP instance need to be create.

In addition, there were also very different and very diverse requirements against the Reference Technology Platform and it was obvious from the beginning that it will be hard to fulfil all of them during the lifetime of the CESAR project. So it is in particular beneficial that the idea of the Reference Technology Platform overcome the lifecycle of CESAR and that it is also adopted and further developed in other projects.

While building the RTP, taking the various different domains and problem spaces into account, it became evident that there is no big single universal tool, usable for complex embedded systems development of safety critical application, which fits all needs and which can be installed with a one-click-installation-wizard. There is also no single meta model which can be used throughout the entire development lifecycle in all domains covered by RTP, although there are some candidate which play in general an important role in systems development like SysML or CMM. There is always a trade-off between having smaller and semantically separated meta models which can be easier maintained but where more semantic transformations are needed or having one or only a few big meta models where everything is mapped into. CESAR has shown that the truth is somewhere in-between. It is very important to be clear about what requirements in particular with respect to involved tools and targeted workflows the RTP instance has to fulfil. So having the RTP as a set of available assets (e.g. meta models, tools) is very important but the same attention needs to be paid to the tailoring process for creating RTP instances.

An important question for the RTP implementation was the question of storing the data produced and consumed during development lifecycle. So the question is basically who is the owner of this data. Is the data store in a logically centralized (but possibly physically distributed) repository, or remains the data with the tool it produces and is each tool responsible for data management. So there is probably no final answer to that, both approaches are possible but most RTP instances realized in CESAR did follow the logically centralized repository as it is supported by ModelBus for example.

One question has been raised several times inside but also outside of the CESAR project: What does it cost to integrate a tool to the RTP? Based on the experiences gained from implementing the RTP, the only valid answer to this question is: It depends! There are always many different ways to integrate a tool to the RTP. So for accessing the cost of integration, it is vital to define what kind of integration is appropriate and needed?

So in summary, whenever a new RTP instance shall be build based on the existing RTP, it is important to make a plan and to clarify quite a number of things beforehand, e.g. how many similar project will run on this RTP instance? What are the involved parties? Which tools do they typically use? What engineering processes needs to be executed? Which level of agility shall be supported?

And there are some more questions like these. Although the creation of the RTP was an ambitious goal, in the end – even in such a heterogeneous environment like a European project – CESAR project has shown that the idea of RTP is working. This is demonstrated by the RTP instances, which have been created by the CESAR partners.

Chapter 8
Interoperability

Parham Vasaiely, Rainer Ersch, Christian Hein, Andreas Keis, Fabien Paganelli, Gerald Stieglbauer, and Ivo Viglietti

8.1 Introduction

Today's most heterogeneous software tool environments and the rising number of different systems engineering methods and processes lead to the need for well-matched tools. A Reference Technology Platform (RTP), like the ARTEMIS CESAR RTP, shall provide a complete systems engineering environment with a set of management or engineering methods and processes, as well as engineering tools. The integration and interconnection of tools is needed in order to support collaboration between all relevant business processes of an enterprise for example interactions with customers, suppliers and partners. Furthermore, the interoperability aspect between tools integrated in such a platform is getting more and more crucial for successful and efficient product engineering.

The main technical challenge in addressing this problem is the lack of open and common interoperability technologies supported by the different tools that generate

R. Ersch
Siemens AG, Munich, Germany

C. Hein
Fraunhofer Gesellschaft zur Förderung der Angewandten Forschung e.V., Munich, Germany

A. Keis · P. Vasaiely (✉)
EADS Deutschland GmbH, Munich, Germany
e-mail: parham.vasaiely@eads.com

F. Paganelli
Sagem Défense Sécurité, Paris, France

G. Stieglbauer
AVL List GmbH, Graz, Austria

Friedrichshafen, Germany

I. Viglietti
AleniaSIA Spa, Turin, Italy

A. Rajan and T. Wahl (eds.), *CESAR - Cost-efficient Methods and Processes for Safety-relevant Embedded Systems*, DOI 10.1007/978-3-7091-1387-5_8,

and provide access to the data, and of the underlying IT infrastructure. Instead they are based on COTS products or custom-built solutions. An interoperability specification (IOS) will guarantee these needs for interoperability and collaboration between tools across the entire engineering lifecycle. However, to provide not only an open, common way to exchange data, but also basic technologies to link that data, in order to achieve full traceability, the CESAR IOS has adopted interoperability technologies proposed by the Open Services for Lifecycle Collaboration (OSLC). In addition, this work will provide a RTP reference architecture, which allows consistent implementations of technology platform. Furthermore the connection and integration of different RTPs based on the same IOS, allow composing a multi domain engineering environment.

8.1.1 Overview

Interoperability can be defined as the ability of two or more systems or components to exchange data and to use the data that has been exchanged [156]. In this work this term specifically relates to the challenges to share data rather than exchange, and to provide integrated information and integrated access to that information. The difference between sharing and exchanging of data is related to the underlying implementation approach. When implementing interoperability to exchange data, the work and information flows supported by the system are predefined, whereas the implementation of data sharing, allows a more generic access which is work and information-flow independent. Integrated information connotes an overall system in which a single view of information is presented, achieved by combining various information sources, whereas integrated access suggests an overall system that provides a single point of access to information, and that combines various individual access mechanisms in order to do so [157].

With respect to software tools and engineering platforms used in product lifecycle, interoperability can be defined as the capability of different tools to share data over an agreed set of data formats and communication protocols. However, the rising number of different technologies related to modelling and programming languages, communication protocols and data formats, makes the identification and use of a common interoperability approach for individual technology providers difficult. Furthermore the high number of different IT infrastructures composed of hard- and software complicates this effort once again. Therefore an interoperability specification (IOS) is needed as a common agreement between tool and technology providers, as well as users. A specification is based on weaker agreement than a standard, however it is required as a first step towards the standardization of an interoperability approach.

The IOS defines the method of communicating and sharing data, which includes common communication protocols and data formats. Today, there several interoperability specifications for many aspects of tool integration are already in existence. However these are either not specific to the needs of an engineering environment used in the domain of embedded systems engineering or not extensively applicable

[24]. The CESAR IOS will guarantee the needs for open interoperability and collaboration between tools across the engineering life-cycle. However the IOS not only provides a common way to share data, but also implements basic technologies to link different data. Moreover the implemented Linked Data approach [75] is an important building block of the IOS, providing a unique characteristic which needs particular attention.

8.1.1.1 Engineering Tool Interoperability

This work focuses on the integration and interconnection of software components like tools and services, further called components. The components may be integrated based on a resource and service oriented architecture like the web or by using an underlying software platform as middleware, which allows the integration of multiple components. The communication and data sharing between components mainly depends on the existing IT infrastructure and the engineering environment architecture. A simple environment is built locally or distributive over a network, but is composed by the same kind of components, which all are using the same data format. This kind of environment is called Homogeneous Tool Environment. However today's engineering environments looks different, they are mainly built distributive over a network and based on different kind of components, which are used within the different product life-cycle domains. This infrastructure is called Heterogeneous Tool Environment.

The following basic scenarios are identified based on the different environments, to introduce the challenges of engineering tool interoperability:

Homogeneous Tool Environment

In this environment, multiple instances of the same tool are installed. The tools may share the same memory or communicate over message passing. Data exchange won't cause data loss or corruption. However, in order to cover the whole product life-cycle the tools used in this environment must understand the same, obviously very powerful and meaningful concepts. The communication can be implemented for example based on the shared memory principles.

Heterogeneous Tool Environment

In this environment, different tools are installed. The tools cannot share the same data since they are using different data formats. This is a typical starting point where an interoperability specification is needed. Starting from the different tools scenario, tool vendors or users start to develop proprietary tool adapters to translate/transform data from their own source format into a destination format. The problem which prevents widespread interoperability among development tools today is not to be

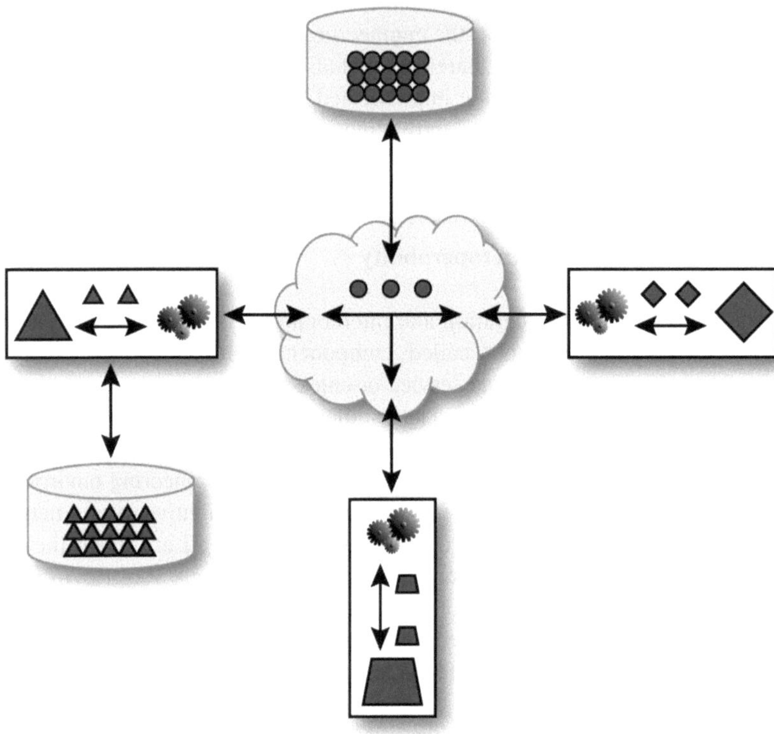

Fig. 8.1 Heterogeneous Tool Environment – tools share information (common interoperability)

found in the realisation techniques, they are sufficiently powerful and efficient to integrate two tools separately. The problem is rather the complexity of large-scale tool-specific realisations in order to allow full interoperability among a number (n) of different tools. If each of them has to be integrated in a point-to-point manner with all other tools, that means from scratch every time, this integration paradigm becomes highly inefficient. For 3 tools, 3 integrations are needed (A with B, A with C and B with C), for 4 tools 6, for 5 tools 10 and for 6 tools 15 integrations are needed. The numbers continue to grow nearly quadratically. Considering the fact that in the CESAR RTP specification for version 1, there are about 85 tools listed which were requested for integration by at least two partners.

According to the formula above, up to 3,570 tool integrations would have to be done to achieve this. Admittedly, this formula describes the worst case where all tools shall to be able to communicate with all other tools. This is often not necessary in practice. However, even if only 10 % of the tool combinations shall be possible, still over 300 tool integrations would have to be made in the scope of the CESAR project alone. To avoid implementing multiple adapters for a tool or to be dependent from one mediation technology vendor a common interoperability approach needs to be applied, as illustrated in Fig. 8.1. This approach requires a common agreement of tool vendors and platform developers on the communication protocols as well as data exchange format. In this approach just one adapter is needed for each tool to understand the common data format or to share its data with others.

In Chap. 4 these and other tool integration patterns are described in more detail based on simple examples.

8.2 The Interoperability Challenge

By analysing the need of interoperability within an integrated engineering environment in the domain of safety critical embedded systems, different aspects of interoperability can be identified. One is the aspect of an open specification. Here the purpose and benefits of an open specification versus a closed specification are identified. Another aspect is the identification of different stakeholders involved in the deployment, maintenance or use of such an engineering environment and their roles. Stakeholders are system or software engineers, software platform developers and tool or technology provider. Each stakeholder will address different needs to the IOS with respect to its requirements and responsibilities within the environment. Finally, the integration depth of two tools as a noteworthy property of interoperability will be discussed.

8.2.1 Open Interoperability Specification

As mentioned in Sect. 1.3, an IOS can be seen as a common agreement between tool and platform providers and users which are using these tools in their environments. There are two forms of specifications with respect to openness and a common agreement, open or closed specifications.

- A *closed interoperability*: A closed interoperability specification, also called post facto interoperability, is mainly driven by one dominant party, e.g. a leading engineering tool or platform provider. The used data formats and protocols are not negotiable and in most cases adapted to be highly efficient to the tools and services this vendor is providing.
- An *open interoperability*: An open interoperability specification is driven by a broader interest group including representatives from vendors, industrial companies as users, academicians and others stakeholders. This approach not only supports the reliability of interoperability and interchangeability between suppliers, purchasers and users regardless of the source, but also enables a more independent, reusable and future-proof specification. For example this allows other tool integrators to leverage a multitude of existing assets by conforming to an open specification. In addition the acceptance of an open specification by third party vendors is distinctly higher. Another important factor for openness is the fact, that a broader number of technologies can be respected.

In conclusion, an open specification is more valuable to a broader number of parties than a close one. However, the time to achieve a common agreed solution obviously is significant higher than a solution driven by a close group of partners.

8.2.2 Implementation-Independent Specification

The need for an implementation-independent interoperability specification is strongly coupled to the need of an open specification, as mentioned above. An implementation-independent IOS can be used by different integration scenarios to apply interoperability. For example:

- Users and tool providers are not forced to use respectively integrate their tools into a specific software platform to enable their collaboration.
- Integration can be applied in a vendor-independent way, but with respect to the enterprise requirements.
- A vendor can use individual and efficient technologies to implement some specific features on the application or service layer of their tools.
- Technologies used in the RTP are not restricted to specific protocols or programming languages.
- Future implementations of an environment can be implemented using improved or even new technologies.

Conclusively one can say that an implementation-independent specification is preferable but not extensively applicable in terms of basic technologies needed for communication for instance. There is the need to define at least the architecture styles of the environment, some data formats and communication protocols. However, it is important that the technologies are open and at provider-independent available.

8.2.3 Integration Depth

The integration depth increases with the amount of information detail which is specified. As the amount of integrated information detail increases, the amount of different tools which may re-use a common shared concept increases as well. This is because tools may vary in specific details, thus when a specification contains more details, it is more likely to fit for many tools. Allowing tools to share as much information as possible, the degree of re-use will be increased. However the re-use may affect the overall correctness. Therefore verification and validation as well as consistency checks are necessary to guarantee the correctness of the shared information. We differentiate integration depth between three levels:

1. **Black-Box**: Elements are integrated only in a minimalistic manner by their type, name and eventually some additional description. The content or details of elements is hidden and not considered for integration. This is similar to a

black box view on a component in engineering, which shows only the input and output of a component but hides its implementation. The main goal is the generic integration of many very different elements, but re-using their content is not relevant. A typical scenario is traceability.
2. **Grey-Box**: For grey-box integration depth, more details of elements are considered for integration. However, the integration still implies considerable loss of information and some details of elements are not integrated. The goals are both the integration of different elements and the re-use of their information content. That means the amount of integrated information increases at the cost of the variety of elements which can be integrated. A typical scenario is the integration of tools from different domains which only have to exchange basic data structures to ensure a seamless development process.
3. **White-Box**: Elements are integrated with close to 100 % of their information content. But deep integration is very tool-specific and often each pair of two tools requires its specific white-box integration concept. The goal is to integrate as much information as possible and completely sacrifice the re-usability of the interoperability concept. A typical scenario is data migration or the collaboration of different tool types inside the same domain.

8.2.4 *Business and Engineering Work and Information Flows*

Work and information flows are based on business processes coming from use cases, as well as scenarios and engineering methods or processes, used in the domain of safety critical embedded systems. The IOS needs to apply a generic interoperability approach to the possibility of implementing and applying different methods and processes

8.2.4.1 Work Flows

A work flow with regard to engineering tools mainly affects the purpose and way how tools are used together, sequentially or randomly. In addition the work flow also affects the direction of information flow, see Sect. 2.7. Figure 8.2 illustrates two different work flow scenarios, which are applied over the same environment. Obviously the user should not be forced to reinstall, reconfigure or adapt tools, with regard to any change on the work flow, since this would affect the dynamic of enterprises to react on situations. Further, any intervention to an environment also has an impact on costs, for example caused by down-times during the deployment.

In the "Work Flow Scenario 1" (on the left), a simple Top-Down development approach will be presented, starting by project or requirements management over to design and simulation of systems. The project management and requirements management tool can be used as initial tools in the development work flow, since both are normally the tools to start with. The CM Service assigns change requests and new work items to the project. In the design tool the engineer is doing the actual

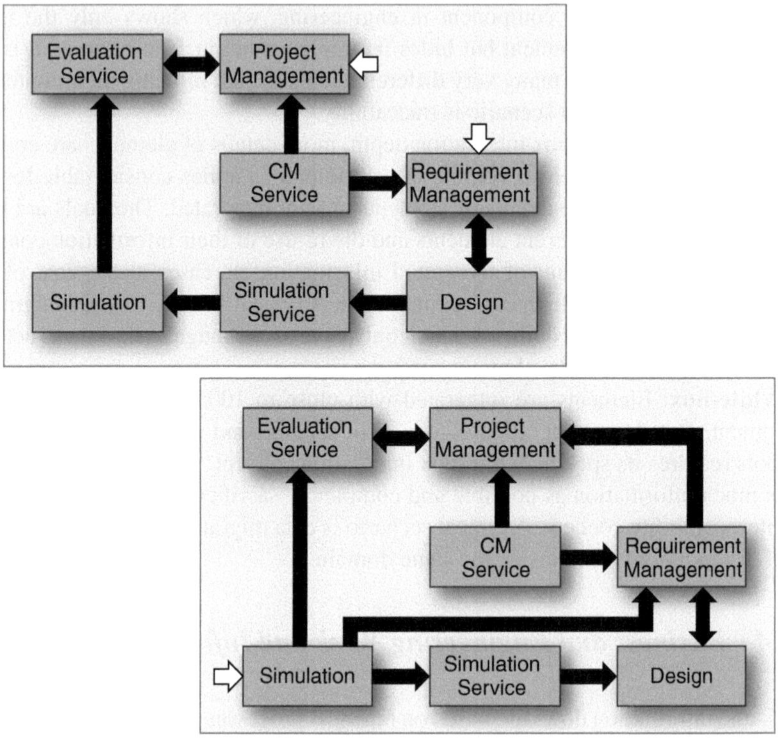

Fig. 8.2 Business and Engineering Process Work Flow Scenario 1 (*left*) and Scenario 2 (*right*)

development. After creating the design and system structure a simulation service filters all information which are related to the behaviour of the system and provides this information for a simulation tool. Using this tool an engineer can simulate a first prototype of the system. Here this simple work flow will end.

In the "Work Flow Scenario 2" (on the right), one have the same order of tools but a complete different work flow. In this scenario a simple Bottom-Up development approach will be presented. In some cases engineers have to use already existing system components, for example when trying to improve a system. Anyway the engineer now simulates the components and then provides the results and changes to the design team as well as the evaluation service. In this work flow a requirement may be affected and a new work item will be created in the project management tool.

With the two different work flows using the same infrastructure, one can realize the importance of a dynamic adaptable environment.

8.2.4.2 Information Flows

As mentioned above, work flow changes are caused by updated business requirements or processes. But also tools and services fall victim to change, for example if

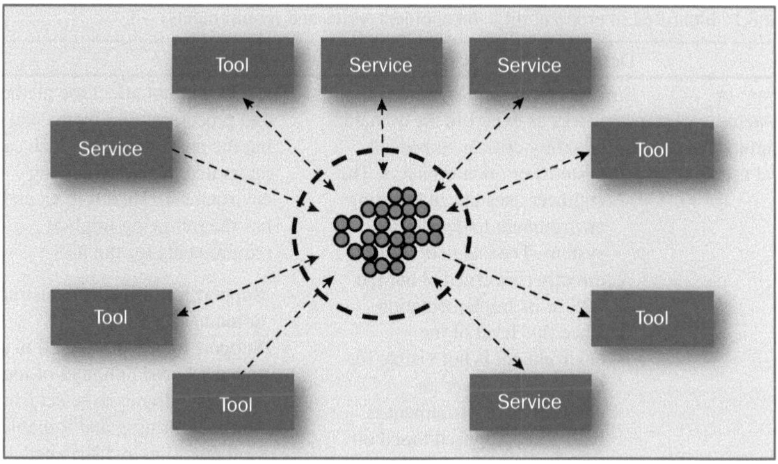

Fig. 8.3 Business and Engineering Independent Information Flow

a tool will be updated or a bundle of tools may be replaced by just one. As a result the RTP Instance should provide a process which allows the user to exclude not needed or add new components to the RTP Instance, when deploying or at runtime. This enables the deployment of an efficient and light weight environment. However the exclusion or integration of components shall not break existing or avoid new work flows, quite the opposite, produced information shall be continuously available, enriched and improved in terms of quality and consistency. This process can be guaranteed by applying the integrated information approach [145]. Integrated information connotes an overall system in which a single view of information is presented, achieved by combining various information sources. As a result overall information flow is bi-directional and not necessarily coupled to a specific work flow. Figure 8.3 illustrates such a business and engineering work-flow independent information flow, which needs to be applied by the IOS in order to meet the needs for integrated information. Components, such as tools and services, can produce or process information (illustrated as circles) in bi-directional way. The circle shall illustrate that information will be provided or shared with others.

8.2.5 Interoperability Stakeholders

As mentioned in Sect. 1.3, the level of interoperability and the needs of a specification are defined by different stakeholders. Each stakeholder has its own requirements with respect to its role in the engineering environment and product lifecycle. These requirements are also coupled to the various RTP components which need to be integrated. The following table lists the stakeholders and their main requirements with regard to interoperability (Table 8.1).

Table 8.1 Identified interoperability stakeholders (roles and requirements) – 1

Role	Description	Requirements
Software or systems engineer as end user	Software or systems engineer works as expert in the domain of safety critical embedded systems or other domains. The engineer uses the engineering environment to develop a system. The end user is not directly concerned about the IOS or its implementation, since this level of the environment is not visible for this role. However an engineering environment is built and composed based on the end user requirements and use cases	The IOS will not affect the methods and processes used by the end user but the possible tools which can be integrated in an engineering environment. Therefore an end user has the following implicit requirements for the IOS – Support for different industrial domains – Support for open standard in order to have a broad number of tools and technologies to select from – High Performing and Reliable – Collaboration in distributed environment
IT systems administrator	IT systems administrator maintains and operates computer system and/or network. The duties of a system administrator are wide-ranging, and vary widely from one organization to another. Systems administrators are usually charged with installing, supporting and maintaining servers or other computer systems, and planning for and responding to service outages and other problems. The IOS affects the admin by setting restrictions to usable technologies and data formats	The IOS affects the admin by setting restrictions to usable technologies and data formats. This may affect the possible tool integration scenarios which need to be composed. Therefore this role has the following implicit requirements – Support for a broad number of tools and technologies – Support for open available technologies – Support for a scalable and performing platform – Distributed environment

8.3 The CESAR Interoperability Specification

The CESAR IOS will guarantee the needs for open interoperability and collaboration between tools across the engineering lifecycle. It supports a powerful approach on the interoperability syntax level, which not only allows a common way to exchange information, but also implements basic technologies to link different information to achieve full traceability. Here the adopted Linked Data approach by the OSLC [171] is an important building-block of the IOS, providing a unique characteristic, which deserves particular attention. This allows the IOS to support the following key features when applying interoperability:

- Traceability of Engineering artefacts as well as work items.
- Visibility into all aspects of the implementation process (performance and historical trends)
- Control through policies and processes

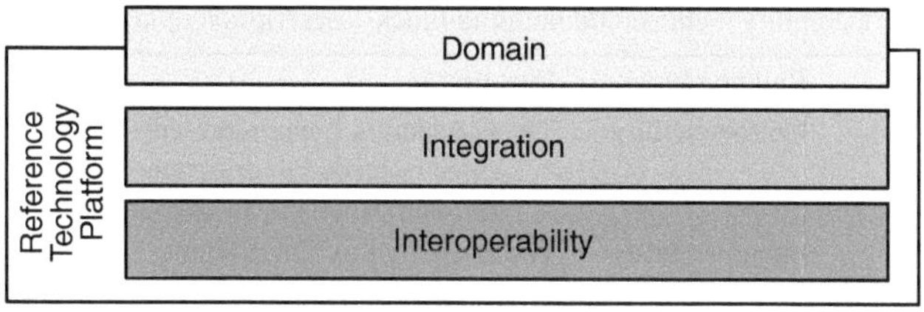

Fig. 8.4 The RTP reference architecture building-blocks

Table 8.2 RTP reference architecture: description of the three main architectural building-blocks

Building-block	Description
Domain	This building-block contains elements which shall represent enterprise or project specific use cases and requirements with regard to management or engineering processes and workflow directives. In other words, it represents the business logic. The Domain building-block is partially part of the RTP, since it contains also engineering pattern, methods and processes, which needs to be applied during the different engineering lifecycle phases. This building-block is not directly concerned by the IOS and will be discussed briefly in the future RTP Chap. 7
Integration	This building-block contains elements which are the actual components of an engineering environment, like tools, services or software platforms. The components need to be integrated into the environment to allow interconnection and collaboration among themselves, like service or information sharing. This building-block is not directly concerned by the IOS and will be discussed briefly in the future RTP Chap. 7
Interoperability	The Interoperability building-block represents technologies needed to implement the actual interconnection and collaboration between tools. The used communication or data exchange technologies are inspired by the Web and the OSLC. This building-block is concerned by the IOS and further defined in this chapter

However, also the interoperability semantic has been respected, by applying a basic semantic layer to the IOS. This layer adopts specifications developed by the different OSLC working groups, and leaves open the possibility to adopt other domains and their specifications, which are not covered by the OSLC working groups yet. This chapter presents the IOS defined by the CESAR project with respect to the requirements and state of the art technologies, presented in Chap. 3. Further the IOS will be presented as an architectural building-block containing different layers, which again can be integrated into an overall RTP architecture. The architecture describes the organisational structure of the IOS, which can be divided into syntax as well as semantic layers. However, first an overview of RTP building-blocks shall help to classify the IOS into the broader picture. Different layers of an engineering environment need to be considered. As a result, a set of corresponding RTP building-blocks can be identified. Figure 8.4 presents three main architectural building-blocks, including the Domain, Integration and Interoperability blocks, further described in Table 8.2.

Table 8.3 Interoperability architectural building-block – description of layers

Layer	Building block	Description
Foundation	Interoperability	The Foundation Layer represents the architecture style, which is needed to design modules and their communications in an environment
Basic Syntax	Interoperability	The Basic Syntax Layer defines simple communication protocols as well as data formats, used to share information and provide services between tools
Extended Syntax	Interoperability	The Extended Syntax Layer defines the structure of information which is represented in a common data format and needs to be shared between tools
Interoperability Semantic	Interoperability	The Interoperability-Semantic Layer is the first and very basic semantic level in the architecture. It allows the understanding, consequently the semantic interpretation and processing, of information in the different tools and platforms

8.3.1 Basic Design Principles

The basic design principles of the IOS are based on the World Wide Web and its Linked Data approach as defined by the four rules of Linked Data by Tim Berners-Lee [19]:

1. Use URIs as names for things
2. Use HTTP URIs so that people can look up those names
3. When someone looks up a URI, provide useful information, using the standards (RDF, SPARQL)
4. Include links to other URIs. So that they can discover more things

8.3.2 The Interoperability Architectural Building-Block

In this chapter, the IOS will be discussed and illustrated as Interoperability architectural building-block containing different layers, which again can be integrated into an overall RTP architecture. The architecture presents the organisational structure of the IOS and is based on a layer approach. The layering is strict, which means that layers from above cannot brake through or bypass lower layers, this shall support consistent implementations. Furthermore the architecture represents the system components and their relationship to each other. The relationship nature of a strict layered architecture can be described as follows (Table 8.3):

- Lower layers provide services, protocols, technologies for upper layers.
- Upper layers use services, protocols, technologies of lower layers.
- Services, protocols, technologies defined on the same layer can use each other.

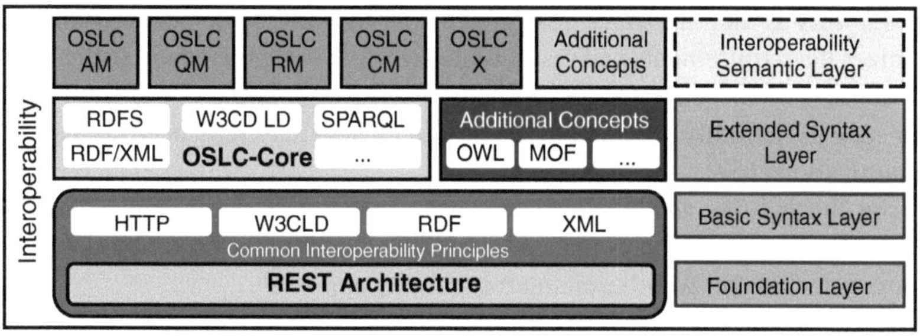

Fig. 8.5 Interoperability architectural building-block – reference architecture

Figure 8.5 presents the Interoperability architectural building-block as standalone, which allows its platform-independent implementation and reuse. Each of the four layers now are implemented using specific technologies and methods, coming from Chap. 4. The layers and their technologies are discussed in more detail in the chapters below.

8.3.2.1 The Foundation Layer

An architecture style is needed to design modules and their communications in an environment. The Foundation Layer represents this architecture style as lowest layer of the overall architecture. One important CESAR requirement of this layer is the agreement on reusing existing architecture principles already applied by the Web, such as the Service Oriented Architecture (SOA) or the Representational state transfer (REST). The decision between SOA and REST as architecture style is coupled with requirements set by systems engineering environments and their users. CESAR has identified that nowadays most engineering tools are installed and used locally as rich client applications, rather than web based services and thin clients. Therefore the value of reusing and sharing of resources (information, documents), which are produced and used during the engineering lifecycle, in a common way is much higher than the common representation of engineering tools as services. However, the latter is also important, in order to guarantee consistency while sharing information and collaborating between tools.

SOA sees services and messages as central elements of an environment. Whereas REST is resource-oriented, focusing on information and documents and their links or relationships.

In REST, a tool exposes its data and services through a resource-oriented Web service, in order to make its data and services available to the widest possible range of clients. The intent of the Web service is to provide a stable long-term web-based interface for directly accessing the facilities and data offered by the tool. These web services use an arrangement of URIs, HTTP methods, and standard representation languages such as XML and RDF, that work like the rest of the Web.

However this is not the only reason why REST has been considered. It also minimizes the requirements and restrictions given by SOA. For example the protocol is stateless, which means that a client state is carried in the client itself and the server state is explicitly reflected in the resources, whereas SOA needs transactional communication [157].

Given the reasons above, REST has been selected as the default architectural style, used as the basic interconnection principle, allowing information sharing on a business oriented and web-based interoperability approach.

8.3.2.2 Basic Syntax Layer

Sharing of information requires forms of communication. The Basic Syntax Layer defines simple communication protocols as well as data formats, used to share information and provide interfaces between tools. This layer can be seen as a set of complementary technologies used to achieve communication and collaboration on a very basic level, therefore the Interoperability building-block has implemented a second and thus extended syntax layer, see Sect. 6.2.3. However, the technologies of this layer are final to guarantee consistent interoperability and its implementation. The following technologies can be seen as the basic syntax for the interoperability:

- *HTTP*: Since the CESAR IOS, like the OSLC, follows the architecture style of the Web and the REST, HTTP is used as the application protocol. HTTP allows clients and servers to provide a uniform interface. A uniform interface between clients and servers simplifies and decouples the architecture, which enables each part to evolve independently. In the context of REST, the set of operations supported by the web service using HTTP methods are GET, PUT, POST, or DELETE.
- *RDF*: A resource-oriented architecture requires an efficient data model to present resources and their properties. The OSLC data model bases on the Resource Description Framework (RDF). OSLC requires RDF representations, but uses just a small number of RDF concepts. By analysing the OSLC approach more carefully one can see the advantages of using the industry standard RDF as data model. First, one avoids defining a new data model but accommodate different schemas [133]. However, more important the RDF model allows to build again on the Web and furthermore to achieve and create Linked Data. The Linked Data approach fulfils the initial requirements of linking different information in order to achieve for example full tractability [75]. Therefore RDF is the common data model in the CESAR IOS.
- *W3C Linked Data*: The linking of resources not only allows expressing relationships among information, but to explore further information and finally achieve a higher level of traceability [160]. Since the CESAR IOS has implemented RDF as common format for its information, consequently the W3C Linked Data approach can be applied to achieve and create Linked Data.

Table 8.4 CESAR Namespace URIs

Namespace URI	Description
http://www.cesarproject.eu/ns/cmm-type/	Official resource types defined by the CESAR project based on the CMM specification
http://www.cesarproject.eu/ns/ext-type/	Non official resource types defined by CESAR partners as extensions to the CMM specification
http://www.cesarproject.eu/ns/resource/	Instantiated OSLC resources based on the defined types

- *XML*: As mentioned above, OSLC uses a simple model of resources with property values intended to be consistent with the RDF data model. OSLC also builds upon the Extensible Markup Language (XML) namespace mechanism [59]. When specifying a resource or a property, OSLC Specifications define its type as a URI which can be decomposed into a namespace URI and a name. OSLC abbreviate type URIs as Prefixed Names, which are represented in XML as QNames. The namespace used for resources defined in the CESAR project is defined as follows:
  - Namespace URI: http://www.cesarproject.eu/ns/. . .#
  - Default Prefix: cmm-type (core meta-model, [CESAR, D_SP1_R3_3_B_M3])

  When defining a Resource type, a specification must provide the following information:
  - Name (as String): Name of the resource which must be valid as the Local Name part of a QName.
  - URI (as URI): The URI of the resource definition. Per the rules of Prefixed Names, this URI is formed by appending the Name to the end of the Namespace URI in the specification that defines the resource. For example, the resource named "Requirement" gets the Type URI of http://www.cesarproject.eu/ns/cmm-type#Requirement.

Table 8.4 (CESAR Namespace URIs) provides a list of namespaces used in the context of the CESAR project. OSLC and partners specific namespaces are not listed.

8.3.2.3 Extended Syntax Layer

In addition to the simple communication protocols and data formats defined in the Basic Syntax Layer, further structuring of information and query languages are required to allow efficient information sharing. The additional structure allows tools to parse and interpret information more efficiently, whereas a query language is used to express queries across diverse data sources. The Extended Syntax Layer has mainly adopted technologies implemented by the OSLC-Core [123]. However extensions are possible, if however based on the basic syntax technologies. The following technologies are instances of the extended syntax layer of the IOS:

RDF/XML

As mentioned in Sect. 4.2.2, one intention of using RDF is the possibility to represent information and their relationships as graphs. The combination of XML and RDF defines a syntax specification for RDF graphs based on XML. The CESAR IOS proposes the resource representations in RDF/XML, in order to define a machine readable form of RDF with Notation 3 as a more human-readable form.

RDFS

An RDF Schema is a set of classes with certain properties based on the RDF, providing basic elements for the description of ontologies, intended to structure RDF resources. RDFS constructs are the RDFS classes, associated properties and utility properties based on the vocabulary of RDF. This allows the definition of user defined resource types and properties used for example in the interoperability semantic layer. For example, one may define an RDFS vocabulary about "Requirement" that includes a class "Requirement" and "SystemRequirement". The "SystemRequirement" class is a sub class of class "Requirements". This hierarchical dependency is expressed using "rdfs:subClassOf". Now one can create a requirement resource of which all individual system requirements are members or "SystemRequirement" expressed with "rdf:type". They may also define a property "id" and "derived", thereby allowing people to publish RDF descriptions of their own requirements using these terms. Although RDFS includes semantic elements like a pre-defined vocabulary, in the context of the interoperability it can be seen as a support language to express ontology and semantic.

Authentication

OSLC Services use standard web protocols for authentication. OSLC Services can use HTTP Basic Authentication, OAuth or both. However in OSLC the application of authentication is optional, whereas the industry partners in CESAR agree that the protection and security of information deserves much higher attention. In addition the Linked Data approach makes it easy to link and reach resources or even to crawl through different sources [LinkedData]. Furthermore a consistent implementation of such authentication mechanism is necessary, since today enterprise environments are build on enhanced IT security standards. Figure 8.6, illustrates the different IT security zones, which break links between resources:

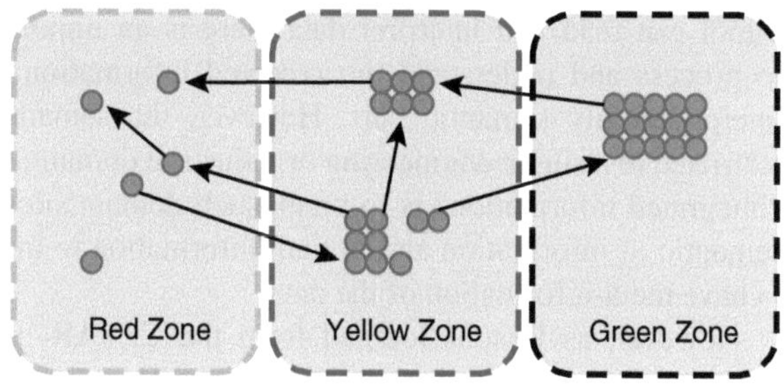

Fig. 8.6 Typical security zones of an enterprise environment

- *Red Zone*: An enterprise intranet, which is protected by firewalls. Access to this zone from outside is undesired. Resources in this zone can be internal requirements or blue prints of products.
- *Yellow Zone*: Also called demilitarized zone (DMZ), allows restricted access to information from internal sources. However collaboration to and from the outside is possible and allowed in a restricted way. Resources in this zone can be requirements for supplier or component interface descriptions.
- *Green Zone*: Resources of this zone can be accessed without authentication. Resources in this zone can be public product information or elements used for demonstration purposes.

The CESAR IOS recommends the consistent implementation of HTTP Basic Authentication and OAuth, to protect resources against unauthorized access. Therefore the requirement level of the authentication is set from "MAY" as proposed by the OSLC to "SHOULD".

Additional Concepts

The OSLC-Core covers many requirements given to the interoperability syntax. However it is obvious that OSLC cannot cover every aspect, since the specification needs be more generic rather than specific. Concepts which are not covered by OSLC-Core have to be specified and implemented dynamically as additional or specific concepts, by respecting the layered architecture. One addition concept would be for example OWL. Like RDFS, OWL provides vocabularies for describing conceptual models in terms of classes and their properties. OWL may be used in the different semantic layers to represent ontologies used to define CESAR types or resources and their properties. Although OWL includes semantic elements like a pre-defined vocabulary, in the context of the interoperability it can be seen as a support language to express ontology and semantic.

8.3.2.4 Interoperability Semantic Layer

As mentioned above, tool interoperability can be divided into two parts, interoperability syntax and semantic. The syntax allows tools to exchange information by defining the communication protocols and data formats. However, even when a tool can read and interpret data there is an important part missing to be able to process and understand the received information. This is the purpose of the interoperability semantic part. However, the semantic must not be specific or restricted to a single engineering or industrial domain, since the purpose of the term "integrated information" is to break such domain silos. In order to understand the semantic of information and to map information to internal concepts, a tool needs to have meta-information of the data.

However, as lessons learned from the CESAR project as well as individual partner experience, the application of one common meta-model for all domains and aspects of the engineering lifecycle is not applicable due to the nature of complexity of artefacts, processes and methods used on project, company or domain level. This insight leads to the need to divide a single common meta-model vertically in different engineering domains or horizontally into different abstraction/complexity levels. For the IOS both approaches have been applied. Semantic concepts are divined on engineering level like Requirements Management or Quality Management, as well as into different abstraction and complexity levels like tool layers, application as well as industrial domain layers. The CESAR IOS defines therefore multiple layers of the semantic as illustrated in Fig. 8.7 RTP and IOS Semantic Layers.

Layer	Part of	Description
Interoperability Semantic	IOS	The Interoperability-Semantic Layer is the first and very basic semantic level of the architecture. It allows the understanding, consequently the semantic interpretation and processing, of information in the different tools and platforms. More precisely this layer defines common types of artefacts produced or processed in the different engineering domains, such as requirements management or quality management
Application-Semantic	Integration	The Application-Semantic Layer is the second semantic level of the architecture. It enhances the basic Interoperability-Semantic Layer with tool specific information. This additional layer is needed since tools evolve their own ontology when implementing or adopting ontology, new concepts or a (modelling) language
Domain-Semantic	Domain	The Domain-Semantic Layer is the third semantic layer in the architecture. This layer represents the workflows as relations between the methods and processes and consequently between tools. However, this semantic layer can be shaped and refined by defining higher domain levels, like industrial domains

8.3.2.5 Assigning CMM Concepts to the Different Semantic Layers

Since the purpose of the IOS is to be domain-independent and to serve the RTP in many different scenarios of tool integration, the specification of data semantics is represented as basic semantic layer in Sect. 6.2.4 based on Core Meta-Model concepts. The basic semantic is not intended to be used for model transformation,

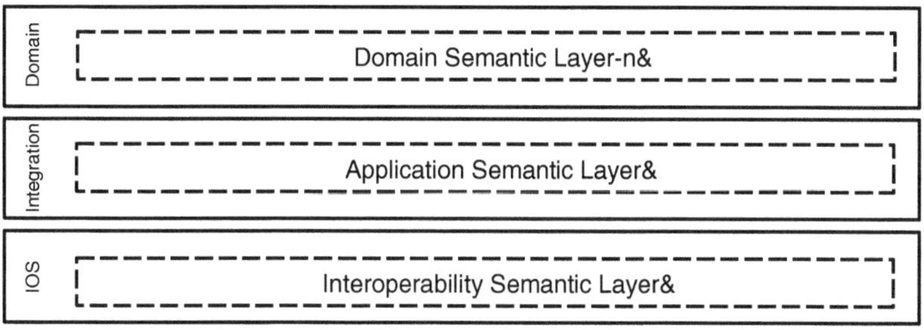

Fig. 8.7 RTP and IOS Semantic Layers

translation or mediation purposes, since the represented semantic is too weak. Today the basic semantic layer is represented mainly by concepts defined by the OSLC. In CESAR the CMM Core contains common modelling concepts which are generally needed such as means for system modelling along the development process, component based design, requirements and traceability as well as verification and validation. These concepts are based on HRC, EAST-ADL, the RMM, the CESAR SP3 data model as well as a real-time extension. The need for these concepts was identified in [CESAR, RR09]. In CESAR the purpose of the core meta-model is comparable to be used as a specification which defines basic concepts for the domain of safety critical embedded systems. However compared to OSLC concepts the CESAR core meta-model is much richer and more precise in the definition of types, properties and their relationships. Here one challenge is to assign the CMM concepts to the different semantic layers of the RTP and IOS. In OSLC "Resource Shapes" are used to describe Types of resources, comparable to Meta-Model languages such as EMF and MOF. In Appendix B an example shape is defined to illustrate the definition of a CESAR System Requirement by reusing OSLC Requirement type, as defined in the OSLC-RM specification.

Additional Concepts for the IOS Semantic Layer

The OSLC specifications are covering basic concepts of semantic used in the context of the IOS. However it is obvious that OSLC cannot cover all aspect, since the specification needs to be more generic rather than specific. Concepts which are not covered by OSLC have to be specified and implemented dynamically as additional or specific concepts in the higher semantic layer and moved down once accepted as basic concepts.

8.4 The Future Reference Technology Platform

The future RTP is a result of lessons learned from the CESAR project. It mainly reflects the need for an easy to implement, scalable, open and common interoperability approach, which can be applied to implement an effective and efficient

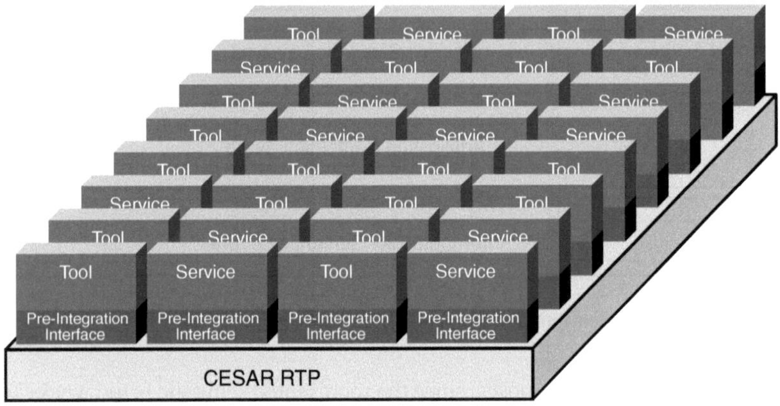

Fig. 8.8 CESAR RTP, set of pre-integrated tools and services

future engineering environment for enterprises. Compared to the interoperability applied to the CESAR RTPv3, the CESAR IOS presented in this work allows the consistent implemented of the future RTP. This chapter will present an outlook to the future RTP, in order to meets the requirements set to a RTP initially. Please note that in this and all following chapters the term RTP is used by referring to the future RTP.

The RTP is composed by a set of heterogeneous tools and services, further called RTP components. Figure 8.8 represents such a set of components, collected in the CESAR RTP. The RTP is not an active or ready-to-use engineering environment, since its components are in a pre-integrated status and inactive in terms of any functionality. The components of the RTP are neither connected to each other nor in a strict order, which allows their randomly compilation and installation based on a respective business or engineering use case.

Anyway, in order to compose components to an actual engineering environment, the RTP needs to provide a set-up or configuration manager. This manager shall allow the selection and composition of components or already existing bundles of components. However, since the decision and selection process needs additional and fundamental understanding of the design of heterogeneous engineering environments, the "Engineering Environment Designer (EE-Designer)" will be introduced as a new role in Sect. 6.2. The EE-Designer supports or leads the traditional IT administrator, see Sect. 2.7, during the deployment and updating phase of the engineering environment.

As mentioned above, the RTP components are pre-integrated, with regard to their interoperability. Figure 8.9 illustrates a RTP component with a "Pre-Integration Interface". This interface allows the seamless integration when deploying the RTP instance based on use cases or user requirements, further described as RTP tailoring process in Sect. 6.2. In order to allow a consistent way to share information, after selecting and installing components to an active and ready-to-use engineering

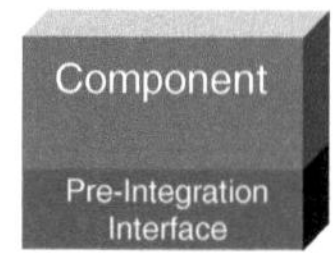

Fig. 8.9 RTP Component in an pre-integrated status

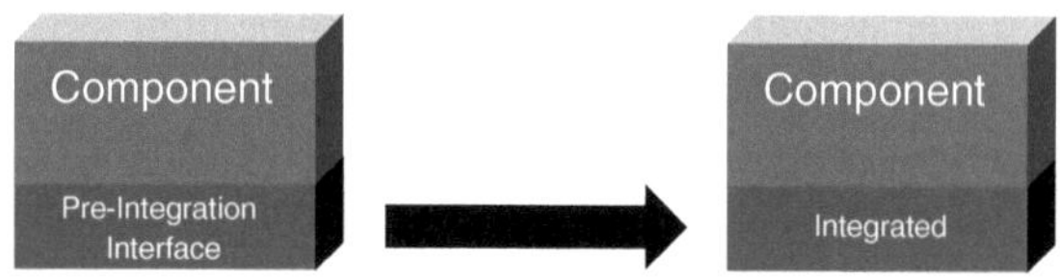

Fig. 8.10 Component installation and activation – change of state from "inactive" to "active"

environment, the interoperability provided by the interface has to base on a common, standardized approach. This common approach is defined by the CESAR IOS, and bases on open and standardized technologies. Therefore the implementation of the IOS for RTP components is done in the "Pre-Integration interface" which is illustrated in Fig. 8.9.

This interface adopts the IOS technologies in order to provide information to a collective of components, as well as process information which are provided by a collective of components. After installing and activating the component, this interface allows consistent information interchange as well work flow control and analysis. The activation process shall initiate the component with regard to all needed information, such as a unique identifier, available network configurations and so on. After the activation phase the component changes its state from "inactive" (Pre-Integrated Interface) to "active" (Integrated), as illustrated in Fig. 8.10.

The following chapters describe the RTP instantiation process which results in an actual ready-to-use engineering environment. The selection and composition of components will be supported by a so called RTP tailoring process. The RTP has to provide this process to the EE-Designer. In addition a technology-independent reference architecture of the RTP will be discussed. The architecture highlights the organizational structure of the RTP and IOS which is needed to guarantee the integration of components, without presenting specific technologies.

8.4.1 Future RTP Instance

Since the RTP components are in a pre-integrated status and inactive in terms of any functionality, the RTP itself is not a ready-to-use engineering environment. The reason is that the RTP is a composition of disconnected components. Only a component selection and composition process allows providing an effective, light weight and ready-to-use engineering environment. The RTP Instance is the result of such a selection and composition process.

A RTP instance, as actual or part of an engineering environment, will fulfil the needs of enterprises and engineers in terms of the following characteristics:

- *Effective and Efficient*: The intended purpose of the created environment is to fulfil a specific use case and requirement, which are common in the domain of safety critical embedded systems engineering.
- *Light Weight*: The environment will be light weight, because components which are not needed to fulfil a use case or requirements can be excluded.
- *Integrated*: The created environment can be integrated in the enterprise IT infrastructure as standalone or as part of an existing engineering environment, which is based on the same interoperability specification.
- *Flexibility*: The seamless adaptation of an existing environment, caused by new or changing requirements, allows its more efficient maintenance, with regard to time and cost.
- *Multifunctional*: Obviously identical tool instances shall be used within different scenarios. More specific, the components of an environment shall be used by engineers within multiple work flow scenarios, coming from different use cases. This shall enable the reuse factor as well as save cost, in term of time and money.

In order to represent the characteristics above, the RTP Instance has to respect the enterprise specific work and information flows as decried in Sect. 2.7. The implementation not necessarily requires a common repository for collecting and storing the information, further a loosely coupled environment can be implemented in which components are providing services to request information in a common way. This characteristic connotes an overall system that provides a single point of access to information, and that combines various individual access mechanisms in order to do so [157]. Again this enforces the overall information flow to be bi-directional and not necessarily coupled to a specific work flow or tool chain, which leads to a more dynamic and future proof engineering environment.

8.4.2 The Engineering Environment Designer

The systems EE-Designer is the actual developer of an RTP instance. The EE-Designer adapts and tailors the RTP to meet the business use case or requirement. Since the IOS is part of the RTP and responsible for the tool interoperability, it may affect the possible tool integration scenarios which can be composed by the EE-Designer by setting restrictions to allowed formats and protocols.

8.4.2.1 RTP Tailoring

In order to meet the requirements described in Sect. 6.1 and to support the EE-Designer in the component selection and composition process, a previous activity of components preselecting and adapting of the RTP has to be applied. The so

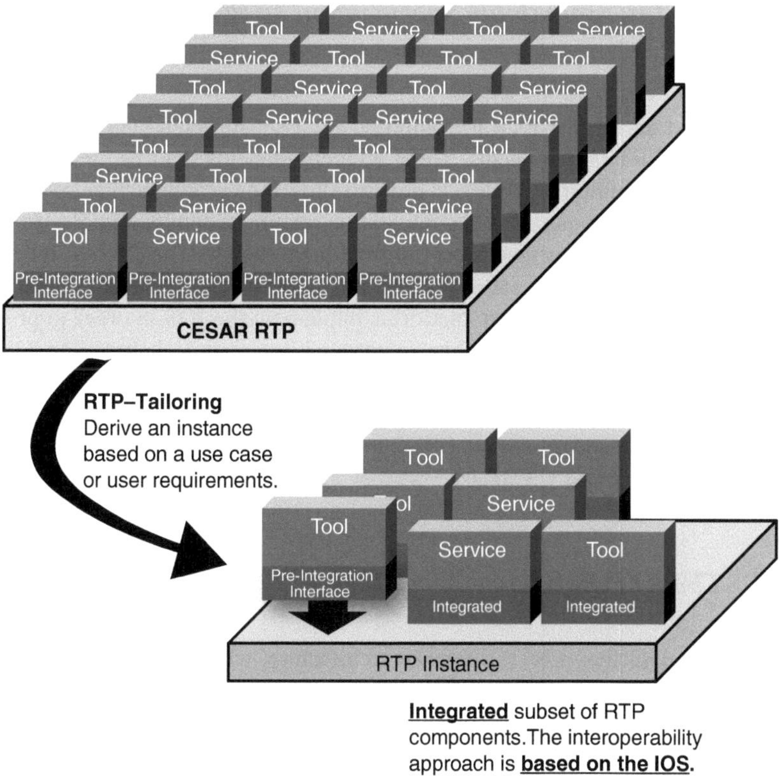

Fig. 8.11 RTP tailoring process, an RTP instance derived by an use case or requirement

called RTP tailoring process shall provide techniques and mechanisms to identify needed RTP components using use cases and requirements as input. The result is a ready-to-use engineering environment, which can be used in different work flow scenarios.

However RTP Tailoring is not part of the IOS, but both topics are tightly coupled together, since they are part of the RTP and share same requirements. Therefore it makes sense to specify and illustrate areas where the two technologies need to work together.

Figure 8.11 illustrates an abstract tailoring process. The process uses requirements and business use cases as its evaluation basis. The IOS has to guarantee and enable the seamless integration of the selected components regardless of the chosen work flow, installation order or even completeness of tool chains.

8.4.3 The Interoperability Specification as Part of the Future RTP

As mentioned above, the RTP is composed by a set of heterogeneous tools and services, called RTP components. These RTP components shall be pre-integrated, with regard to their interoperability, in order to allow their seamless integration when deploying an RTP instance. The pre-integration requires a common interoperability approach in order to allow consistent information exchange. The CESAR IOS shall define this common interoperability, based on open and standardized technologies. Components as well as software platforms implement the IOS in order to provide a pre-integrated interface. Since the interoperability approach bases on OSLC, components can adopt OSLC specifications as their native language or implement an adapter from their proprietary language to OSLC. In any case, a component needs to implement and provide an OSLC Interface. Further the interfaces are OSLC ServiceProviders or OSLC Clients.

8.4.4 The RTP Reference Architecture

This chapter describes a RTP reference architecture, where the IOS is embedded in as a building-block. The architecture is a refinement of the main building-blocks presented in Fig. 8.4. The three main architectural building-blocks of the reference architecture are the Domain, Integration and Interoperability blocks, further described in Table 8.5. The architecture presents the organizational structure of the RTP as well as IOS, which is needed to guarantee the seamless integration of components. Like the IOS architecture it is based on a layer approach. Please note, that the RTP reference architecture presented in this chapter and the described building-blocks are related to the tool integration and interoperability problem of the RTP. Other parts of the RTP can be represented based on different architecture principles and designs.

8.4.4.1 A Technology-Independent Reference Architecture

In order to discuss the roles and expected purpose of the different identified RTP architecture layers, Fig. 8.12 presents a technology independent reference architecture. The reference architecture classifies the RTP-related architectural building-blocks into nine layers, as shown below.

The layers of the three building-blocks are described in Table 8.6, and in more detail in Chap. 6.

Table 8.5 Identified interoperability stakeholders (roles and requirements) – 2

Role	Description	Requirements
Tool and service provider	A tool is used by a systems engineer in the domain of safety critical embedded systems. Using a tool the systems engineer produces information, like design elements, requirements or test cases. An instance of such a tool can be a design or simulation tool. A service is processing the information, produced by users and their tools, but can also enhance or create new information. An instance for such services can be traceability, impact analysis and transformation or information repository. A tool and service provider will adopt the IOS in order to allow the integration of its product	The IOS will not affect their internal way of information handling and processing. The tools are affected regarding their data exchange formats, communication protocol and their way of publishing and providing information. Therefore a tool or service provider has the following explicit requirements – Standardized and open available data exchange and communication protocols – Concrete data structure for tool collaboration – Minimum effort of integration – Easy adapter or interface implementation – Future-proof architecture and concepts
Platform developer	The Platform Layer can be seen as a middleware which allows the integration of tools and services based on common interoperability technologies. The platform has to adopt the underlying layers in order to provide interfaces to tools. Furthermore the integration platform provides the basic platform of an engineering environment. The environment may be composed by many different platform providers. It may be a loosely coupled technology (Network) or tightly coupled technology (Soft-, Hardware Platform)	The IOS is mainly affecting the interoperability with respect to communication and data processing. Therefore the requirements are similar to the tool or service provider requirements but more distinctive – Standardised, reliable and scalable interoperability platform architecture – Standardized, open and free available data exchange and communication protocols – Minimum effort of integration support – Easy adapter or interface implementation – Future-proof architecture and concepts

8.4.4.2 The Domain Architectural Building-Block

Today's most heterogeneous systems and the rising number of partners involved in the engineering process leads to many different possible and valid ways of developing, managing or maintaining safety critical embedded systems. This building-block, illustrated in Fig. 8.13, contains elements which shall represent enterprise or project specific use case and requirements with regard to management and

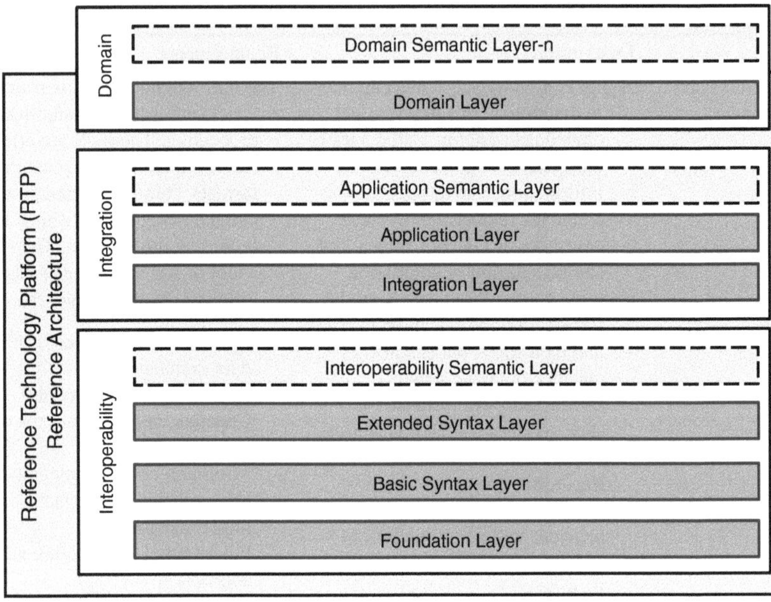

Fig. 8.12 RTP reference architecture – a technology independent view

engineering processes and work flow directives. The following two layers have been identified in Chap. 6 for this building-block:

Domain Layer

Represents methodologies and processes used in the different domains involved in the systems engineering lifecycle. These methods and processes are used by domain experts and other stakeholders to develop, manage or maintain safety critical embedded systems. Examples are for instance the different development processes like Waterfall, V-Model, Unified Process or systems modelling process like SYSMOD. The RTP shall support different domains as well as processes without forcing users to adapt their style to a specific one.

Domain Semantic Layer

This layer represents the workflows as relations between the methods and processes and consequently between components. However, this semantic layer can be shaped and refined by defining higher domain levels, like industrial domains. For instance, the engineering process V-Model may be used differently in the Aerospace than in the Automotive or Healthcare domain. The RTP shall support different levels of domains based on industrial enterprise needs. A work flow can represent and implement the different enterprise or project specific approaches. It is based on

Table 8.6 Technology-independent description of the RTP and IOS Architecture Layers

ID	Layer	Building block	Description
1	Foundation	Interoperability	The Architecture Layer covers common architecture principles which are needed by any IT infrastructure
2	Basic Syntax	Interoperability	The Communication Layer defines communication protocols as well as data formats, used to exchange information between tools and services
3	Extended Syntax	Interoperability	The Syntax Layer defines the structure of information which are represented in a common data format and are needs to be shared between tools
4	Interoperability semantic	Interoperability	The Interoperability-Semantic Layer is the first and very basic semantic level in the architecture. It allows the understanding, consequently the semantic interpretation and processing, of information in the different tools and platforms
5	Integration	Integration	The Integration Layer can be seen as a middleware which allows the integration of tools and services based on common interoperability technologies
6	Application	Integration	The Application Layer covers the engineering tools and services provided by the RTP
7	Application-Semantic	Integration	The Application-Semantic Layer is the second semantic layer in the architecture. It enhances the basic IOS-Semantic Layer with tool specific ontology
8	Domain	Engineering domain	The Domain Layer, represents methodologies and processes used in the different domains involved in the systems engineering lifecycle
9	Domain-Semantic	Engineering domain	The Domain-Semantic Layer is the third semantic layer in the architecture. It represents the workflows as relations between the methods and processes and consequently between tools. This semantic layer can be shaped and refined by defining higher domain levels, like industrial domains

business processes coming from use cases, as well as scenarios and engineering methods or processes, used in the domain of safety critical embedded systems.

8.4.4.3 The Integration Architectural Building-Block

The actual implementation of an engineering environment, using hard- and software components, is done in the IT infrastructure of an enterprise. However the installation of RTP components in such environment needs specific attention with regard to the used integration technology. More specific, a decision of what middleware

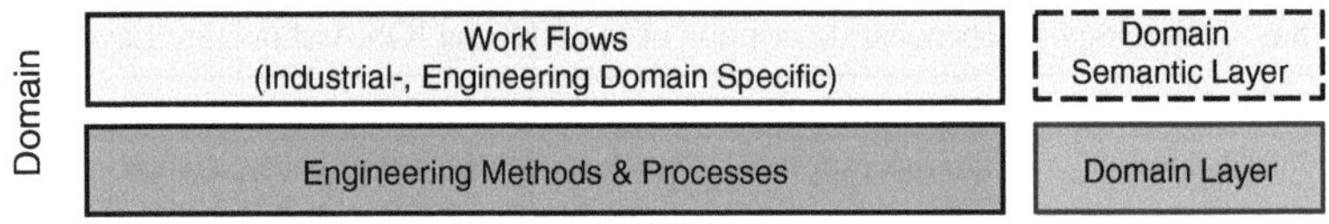

Fig. 8.13 RTP reference architecture: domain architectural building-block

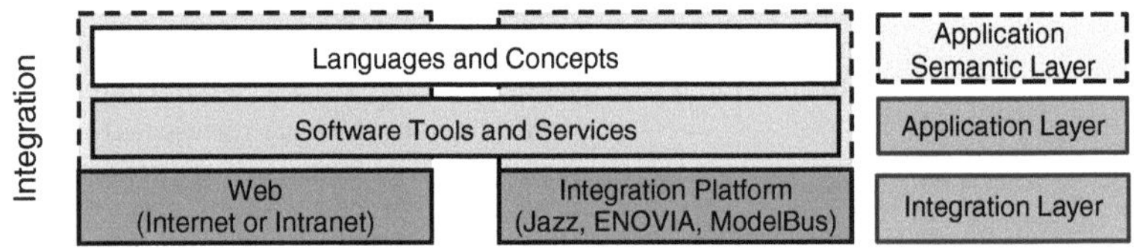

Fig. 8.14 RTP reference architecture: integration architectural building-block

shall be used needs to be taken. The platform can be seen as the backbone of the environment whether it is implemented as a loosely coupled environment using network technologies, like client an servers, or tightly coupled using a software platforms. The platform needs to provide services to software applications beyond those available from the operating system. Thus the platform can be seen as a middle which is obviously not part of an operating system, not a database management system, and neither is it part of a particular software application. The platform makes it easier for software developers to perform communication and data exchange, by providing specific set of technologies and methods to do so. This building-block, illustrated in Fig .8.14, contains elements which are the actual components of an engineering environment. The following three layers have been identified in Chap. 6 for this building-block:

Integration Layer

This layer can be seen as a middleware which allows the integration of tools and services based on common interoperability technologies. The platform has to adopt the underlying layers in order to provide interfaces to tools. Furthermore the integration platform provides the basic platform of an engineering environment. The environment may be composed by many different platform providers. It may be a loosely coupled technology (Network) or tightly coupled technology (Soft-, Hardware Platform).

Application Layer

The Application Layer covers the engineering tools and services. The term tool or service is used interchangeably in this paper. Further the term tool is used for software tools and services specific to the systems engineering domain. A tool is used by a systems engineer in the domain of safety critical embedded systems. Using a tool the systems engineer produces information, like design elements, requirements or test cases. An instance of such a tool can be a design or simulation tool. A service is processing the information, produced by users and their tools, but can also enhance or create new information. An instance for such services can be traceability, impact analysis and transformation or information repository. Tools and services need to adopt the CESAR IOS in order to be integrated into an engineering environment based on the Integration Layer.

Application Semantic Layer

This additional layer is needed since tools evolve their own ontology when implementing or adopting ontology, new concepts or a (modelling) language. As described in Chap. 1, many concepts like modelling languages are not described in a formal way. This leads not necessarily to misinterpretations when implementing the language, but obviously it does not support a consistent implementation. Tool providers can implement just parts of the language or combine it with other language concepts to fill gaps. In any case, even when the modification of an existing or standardized ontology is not as the inventor intended, it is a valid and in often applied method to build effective and efficient languages specific to the enterprise or engineer needs.

Integration Scenarios

As mentioned in Chap. 1, the integration and interconnection of software components like tools and services, can be applied on different approaches. The components may be integrated based on a resource and service oriented approach like the web or by using an underlying software platform as middleware, which allows the integration of multiple components. However in both scenarios the interoperability may base on the same specification, but change the behaviour of tools when sharing information. In both approaches the information is presented as HTTP Resources as expected by the IOS. The following section descries two approaches considered for the RTP Integration Layer and their main differences.

- *Web based Environment*: In this environment, tools implement individual web services, further called service providers, in order to share information with other tools. A service may be used to query or create new information. One powerful consequence of this is that integrations are "loosely coupled" and independent of the target tool. For example, if one part of your testing organization is using Tool A from one vendor while another part is using Tool B from another vendor, your requirements analysts can link their requirements to test cases in exactly

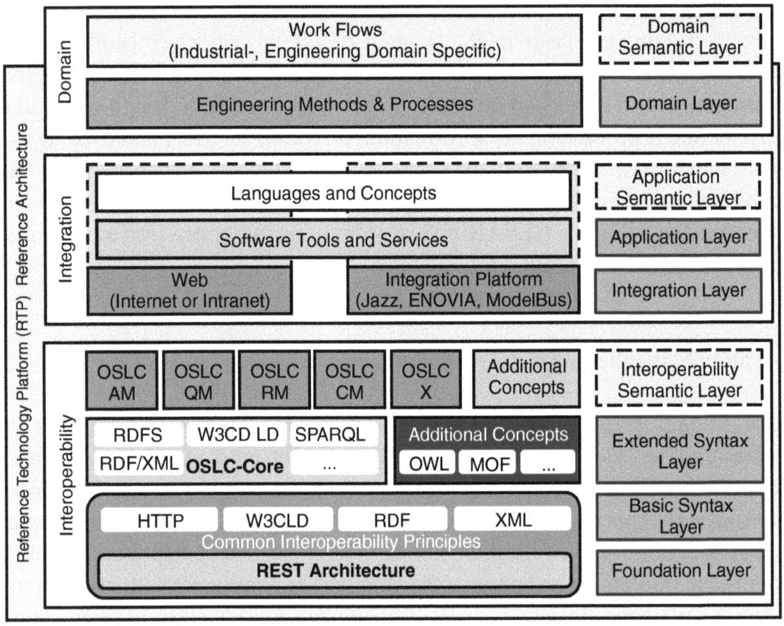

Fig. 8.15 The full RTP reference architecture

the same way, regardless of the test team with which they collaborate. And if the requirements analysts decide to switch to a different requirements tool, it does not disrupt their collaboration with the testers, provided that the new tool also implements the OSLC interfaces. This service and resource oriented approach allows the availability and sustainability of implementations.

- *Integration Platform*: In this environment, a tool provider or user need to implement an individual tool adapter, proprietary or OSLC compliant, to be used as integration interface to a software platform. The software platform then will guarantee the exchange of information between its different integrated components. Software platforms may provide a SDK to support users to implement such an adapter. The implementation effort for a tool is obviously lower since the platform take over many tasks such as service provider or repository functions.

8.4.5 The Full RTP Reference Architecture

Finally, Fig. 8.15 presents the full RTP reference architecture, including all building-blocks and their layers.

8.5 Conclusion and Future Work

The CESAR Interoperability Specification (IOS) is a robust and relevant technical foundation to implement common interoperability. It is the result of concrete lessons learned and practical experience gained in CESAR on the basis of the many experimentations and evaluations of different interoperability technologies and different kind of tools and tool chains as practiced by industrial users. The tool integration technologies are based on open and standard Web technologies like REST, RDF and HTTP. Furthermore the IOS has adopted the W3C Linked Data Basic Profile to link information, in order to support the following key tool integration and interoperability features:

- Traceability of all artefacts
- Visibility into all aspects of the implementation process (performance and historical trends)
- Control through policies and processes

Chapter 9
Evaluation of CESAR: Pilot Applications

Gianfranco Fenu, Eric Armengaud, Olle Bridal, Salvatore De Martino, Øystein Haugen, Odile Laurent, Ida Petrone, François Pouzolz, Thomas Söderqvist, and Tormod Wien

9.1 Introduction

Definition The term *Pilot Application* (PA) refers to the development process of an example application, which has the intention to serve as a source of inspiration in terms of requirement elicitation and evaluation of the *Reference Technology Platform* (RTP) (extracted from [155]).

As described in previous chapters,[1] the main intents of the CESAR project may be summarized as:

[1]In particular, the reader may refer to Sect. 1.4

G. Fenu (✉)
Università degli Studi di Trieste, Trieste, Italy
e-mail: fenu@units.it

E. Armengaud
AVL List GmbH, Graz, Austria

O. Bridal · T. Söderqvist
Volvo Technology Corporation, Gothenburg, Sweden

S. De Martino
Ansaldo STS S.p.A., Genoa, Italy

Ø. Haugen
Stiftelsen SINTEF, Trondheim, Norway

O. Laurent · F. Pouzolz
AIRBUS Operations SAS, Toulouse, France

I. Petrone
Ansaldo STS S.p.A., Genoa, Italy

T. Wien
ABB AS, Oslo, Norway

A. Rajan and T. Wahl (eds.), *CESAR - Cost-efficient Methods and Processes for Safety-relevant Embedded Systems*, DOI 10.1007/978-3-7091-1387-5_9,

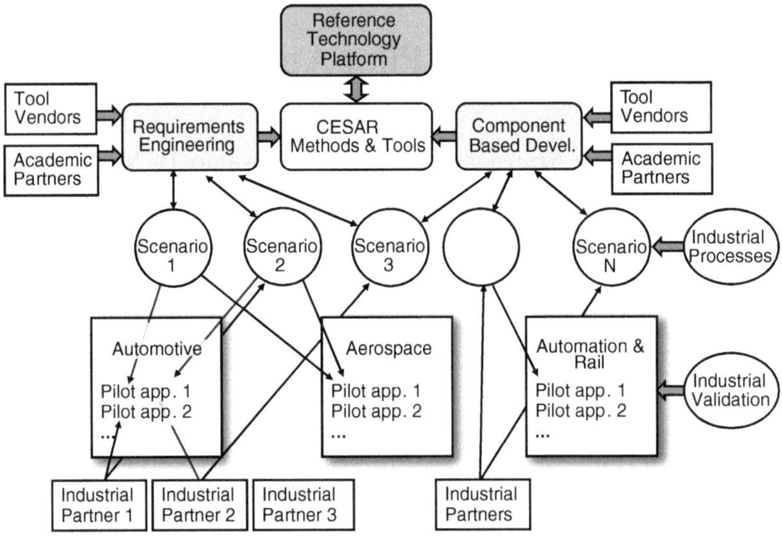

Fig. 9.1 Information flow and role of PAs

1. **Guaranteeing strong industrial take-up** of technical innovations;
2. **Covering requirements of multiple application domains** with minimal R&D costs;
3. **Assuring technical coherency and maturity** of technical innovations of the project.

With the aim to achieve these goals, the following concepts have been derived:

- Business requirements and a representative set of targeted applications, henceforth called *Pilot Applications*, jointly enabling the demonstration of business objectives in the automotive, aerospace, automation and rail industrial application domains;
- Steps in industrial processes with high potential for improvements towards achieving business requirements on targeted use cases (henceforth called *Scenarios*[2]);
- Cross–domain harmonized technical requirements covering all business requirements and initial scenarios.

The role of PAs is described in Fig. 9.1: in particular

- Using the Pilot Applications and Scenarios, the industrial partners provided requirements – in terms of tool integration, interoperability, requirement engi-

[2]*Scenario*: a selected and limited part of a Pilot Application with the purpose of highlighting an important aspect with respect to the development and evaluation of the RTP (extracted from [155]).

neering, component-based design needs – to the partners, responsible of the development of the RTP platform;
- The industrial partners received the RTP to evaluate, together with guidelines for the automated construction of customized system development platforms;
- They performed evaluation and integration of the methods provided; moreover they passed feedback to the development team of the RTP.

This cycle has been performed three times during the whole project.

9.1.1 Structure of the Chapter

This chapter is focused on the pilot applications and their results in terms of evaluation of the RTP. As stated above, there were four application sectors, addressed in the project: automotive, aerospace, automation and rail domain. The partners, involved in the application domain, proposed 23 pilot applications, explaining in many scenarios their needs and desiderata with respect to the RTP. Obviously, we do not intend exploring each scenario of every pilot application in this chapter. Indeed, for each domain one pilot application has been chosen as representative of the standard practice in each domain and, most important aspect, it is illustrating how, in which steps during the project process, and how much improvement may be obtained applying the RTP.

The rest of this chapter is devoted to illustrate these PAs. In particular in Sect. 9.2 the main features of these example applications are presented, Sect. 9.3 is focused on describing how a picture of the state-of-practice has been obtained and how the Pilot Applications have been used to explain and motivate the requirements on system engineering and tools integration. Finally the processes of tailoring and evaluating the RTP by means of example applications are described in Sect. 9.4.

9.2 The Pilot Applications

For each sample application, we report the scope of the application in the project, a description of the system (or the industrial process), and some details regarding the development process. Of course, we do not describe how effectively such kind of application projects are performed by the industrial partners, involved in the project. Indeed, we prefer pointing out, in this section, the main steps in developing each project, taking into account issues, requirements and technical needs (in the next Sect. 9.3), and describing how the proposed platform can be adapted to a specific industrial domain, in order to better satisfy particular industrial needs (in Sect. 9.4).

Moreover, with the aim to support the reader, in the case he/she would like to obtain more information on a particular topic or he/she is not so acquainted with standards, concepts, practices in a particular industrial domain, we insert appropriated references to standards, state-of-the-art and relevant topics in each application domain.

9.2.1 Automotive Pilot Application: Brake–by–Wire System

The automotive pilot application is a Brake-by-Wire (BBW) system for trucks, i.e. an electro-mechanical[3] system. The system is moreover augmented with ABS functionality.

Scope. The scope of this pilot application is the entire development cycle for a Brake–by–Wire system, including development activities that are typically performed by vehicle manufacturers as well as those carried out by suppliers.

9.2.1.1 System Description

A BBW system monitors a number of physical quantities including brake pedal position and wheel speeds. Based on these inputs and taking into account any brake requests from functions such as for example Adaptive Cruise Control, the system controls its brake actuators at the wheels. A schematic view of a possible and simplified BBW system layout is shown in Fig. 9.2. The following abbreviations are used in the figure:

- WSS – Wheel speed sensor
- ECU – Electronic Control Unit
- HMI – Human-Machine Interface
- comm. – Communication

Note that Fig. 9.2 is heavily simplified: indeed it does not show how the trailer brake interface is integrated with the tractor's brake system. This integration and interfacing may be obtained in different ways and therefore represents a design decision that would not be appropriate to include in this conceptual overview. It should also be noted that the Figure only shows four wheels being braked whereas in reality there could be a different number of braked wheels.

The system boundaries are considered to be:

- Sensing of driver actions (sensors are considered to be inside the BBW system but the physical pedal is considered to be outside);
- Sensing of wheel speeds ("ABS" sensors are considered to be inside the BBW system);
- Electromechanical actuators for braking (the boundary is either the electrical outputs from ECUs to electromechanical actuators or within these actuators so that the electrical part of the actuator is considered to be inside the BBW system);
- Communication network for interacting with other systems (transmission of info to driver, transmission of requests to other systems, reception of requests from other systems, etc.). The boundary is the physical connection to the

[3] In many of today's trucks an electro-pneumatic system is applied, instead of a BBW system.

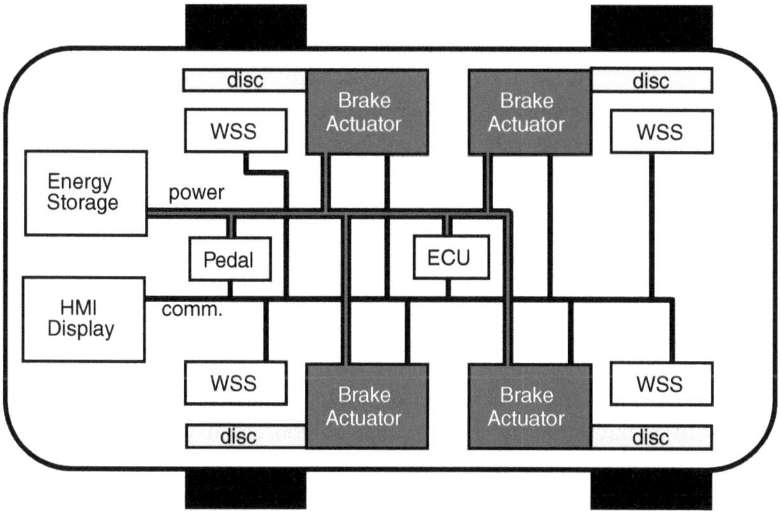

Fig. 9.2 A schematic overview of a BBW system and its environment

communication bus. However, the BBW-specific bus (in blue in Fig. 9.2 above) is considered to be a part of the BBW system;

9.2.1.2 Development Activities

Here are reported the development activities, related to the BBW PA, with a short description of each step/activity.

- Hazard identification: identification of potential hazards of the Brake-by-Wire application.
- Hazard classification: classification of hazards in terms of their degree of "unwantedness".
- Requirements elicitation: elicitation of requirements including collection of stakeholder 'needs' and 'wants'. The requirements can be of the categories:
  - Functional requirements: e.g. "The driver's brake request shall be translated into a determined brake force that is proportional to the magnitude of the driver's brake request.";
  - Non-functional requirements: e.g. "The probability of a non-fulfillment of a functional requirement within the first 5 years of use shall be less than 0.01.";
  - Constraints: e.g. "The software archtecture shall conform to the AUTOSAR release 4.0 specification."

- Requirements specification: the specification of requirements, of any type, at any relevant abstraction level (for example the EAST-ADL[4] levels Feature, Analysis, Design).
- Requirements refinement/breakdown: the refinement of requirements, from higher (user-oriented) levels to more detailed system-internal levels, as the development progresses down the left side of a typical ‘V model’ for the system development.
- Verification between different levels of specifications: verification that requirements and design decisions at one level are consistent with higher level requirements (in the EAST-ADL world, this activity is primarily applicable to the Analysis architecture and the Design architecture, which need to be verified against the Vehicle/Feature level and the Analysis level respectively).
- Hazard occurrence analysis: hazard occurrence analysis, including occurrence analysis of failures that are not hazards, is concerned with investigating the potential causes of hazards and identifying the causal relationships between these causes and these hazards. The analysis may be qualitative (i.e. relationship-oriented) or quantitative (i.e. probabilistic). Typical methods include FMEA, FTA, Reliability block diagrams and Markov analysis.
- Requirements management: overall management of a requirement repository, allowing requirements to be studied from various angles and allowing requirements changes, requirement tracing, generation of documentation, etc.
- Requirement tracing:
  - Tracing a requirement or a specific solution in the system design to a requirement at a higher abstraction level, from which the lower-level requirement or solution has been derived
  - Tracing a requirement to another requirement at a lower level or to a specific solution in the system design, when the lower-level requirement or the specific solution is derived from the original requirement
- Change impact analysis: a part of the change management process, and is concerned with identifying the effects of a proposed change of either a requirement or a design solution. The underlying reason for a change could be an improvement of the system or a correction of an existing flaw.
- System modelling: modelling of “anything” (physical structure, logical structure, behaviour, etc.) at any relevant abstraction level from vehicle-level to detailed design, for example the EAST-ADL levels Feature, Analysis, Design, Implementation.
- OEM/supplier interaction: the responsibility split and interaction between vehicle manufacturer and suppliers (possibly at different levels such as Tier 1, Tier 2, etc.) is an important aspect of any automotive development project.

[4]EAST-ADL is an Architecture Description Language (ADL), aligned with the more recent AUTOSAR [11] automotive standard; the reader may refer to [46] for details.

- Simulation: the simulated entity could be the system's environment, the system itself, or both. Examples include Hardware-in-the-Loop and Software-in-the-loop simulation. Simulation may of course be performed at different levels of abstraction, from simulation at overall system level to simulation of hardware and software components.
- Detailed design: a collective term for all hardware and software development where low-level requirements are transformed into a design and finally implemented.
- Requirements validation: ensuring that requirements are complete (i.e. cover all possible scenarios), consistent (i.e. that each requirement does not contradict itself or any other requirement) and non-ambiguous (i.e. that a requirement is clearly expressed and not open to interpretation).
- Verification by testing: testing of a system in order to check whether it fulfils its stated requirements. This activity includes the specification of the testing, which may involve the automatic generation of test cases. The activity is intended to cover testing at different levels, from unit testing, via integration testing to complete vehicle testing.
- Verification by analysis: Analysis of a system in order to check whether it fulfils its stated requirements. This activity includes formal verification of software as a special case.

Standards

Legal requirements on brake systems are defined in UN ECE Regulation No. 13 (Uniform provisions concerning the approval of vehicles of categories M, N and O with regards to braking) [163]. In addition, the quality management standards ISO 9001 [88] and ISO/TS 16949 [89] are applicable.

9.2.1.3 Development Cycle

Figure 9.3 shows, by means of a V-Cycle chart, a summary of the mapping of the activities of the pilot application (development activities) and the activities that have been identified by the automotive partners as part of the phases of a typical development cycle in this specific industrial domain. All the activities in the chart are considered to be addressed by one or more development activities of the pilot application.

9.2.2 Aerospace Pilot Application: Simplified Doors and Slides Control System

Scope. Significant A/C accident and incident occurred due to uncontrolled decompression. In some of them doors, which were not fully closed latched and locked,

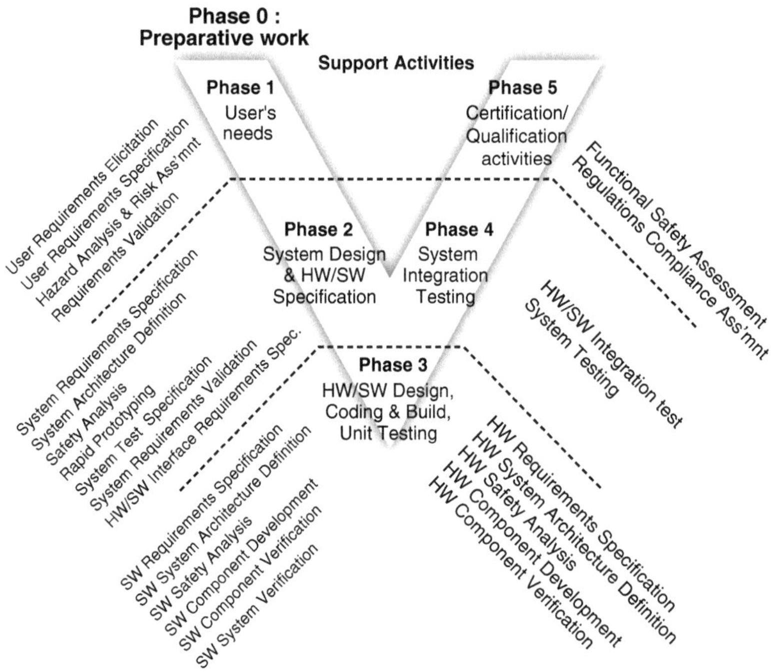

Fig. 9.3 V-Cycle with main activities covered by BBW pilot application

opened in flight leading to an explosive decompression. The 3rd March 1974 an A/C encountered an explosive decompression due to the loss of a cargo door in flight. The A/C crashed in Ermenonville making 346 fatalities. The Cargo door was not fully latched and locked. The differential of pressure between the pressurised cargo bay and the atmospheric pressure rose during the climb leading to an inadvertent opening (for details, the reader may refer to [25]).

To solve this issue, airworthiness authorities set regulations. One of them concerns pressure prevention means, which forbid pressurizing the A/C as long as all doors are not fully latched and locked. Pressurisation prevention shall be failsafe with a probability of failure lower than 10^{-9} per flight hour. The aim of this PA was to develop and capture functional, system solutions according to CESAR tools and methodology starting with this partial specification.

9.2.2.1 System Description

The "Simplified Doors and Slides Control System" (SDSC) clearly it's a simplified representative of avionics systems, but is not a full description of a real one. Nevertheless, the solution of architecture for the simplified doors and slides control system is respecting the safety objectives. Moreover, this system is a good

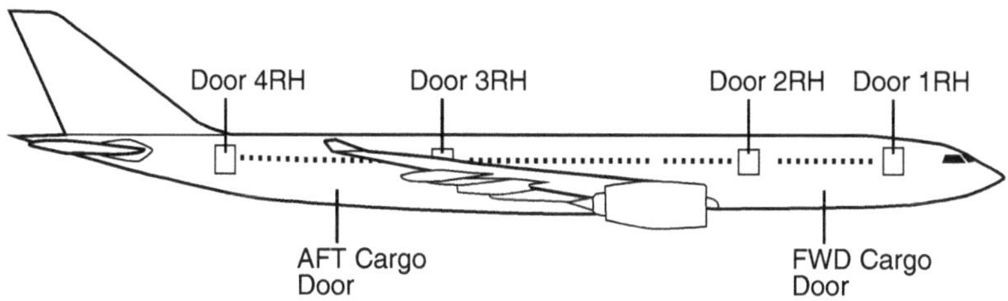

Fig. 9.4 An example: aeroplane doors managed by the SDSCS.

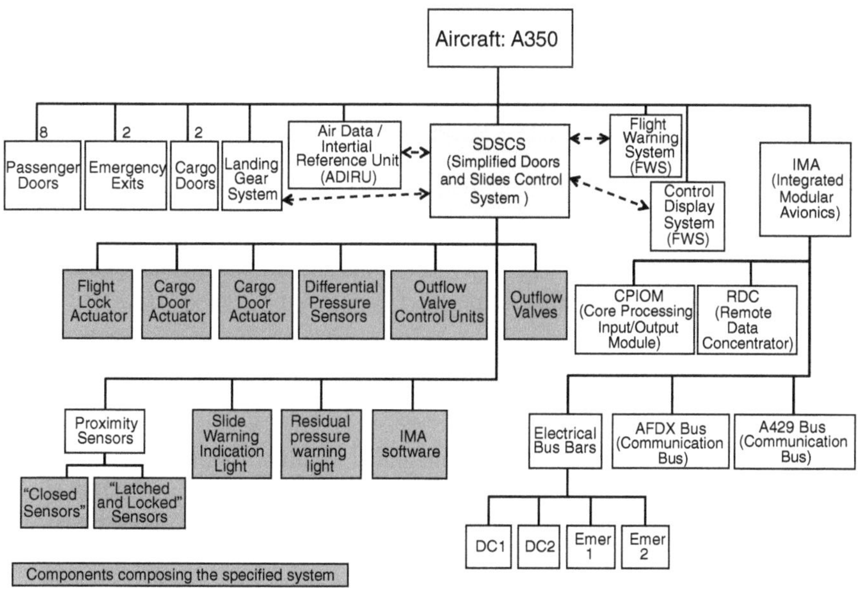

Fig. 9.5 The structure of the SDSCS system, for an aircraft with eight passenger doors, two emergency exits and two cargo doors

candidate for reuse (with aircraft specific definitions and operating conditions) and consequently fits quite well the component based development approach.

The reader may refer to Sect. 9.4 for a brief description of the Product-Line toolchain applied to the SDSC system and to Chaps. 3–5 for examples using this PA.

As illustrated in Figs. 9.4 and 9.5, main components of a SDSCS are:

- The doors to manage, for example eight passengers doors, two emergency exits and two cargo doors;
- Proximity sensors, which are used to monitor the external doors and emergency exit status. In particular, each passenger door, emergency exit and cargo door is controlled by "closed" sensors and "latched and locked" sensors.

- Slide warning indication light for each passenger door and emergency exit, that is activated when the escape slide lever is in armed position and the initial movement of the door inner handle is detected.
- Residual pressure warning light for each passenger door, emergency exit and cargo door. These doors with external opening are equipped with a red flashing indication light to alert the door operator about residual cabin pressure on ground. The passenger doors are also equipped with a buzzer which is activated when the aircraft is on the ground and a residual cabin pressure is detected and the associated slide is disarmed.
- Flight lock actuator for each passenger door and emergency exit, which prevents from an unintentional opening in flight.
- Cargo door actuator for each cargo door, which allows opening and closing the cargo door.
- Differential pressure sensors that permit to compute the differential cabin pressure.
- Outflow Valve Control Units and outflow valves that manage the pressurization of the cabin.

A similar system is, in fact, present on most of the aircraft, but the number of doors to monitor depends on the aircraft type.

Architecture Constrains

The SDSCS is implemented on IMA (Integrated Modular Avionic) system, which is based on the AFDX (Aircraft Full Duplex) network[5] on which CPIOM[6] (Core Processing Input/Output Module) and RDC[7] (Remote Data Concentrator) are connected. The maximum CPIOM allocated to this system are 4. The maximum RDC allocated to this system are 14.
The door-mounted equipment of the SDSCS (proximity sensors, indication lights, flight lock etc.) is connected to the RDC which provides power supply for the sensors. The SDSCS application is implemented on redundant CPIOMs for safety reasons. The CPIOM receives the sensor data from RDCs via AFDX and communicates also with other system via AFDX and with discrete signals. In backup ARINC 429 bus can be used (one emitter N receiver link).

---

[5] AFDX Network (Aircraft Full Duplex): this network is based on deterministic Ethernet switches, which allow to implement 100 Mb/s full duplex communications.

[6] CPIOM (Core Processing Input/Output Module): computation node, which can host different functions. Main communication media for CPIOMs are AFDX links. An ARINC 429 bus (It's an application-specific standard for aircraft avionics; see [6–8]), discrete inputs and outputs are available for backups.

[7] RDC (Remote Data Concentrator): Can be seen as a remote transceiver installed in the A/C close to the sensors, actuators, or equipments to be acquired. RDC can acquire Analog, Discrete, and ARINC 429 signals and convert the signals to AFDX.

Power supply for this use case are based on a four electrical bus bars (namely DC1, DC2, Emer1, Emer2). Overall architecture, number of occurrences to take into account, links between components are choices to be made by the system architect and designers.

Functions Description

The aim of the SDSCS may be described as follows:

- Monitoring of the doors status and the slide armed status of the evacuation slides: proximity sensors are used to monitor the status of each door. Each doors status is the result of
  - A vote between the different occurrences of sensors (vote between close information, vote between latched & lock information)
  - A logical "AND" between the result of each sensors occurrences vote (closed and (latched & lock))

  The SDSCS transmits the door status data to the flight warning system (FWS) and ECAM (Electronic Centralized Aircraft Monitoring). The ECAM system provides the door status information of all the doors on the flight deck.
  The SDSCS activates the Slide Warning Indication Light (located in the door window) when the escape slide lever is in armed position and the initial movement of the door inner handle is detected in order to make the operator aware of the slide armed status and that consequently the emergency slide is going to deploy. The slide status indication function monitors the armed/disarmed status and it is indicated on ECAM. The ECAM system provides necessary slide armed/disarmed information of all applicable doors to the flight crew in the cockpit.
- Control of the passenger doors and emergency exits locks actuators: each passenger door and emergency exit is protected against unintentional opening during flight by a flight lock. The flight lock engages before take-off, it unlocks in cruise and during descend and approach the flight lock is in lock condition again until the aircraft is on ground.
  To determine the on-ground and in-flight status, the SDSCS application is using the landing gear signals, the engine running signals, the computed air speed signal from 3 ADIRUs (Air Data/Inertial Reference Unit) and the emergency evacuation signal. In case of no signal the default is always the ground status and the flight lock unlocks to ensure evacuation is possible in an emergency case.
  In addition an override function is implemented that uses the discrete signal from the Differential Pressure Sensor (DPS), where two DPS are used, one for each aircraft side. The flight lock unlocks, if the DPS detects a differential cabin pressure below 2.5 mbars.
- Control of the cargo doors actuators: a hydraulic actuator is used to open and close the cargo doors. The control logic of the hydraulic actuator is implemented

in the SDSCS. The latching and locking shaft is driven manually by the cargo door handle.
Open/Close sequence: the cargo door will unlock and unlatch when the cargo door handle is moved to open position manually. When the sensors detect unlatched position and the toggle switch is in open direction a servo valve will be activated and the hydraulic pump will be initialized. The hydraulic pump powers the hydraulic actuator and the door moves to open direction. The "close" direction follows the identical principle vice versa. Releasing the toggle switch in either open or close position during operation causes immediate stop of the cargo door movement.
- Management of the residual pressure: passenger doors and cargo doors are equipped with a red flashing indication light to alert the door operator about residual cabin pressure on ground. The SDSCS detects a differential cabin pressure by monitoring the signal from two Differential Pressure Sensors. The SDSCS illuminates the residual cabin pressure indicator lights for max. 45 min and an aural warning at the door is activated if the aircraft is on ground and if the corresponding slide is in disarmed condition for the passenger door and emergency exit. The in-flight/on-ground data is provided to the SDSCS via two discrete signals from the Landing Gear System.
- Inhibition of cabin pressurization: two Outflow Valve Control Units (OCU) manage the pressurization of the cabin. Each OCU controls one outflow valve and to allow pressurization both outflow valves have to be closed. The SDSCS sends an "all doors closed, latched and locked" (CLL) status signal to the two Outflow Valve Control Units (OCU) to allow pressurization only if:
  - All passenger doors are closed, latched and locked **AND**
  - All cargo doors are closed, latched and locked **AND**
  - Emergency exits are closed, latched and locked.

  In case of an unsafe door or defined failure case is detected by the SDSCS the pressurization of the Aircraft is inhibited.
- Interactions with other systems: the SDSCS system interacts with many avionic systems. In particular
  - The SDSCS sends to the OCUs the CLL status via the ADFX and directly through two discrete signals;
  - The status of the doors are displayed in the ECAM and managed by the FWS;
  - The Air Data/Inertial Reference Unit (ADIRU) provides the air speed system;
  - The Landing Gear System (LGS) provides the landing gear extension and retraction signals.

9.2.2.2 Safety and Operational Constraints

The ACJ 25.783 [58] describes the regulations for aircraft doors. The rules, taken into account in the SDSCS, are described below. Each door that could be a hazard if it unlatches must be designed so that unlatching during pressurised flight from the

Table 9.1 Safety objectives for the SDSCS system

Failure condition	Safety objective
Incorrect activation of pressurization when any door is not fully closed, latched and locked	$1.0 \cdot 10^{-9}$
Loss of residual pressure indication on ground in case of residual pressure	$1.0 \cdot 10^{-7}$

fully closed, latched, and locked condition is extremely improbable. This must be shown by safety analysis. The system shall manage:

- A latching function and a locking function: there must be means to safeguard each door against opening during flight due to inadvertent action by persons. In addition, precautions must be taken to minimise the possibility for a person to open a door intentionally during flight. If these precautions include the use of auxiliary devices, they must be designed so that a single failure will not prevent more than one exit from being opened.
- Pressure prevention means: there must be a provision to prevent pressurisation of the aeroplane to an unsafe level if any door subject to pressurisation is not fully closed, latched, and locked. The provision must be designed to function after any single failure, or after any combination of failures not shown to be extremely improbable. The provisions for preventing pressurisation must monitor the closed, latched and locked door conditions.
  It should not be possible to override the pressurisation prevention system unless a procedure is defined in the Master Minimum Equipment List (MMEL) that confirms a fully closed, latched and locked condition. In order to prevent the override procedure from becoming routine, the override condition should not be achievable by actions solely on the flight deck and should be automatically reset at each door operational cycle.
- An indication function: it must be a visual means on the flight deck to signal the pilots if any door is not fully closed, latched and locked. The visual means may be a simple amber light or it may need to be red warning tied to the master warning system depending on the criticality of the door. The door closed, latched and locked functions must be monitored, but only one indicator is needed to signal the door closed, latched and locked condition. The probability of erroneous closed, latched and locked indication should not be greater than 10^{-5}.

The means must be designed such that any failure or combination of failures that would result in an erroneous closed, latched and locked indication is improbable for:

- Each door that is subject to pressurization and for which the initial opening movement is not inward
- Each door that could be a hazard if unlatched.

The safety objectives for the aircraft doors management system are grouped in Table 9.1 and the failure rates of the components forming the aircraft doors management system are grouped in Table 9.2.

Table 9.2 Safety objectives for the SDSCS system

Failure mode	Failure rate
Doors sensor loss (close or latch or lock sensor)	$2.0 \cdot 10^{-6}$
RDC loss	$5.0 \cdot 10^{-5}$
RDC erroneous	$1.0 \cdot 10^{-6}$
CPIOM loss	$2.0 \cdot 10^{-5}$
CPIOM erroneous	$1.0 \cdot 10^{-6}$
AFDX switch loss	$1.0 \cdot 10^{-5}$
AFDX switch erroneous	$1.0 \cdot 10^{-7}$
OCU loss (single channel)	$2.0 \cdot 10^{-5}$
OCU erroneous (single channel)	$1.0 \cdot 10^{-6}$

9.2.2.3 Development Activities

Two main scenarios covered the activities related to the SDSCS Pilot Application, namely a functional analysis approach and a multi-physics modelling scenario. For each of then we identified a set of development activities.

In particular, the *functional analysis* leads to the following actions (with indication of fully or partially support by tools in CESAR project):

- The capture of the needs: various "Aircraft doors management system" needs from A/C manufacturers, Airports, Airlines, etc.; an example on how to define SDSCS requirements is introduced in Sect. 3.3. Moreover, how to analyse SDSCS requirements is reported in Chap. 3, starting from **page 102**
- The identification and characterization of external (main functions and secondary functions between the system and its environment) and internal functions (specific functions that responds to the needs previously identified);
- Using a library of functions (if it already exists); constructing the need diagram (i.e. the relations between functions and typing of their characterization in the project database).
- The automatic generation of the external and internal interaction schemes;
- Checking characterization consistency, to unlock specific functionalities;
- The definition of the failure modes that are to be monitored: warnings on nodes or function (FMEA) (partially supported in CESAR; refer to Sect. 9.4 for an explanation);
- Generation of a fault tree analysis (partially supported in CESAR; please refer to Sect. 9.4);
- Identification of a list of criteria to take into account during decision phase (refer to Chap. 4 for the description of what was done in CESAR project);
- Simulation of the functional architecture.

- Automatic update of the requirement management database: adding a link to the reports related to each requirement (partially supported; refer to Sect. 9.4 for a brief explanation).

Modern A/C control system combines different types of technologies: hydraulic, mechanic and structure, software ... A *multi-physic modelling* ability improves the repartition of the efforts to provide during the development phase. In particular, using a multi-physics modelling approach allows

- To develop a system architecture for the "aircraft doors management system";
- To link the system architecture to the previous functional analysis modelling (which function from the system architecture fulfils the need identified from functional study);
- To check the consistency of the system architecture regarding the functional analysis study. The reader may refer to Chap. 4 for details on architecture exploration, evaluation and analysis in CESAR. Moreover, an example of CBD and CMM, applied to this Pilot Application, are described respectively in Section and in Sect. 5.5;
- To add properties to each block of the system architecture;
- To develop a second system architecture (different technological choice with no impact on functionality of the system). An example of PL tool-chain, applied to SDSCS, is described in Sect. 9.4;
- To run a simulation of the complete system;
- To simulate the system adding defaults and/or failures (partially supported in CESAR; please refer to Sect. 9.4 for an explanation).

9.2.3 Rail Pilot Application: On Board Unit ERTMS Level 1

Scope In order to meet the future needs of the European Railways, the European railways and the supply industry, with the support of European commission, designed up the ERTMS, the European Railway Traffic Management System [55]. The aim of the ERTMS is providing a unique standard in order to avoid the differences arising from the different national train control systems in Europe improving the competitiveness of the overall European railway system. In this context, the European Train Control System (ETCS) can be considered as a basic component of the ERTMS: the existing national ATP-systems should be replaced by ATP. The ERTMS is made up of three levels, each one describing different uses of the ERTMS as a train control system, going from track to train communications (described in Level 1) up to the continuous communication between the train and the Radio Block Centre (RBC) for the Level 2. The third one, still in a conceptual phase, will introduce a "moving block" technology in order to improve the ERTMS potential.

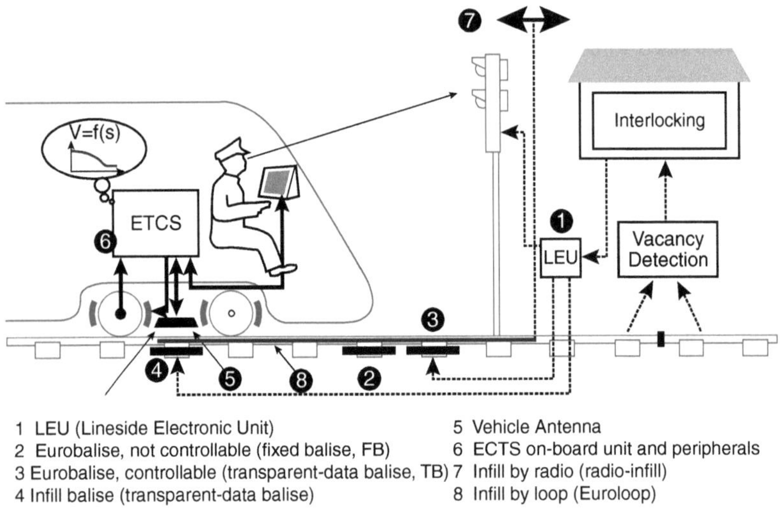

Fig. 9.6 OBU ERTMS Level 1 schematics

9.2.3.1 System Description

The two sub-systems communicate by means of specific dedicated antennas, called eurobalises in literature (Fig. 9.6). Balises, containing pre-programmed track data, are located on the track, in the nearby of the line, at required intervals in order to ensure the communication between the two sub-systems.

The collection of the data is sent to the control centre, that receives the position of all the trains on the line and determines the updated movement authority that will be sent to the balises. This updated information is sent to the train when it passes over the balises, receiving then the newest movement authority: the on-board unit can finally calculate the speed profile and the next breaking point, displaying the information to the driver. For more details, please refer to [22].

The aim of the Rail TPA was to develop *Interoperability tests*, to obtain a complete and consistent interface of the ERTMS Level 1 On-Board Unit. The development process had to be compliant to the CENELEC 5012x standards [48–50], CENELEC 50159 [51] and IEC 61508 [85] standards. The Interoperability tests are standardized by UNISIG (Union Industry of Signalling [166]) and had been developed exploiting the RTP. Therefore, the Rail TPA contributed to the CESAR goals providing the assessment of RTP applicability to perform these tests.

9.2.3.2 Development Activities

The following activities are part of the testing process

- **Test definition**: test specification, based on the system requirements;
- **Test execution**: the test is performed;

- **Test report analysis**: analysis of the test results (comparison of obtained outputs with the expected results etc.);
- **Report document**: the documentations, based on the analysis of the test results;

The report on the state-of-practice pointed out some technical gaps (for a *baseline*, related to efforts and costs, see page 314). In particular, the following gaps have been noticed:

- The test definition was manually performed;
- During the test execution phase, there were interoperability problems due to different proprietary data format from heterogeneous providers (the execution of the interoperability tests aims to verify that different subsystems from heterogeneous providers can work together). In fact, usually either test data and test outputs are expressed in proprietary data formats that are different for each provider. Indeed, the gap among different data formats has not been addressed in CESAR but still exists and it is filled manually;
- The analysis of the test results was manually performed;
- The documentation was manually drawn up.

The proposed actions to be taken, in order to improve the above mentioned phases, were:

- Test definition: implementation of methods, that allow the definition of automatic test. In particular, the methodology was a proprietary testing procedure, developed during the years thanks to a strong experience in testing of complex railway systems. Core of the method concerns the classification of the system input in several categories and the definition of some rules. These rules must be taken into account by the tool for the automatic test generation phase, to discard the combinations not possible for the system. In this way it's possible to reduce the cardinality of the test signals set, without jeopardize its completeness. In detail, the automatic test definition activity concerns

  – Automatic test delineation, analysing the system requirements specifications: for each system requirement a test should be automatically specified, in order to determine whether the system satisfies the requirement;
  – Customizable test generation: it should be possible to define rules that allow the user to generate a customized set of tests;

- Test execution: it's mandatory to follow the approach proposed by UNISIG (subsets 110, 111 and 112) [166], which allows converting

  – Test data input: test data expressed in a standard data format that is common to all providers, should be converted to test data expressed in a proprietary data format;
  – Test data output: test logs expressed in a proprietary format should be converted to a standard log format, common to all providers.

Such kind of conversions – from standard to proprietary format and vice-versa – are necessary to allow different subsystems from heterogeneous providers working together.

- Test report analysis: by comparison of the test result with the expected test output, it should be possible to automatically analyse the test report;
- Test report document drawing up: it is necessary to automatically generate the test report document.

Scenarios

With the aim of emphasizing some features of this application, both during the development and the evaluation of the Reference Technology Platform, three *Scenarios* have been proposed:

- **Interoperability Tests Requirements Analysis**: before the execution of interoperability tests, it is necessary to analyse the tests requirements. In particular, it is mandatory:

  – Verifying that requirements on the ground sub-system are coherent with requirements on the on-board sub-system;
  – Checking the coherence and completeness of requirements on the on board sub-system.

- **Evaluation of On Board Unit Level 1 functionalities**: due to the degree of freedom of the standards, slight incoherence between different subsystems from heterogeneous providers could come true at the interface level. Starting from a new approach proposed by UNISIG, one goal of the Rail TPA was the definition of interoperability tests in order to obtain a complete and consistent interface of the OBU for ERTMS Level 1.
- **Evaluation of new or modified On Board Unit Level 1 functionalities**: with the system already running, some modifications or new functionalities (implying modifications on the external behavior) could be required. In such context, tests should be performed, to guarantee a correct interaction of the new parts with the "old" ones, without degradation in the interoperability aspects.

Development Details

The activities performed throughout the TPA were arranged taking into account the different steps of functional testing activities. Therefore, the following activities had been identified:

1. System Requirements Specification (SRS): system requirements shall be specified in a formal Requirements Specification Language (RSL); in this phase also a simple model of the system has to be provided;

2. Test definition: definition of test scripts starting from system requirements. Then, the scripts are translated in a proprietary data format. Two complementary approaches were taken into account to define the test
   (a) Automatic test specification from the SRS: for each system requirement a test shall be automatically specified, in order to determine whether the system satisfies the requirement.
   (b) Customizable test generation: it should be possible to define rules, which allow the user to generate a customized set of tests. The tool that supports the test definition shall generate tests for all combinations of the possible input variables and should also allow the user to reduce the number of tests, by choosing only the permutations to be considered (i.e., the possibility for generating only definite permutations of these variables and so not all the possible ones).
3. Test specification and test oracle description: the scripts of the identified tests shall be expressed using a flexible formalism (for example using a XML–based language). The test oracle shall be automatically defined by means of the expected verdict from the observed test results, the expected outputs of the test, in order to detect if the actual outputs obtained by the test are correct or not.
4. Test specification in UNISIG formalism: it is mandatory to translate the scripts of the identified tests from the XML formalism in the formalism proposed by UNISIG (subset 076) [166], in order to make the test specifications easily understandable by all actors involved in rail applications (suppliers, customers, assessors).
5. Common UNISIG interface to allow test execution: it is necessary to follow the approach proposed by UNISIG (subsets 110–112) [166], that allows to convert the scripts of the test from a standard format (that is common to all providers) into a proprietary format.
6. Test execution and automatic verification: tests shall be executed and the results automatically analysed, comparing the actual outputs with the expected ones, as defined above (test oracle). However, in this way it is only verified that the system behaves as expected. Indeed, it is not checked if the test raises other unexpected and not requested behaviours and actions. For this reason it is necessary also to verify that all the boundary conditions[8] maintain a coherent status.
7. Common UNISIG interface to allow Log Analysis: it is necessary to follow the approach proposed by UNISIG (subsets 110–112) [166] to convert the log of the test from a proprietary format into a standard format (common to all providers).
8. Automatic drawing up of the test report: the test report document shall contain the test specification and the result of each test.

[8]It means, for example, that if a variable is classified as "not determinant" for a particular test, the result of that test must be the same, whatever is the value of the "not determinant" variable.

9.2.3.3 Development Cycle

The results of a workshop, carried out among experienced test engineers, suggested that the above mentioned activities of test definitions, test report analysis and test report document drawing up could be improved. In particular, the technical gaps identified in each activity and the amount of effort (in terms of person months) spent to develop each activity are illustrated below. The amount of effort put on each testing activity is indicated in percentage points. The sum of percentage points should be 100 %, as it is assumed, that 100 % represents the current overall effort spent in the testing process:

- The first technical gap was identified in the definition phase of the testing process. During the workshop it was stated that the most efforts spent in this phase were due to test definition that is manually performed. This activity took 25 % of the total efforts spent in the testing process.
- The second technical gap was identified in the test execution phase. The workshop revealed that using the "current practice" there were interoperability problems due to different proprietary data format from heterogeneous providers. These problems hindered the execution of the interoperability tests, which aim to verify that different subsystems from heterogeneous providers can work together. In particular, test data, input during the test, were expressed in a proprietary data format, different for each provider. In addition to the test logs, the test outputs were also expressed in a proprietary log format. Test execution took 15 % of the total efforts spent in the testing process.
- The third technical gap was identified in the test report analysis phase. The workshop revealed that most of the efforts spent in this phase were due to the test report analysis, manually performed. It took 50 % of the total efforts spent in the testing process.
- The last technical gap was identified in the document drawing up phase. As result of the workshop, most of the efforts spent in this phase were due to the test report document, that was manually drawn up. It took 10 % of the total efforts spent in the testing process.

Figure 9.7 represents, through a V-Cycle chart, a summary of the mapping of the PA's activities (development activities) and the activities that have been identified by the railway partners. All the activities in the chart are considered to be addressed by one or more development activities of the pilot application.

9.2.4 Automation Pilot Application: Safe Controller Applications in Mechanical Plant

Scope The safety loop (the task to minimise risks associated with the industrial processes that can result in endangering personnel, the environment and systems) in the automation scenario must always be split into a safe product development

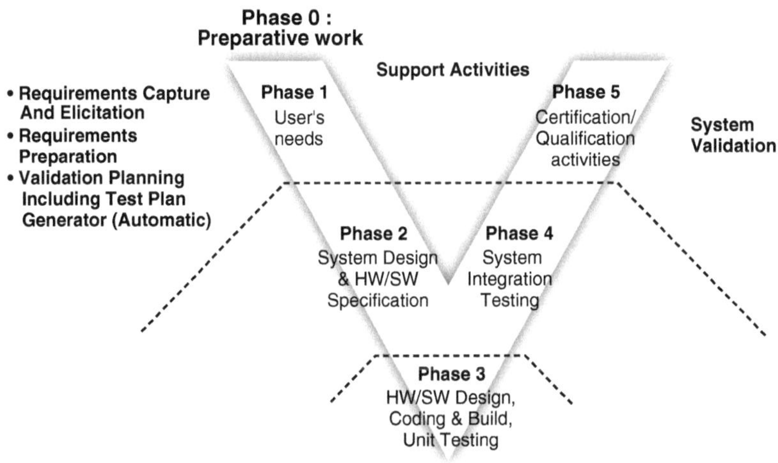

Fig. 9.7 V-Cycle with main activities covered by the OBU ERMTS pilot application

and safe engineering. To clarify this in detail; to introduce safety in a process plant, functional safety compliant devices must be used. Such devices can be safe controllers, safe IO–controllers and safe actuators or sensors. These are provided by different vendors for automation products like ABB or Siemens. The vendor has the responsibility to develop these devices compliant to safety standards (e.g. IEC 61508, see [85]) and to certify them by use of a certification body like TÜV (Technischer ÜberwachungsVerein). Each safety certified device has its own hazards and countermeasures to be integrated into the device to guarantee the functional safety. For example, in the case of a safe PLC, the diagnostics ensure the detection of hardware errors. The development of functional safety compliant devices is part of the work needed to satisfy the safety requirements. With safe devices only, the plant is not safe. Safe devices will be the basis for building a safe process plant, but it requires more safety tasks on the engineering level to achieve a safe process plant in the end. Engineers will have to identify the top hazards. For example, when exceeding a specific pressure in a boiler, measured by a safe pressure sensor, the safe controllers shall open the safe valves, so that the pressure will decrease and no accident will happen. This behaviour is then implemented using IEC 61131–3 (see [83]) in functional safety certified control system and devices.

9.2.4.1 Introduction

As a result of automation, demand for increased production and reduced operator physical effort, control systems of machinery and plant items play an increasing role in the achievement of overall safety. These control systems increasingly employ complex electrical/electronic/programmable electronic devices and systems. Noticeable amongst these devices and systems are electrical power drive systems that are suitable for use in safety-related applications. Examples of industrial applications are machine tools, robots, production test equipment, papermaking machines, conveyor belts, cranes, pumps, fans, etc.

We investigated how safety measures can be added to a conventional power drive system in terms of an optional plug-in safety module and how the safety module can be set up to work in different industrial applications. The purpose of adding safety measures to a power drive system was to achieve risk reduction. A typical case is guard interlocking in order to exclude personnel from hazards where access to the danger zone is only possible when rotating parts have attained a safe condition. Different industrial applications have different safety measures depending on the use and the risk for harming people or material: a hoisting system with a hanging load, e.g., a crane used in a construction area, could for example require an additional braking system preventing the load from falling down and harming people or destroying buildings, while a conveyor belt could require speed or acceleration limiting measures. These could also be measures protecting the device itself, e.g., the motor from overheating thereby causing a fire destroying the motor itself and/or the environment (e.g. buildings), or in worst case causing casualties. The power drive system with the added safety module is denoted as being a safety related power drive system. The International Electrotechnical Commission (IEC) has given directions on how to implement an adjustable speed safety related power drive system. The safety requirements are given in the standard IEC 61800-5-2 [84]. The standard gives direction on how such a system should be designed and implemented (e.g. SIL level to be used) and which safety functions to consider. The safety functions are stopping functions, e.g. Safe Torque Off (STO) removing the power from the motor, or limiting functions, e.g. Safely-limited Speed (SLS) preventing the motor from exceeding the specified speed limit. Some of these functions are mandatory in all industrial applications, e.g., STO, while some depend on the actual application.

9.2.4.2 Safety Module

The Fig. 9.8 presents the layout of the safety related power drive system with the Safety Module, the Motor Controller and the Motor. The Safety Module has several safety I/O's for external safety devices, like gates and indicators. Encoder (tachometer) can be connected option, but also encoderless safe speed is possible for limited applications. Safe brake control is connected directly to Safety Module, but also resources cooperation of regular brake. The Safety Module does not control the Motor Controller, it only monitors it. The only thing the Safety Module can do is to activate STO by releasing the power from the motor and to control the mechanical brake. The Safety Module requests the Motor Controller to execute safety functions, and monitors the outcome. If the Motor Controller does not follow the given commands the Safety Module executes STO. The Safety Module also handles Prevention of unexpected start-up (POUS). The Safety Module cannot be

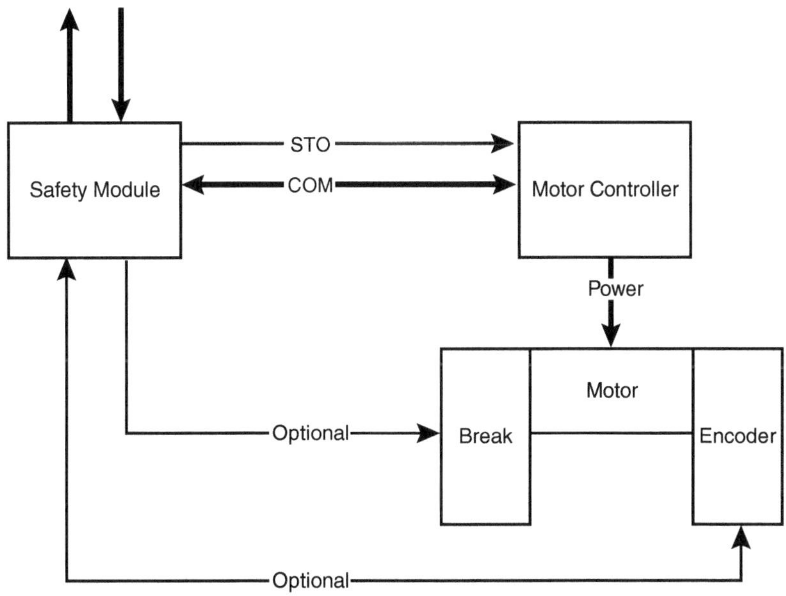

Fig. 9.8 Safety Module system layout

programmed in any way, but it is run-time configurable. Configuration is done by approved personnel using a configuration tool. The configuration parameters are validated and downloaded to the Safety Module and Motor Controller. At start-up the Safety Module performs comprehensive self- and IO-diagnostics and Motor Controller parameter comparison. While running, the Safety Module performs some self- and IO-diagnostics and Motor Controller parameter comparison at regular intervals.

Safety Functions

Safety functions are commanded and monitored by Safety Module, executed by the Motor Controller. Safety functions does not really have execution priorities but an allowance/blocking table. Functions can be simultaneously in execution though stopping functions will just stop the system. The following safety functions are implemented to the Safety Module:

- STO (Safe Torque Off): EN 60204-1 category 0 stop; an uncontrolled stop by removing power from the machine actuators. Safe Brake usage: Brake can be used with STO (configurable). Delay can be set to positive or negative values. (negative delay means the brake is activated before STO).
- SBC (Safe Brake Control): provides a safe output signal(s) to control an external brake(s). Operated only with STO (positive/negative delay)

- SSE (Safe Stop EMERGENCY): can be configured to execute STO OR SS1. Will use harder ramp down than normal SS1.
- SS1 (Safe Stop 1) EN 60204-1 category 1 stop: a controlled stop with power to the machine actuators available to achieve the stop then remove power when the stop is achieved. Can use SAR to reduce speed.
- SLS (Safe Limited Speed) prevents the motor from exceeding the specified speed limit. There will be four configurable SLS-levels. Can use SAR to reduce speed.
- SMS (Safe Max Speed): absolute max speed that motor is allowed to run, monitored all the time.
- SAR (Safe Acceleration Range) keeps the motor acceleration and/or deceleration within specified limits.
- SSM (Safe Speed Monitor) provides a safe output signal to indicate whether the motor speed is below a specified limit.
- Prevention of unexpected start-up. Some of the safety functions use POUS, prevention of unexpected start-up. It means that when the safety function is commanded, it is fulfilled regardless the operator actions, like deactivating STO is not possible before the motor is stopped. The following safety functions have POUS: STO, SSE, SS1, SLS (if the starting speed is higher that the tar-get). Restart or releasing the control from Safety Module after the function is fulfilled can be configured to happen in two different ways: automatically or manually. In the first case control is released after a time delay, where in the second case a manual button directly connected to the Safety Module has to be installed.

9.2.4.3 First Scenario: Component Based Modelling

This scenario dealt with component based modelling, certification of components and reuse of already safety certified components. A considerable effort could be saved, if the safety system is split into modules with minimum dependency. During the project, formal methods for this have been developed and tested. The research issues (i.e. modelling and component composition) was focused on collaborations and functionalities of the components of the safety related power drive system. Topics addressed were: architectural design, modelling and component composition, reuse of components, simulation.

9.2.4.4 Second Scenario: Variability Modelling

A safety related power drive system is used in a vast amount of industrial applications, such as machine tools, robots, production test equipment, papermaking machines, conveyor belts, cranes, pumps, fans, etc. The use of the system in various industrial application areas with different requirements points in the direction of having an efficient way of handling the application requirements and to be able to configure the system accordingly. The recommended way of setting up the power drive system is to identify the possible hazards, assess the risk on

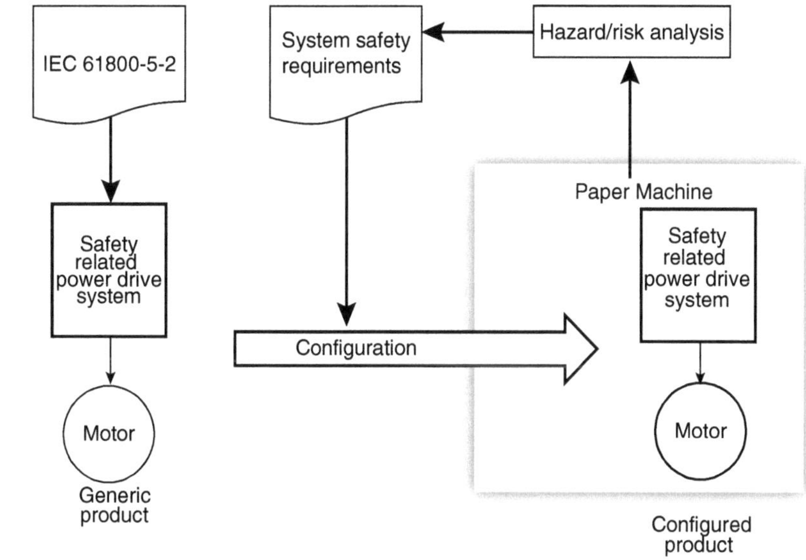

Fig. 9.9 Modelling of a safety related power drive system for a paper machine

people and material and elicit the safety requirements. We used a paper machine as an example.

Hazards

Hazards for a paper machine can be everything from mechanical (e.g., running pinches on rotating rolls, reels and cylinders; outer rolls; knives, edges of wires or sharp edges of machinery frame causing a person to be drawn in or trapped, amputation or cutting injuries end even death) to thermal and noise hazards, causing e.g., hearing loss or fire (thermal) (Fig. 9.9).

Requirements

Having identified the hazards and potential risk one can state the system safety requirements. Typical requirements are:

1. The system shall be provided with an emergency stop system.
2. The emergency stop function shall act on all hazards of the entire machine including the machine pulper and measuring device.
3. Steam and condensate system and burners shall be included in the emergency stop function.
4. The emergency stop system shall act on the whole machine.

5. The machine shall be fit with braking devices.
6. The braking devices shall be operated in such a way that the stopping time is as short as possible.
7. Devices for the prevention of unexpected start-up of the sectional drives shall be provided
8. The system shall be SIL2 compliant

The first requirement states that the paper machine shall be provided with an emergency stop system, so we selected to use the Safety Module with the STO functionality (mandatory function). From the fifth requirement "... braking devices" we selected the SBC function. Nothing was stated about the stopping time except that it shall be "as short as possible". This could indicate that we apply the brakes just before or at the same time the STO is activated. We here selected −500 ms indicating that SBC will be activated just before STO activation. SIL2 compliance (last requirement) was provided with no tachometer installed on the motor, so we omitted this part.

Topics addressed were:

- How to use ontologies at different levels (hierarchy) to check for consistency and completeness of requirements;
- Requirements and ontology variability;
- Variability modelling and system configuration;
- Product Line Engineering.

9.3 Requirements on the Reference Technology Platform

As briefly described in Sect. 9.1, the interaction between domain SPs and technical SPs started deriving a representative set of targeted applications and requirements, related to such example applications, to the standard practice in each industrial domain and to the challenging project goals. In total, 23 pilot applications were defined and it resulted to the identification of more than 600 end-user requirements. Moreover, a "picture" of the current practices in the industrial domains was needed, to consolidate a common baseline and to identify different approach for similar tasks, gaps and possible improvements.

9.3.1 State of Practice

At the beginning of the project the technical SPs and the Task Force Safety& Diagnosability (TF-SD) elaborated a questionnaire, with the aim to capture the current practices of the industrial partners, i.e. the context that the RTP must accommodate, complement or even replace for some parts. Each industrial partner filled in the questionnaire and the results were collected and harmonized in three different surveys, one for each industrial domain. The harmonized information was then analyzed by the technical SPs and by the TF-SD staff, allowing to identify the commonalities, in terms of methodologies, tools, gaps, together with the variations and the differences in the development of an embedded system in the diverse domains.

The questionnaire covered several topics, as explained below. For each topic, state-of-practice details were described concisely (generally, names of well known methods, techniques, tools, etc. should be sufficient for understanding), together with the following requested information:

- **Rationale**: "Why does the user use the item?" with three possible answers:
  - Imposed by Law (regulated domain);
  - Imposed by contract with customer;
  - Own choice or decision, because it is considered good practice;
- **Benefits** and **drawbacks** (or limitations);
- **Requirement**: if the user considered that the corresponding item should be supported by the RTP, the identifier of the corresponding user requirement was indicated here.

The topics addressed in the questionnaire were

- Applicable standards, norms, regulations:
  - Applicable standards, norms, regulations related to engineering, safety, product lines;
  - Constraints imposed by standards, norms, regulations: the constraints imposed on the CESAR-relevant processes, methods and tools – RE engineering, design, V&V;
- Requirement engineering process:
  - Process description: if relevant, detail of the process at system, software and hardware level; requirement capturing, elicitation techniques, analysis methods, requirement documentation, management of changes and requirement evolution;
  - Traceability: the purpose of traceability in the process (e.g. impact analysis, change management, etc.); integration in the RE tools set;
  - Consistency and completeness of requirements;
  - Reuse: what reuse means for each industrial partner at RE engineering level and how each partner reuses RE artefacts;
  - Non-functional requirements: the classes of non-functional requirements that each partner handles and briefly how to handle them;
  - Interoperability: is interoperability within RE process supported across projects/companies?
  - RE languages: how requirements are represented: text, structured text, models ... If requirements specification languages are used, which types

  of languages are used in which phases (formal semantics, executable, supporting synthesis, environment modelling, test case generation); selection criteria for RE languages;
  - RE methods and tools: if relevant, details on the methods/tools at system, software and hardware level; selection criteria for RE methods/tools; details on which tool capabilities are used for which purposes, how many people use the tools, scalability issue, how the tools interact; association of tools to the relevant method(s) and processes;
  - Examples: some examples of typical product requirements/requirement statements according to organizational rules and/or RE guidelines;

- Design process:

  - Process description: identification of the classes of applications that are addressed – e.g., control/command, protocols, . . . ;
  - Languages, methods and tools: which languages are used in which phases (UML, SysML, East-ADL, AADL, . . .); selection criteria for languages/methods/tools; description of which tool capabilities are used for which purposes; how many people use the tools; how the tools interact (exchange formats, etc.); association of tools to the relevant method(s);

- Implementation process:

  - Process description, at system, software and hardware level if relevant;
  - Languages, methods and tools: description of used languages/methods/tools and selection criteria for them;

- V&V process:

  - Process description, with focus on CESAR-relevant issues, i.e. V&V at RE process level, component-based V&V; details of analysis techniques that are used;
  - Languages, methods and tools: which tool capabilities are used for which purposes; how many people use the tools – scalability issue, how the tools interact;

- Configuration management process: process description; methods and tools;
- Documentation process: process description; methods and tools;
- Safety process:

  - Certification issues: certification and qualification requirements applicable in the current practice;
  - Process description;
  - Languages, methods and tools;

- Diagnosability:

  - Design an assessment: the main objectives of diagnosability and on-line diagnosis functions in each application; details on how the on-line diag-

  nosis functions are designed, coded, and distributed within the software architecture;
  - Procedures, methods and tools: methods and tools supporting diagnosability issues; selection criteria for methods/tools;

- Product lines: description of product line engineering activities.

Results and Remarks The questionnaire provided a detailed "picture" of the state-of-practice regarding the development of safety critical systems in the automotive, aerospace, automation and rail domains to the technical SPs, allowing to identify a common baseline and pointing out the different approaches and/or points of view with respect to similar tasks. In particular it can be noticeable that

- The *automotive domain* survey showed that even if the standards or languages (like the ISO DIS 26262, AUTOSAR, EAST-ADL) are the same, the applications in the organizations were different. This was due to the large scale of tools used (sometimes internal tools) and also due to the different roles and responsibilities of the organisations involved (depending if they are OEMs or tier suppliers (equipment or tool providers)). Regarding the safety activities, the methods were often the same, like "Hazard and Risk Analysis", "Failure Modes and Effects Analysis", "Fault Tree Analysis" but a large diversity was registered regarding the tools used.
- In the *aerospace domain* instead there were not substantial differences between the general engineering processes of the different partners. In particular, the processes were similar for companies situated at different levels in the customer/provider relationship (primes, equipment providers ...). In addition, even though the aeronautic domain has stronger certification constraints than the space domain, the actual deployed processes, as captured in the survey, were similar. The implementations of the process by the different companies slightly differed, depending on the chosen tools and methods. However, some means were used by most of the partners (e.g., DOORS for requirement management, Matlab/Simulink and Scade for algorithm development, software compilers ...).
- The state-of-practice contributions of the *rail and automation domains* partners were quite different, due to the different industry domain at which the partners belong. However, there were also some analogies regarding the applicable standards and the tools used for the definition of requirements. All partners related to safety standards (even if with some differences, depending by the domain) and the use of documents (tools such as Word or Excel), for the definition of the requirements, appeared as a common characteristics. Some drawbacks have been noticed by partners mainly regarding the inadequacies of some safety standards and the difficulties in the requirements elicitation and in the V&V process.

9.3.2 Collection, Prioritization and Harmonization of Requirements

After the assembling of a "picture" of the state-of-practice, it became a priority for the project capturing the needs and desiderata of domains SPs. The idea was using the proposed pilot applications as means to explain and motivate requirements on system engineering and tool integration, then harmonizing and prioritizing such requirements and finally using them for the development of the RTP platform. Owing that one of the main goals of the project was to provide common methods and integrated tools, the requirements had been analysed and clustered, exploiting commonalities first of all in each domain SP and then within the different domains (when possible). More in details, each industrial partner elicited requirements, starting from needs as expressed in their proposed scenarios. Each requirement was associated to rationale, priority, and preliminary allocation to categories. These requirements were then consolidated at the industrial domain level, principally merging similar requirements. Finally, a cross-SPs classification of the whole set of resulting requirements was performed in order to facilitate their handling. The result of these steps was a classification of requirements, with respect to:

- Interoperability: Interfaces between different level and tools;
- Traceability: requirement dealing with traceability;
- Requirements engineering: specific requirements regarding RE;
- Design: requirements dealing with the design process;
- V&V: verification and validation requirements
- Configuration management: requirements for configuration management
- Documentation: requirements for tools and processes on documentation
- Standards: requirements that arise from domain norms, standards, law, etc.
- General RTP: requirements with respect to the RTP
- Safety-diagnosablity: requirements regarding safety and diagnosability issues
- Reuse and product lines: requirements regarding reuse and product line engineering

For each requirements the following details were provided:

- Link with the Pilot Application and with the scenario;
- A description of the requirement;
- Rationale for process requirement;
- Category: standard categorization of requirements (compliance to safety or design and modelling standards, interoperability, engineering disciplines, eng. level);
- Process requirement level: direct or derived;
- Domain/subproject;
- Priority: mandatory, highly recommended, recommended;
- Stakeholder: identify technical SP, TF and domain SP;
- Added value: additional value due to the requirement;
- Verification/test: test and verification proposal for the requirement.

Table 9.3 Examples of requirements in the automotive domains

Description	Rationale
Requirements elicitation shall be supported	The RTP should therefore include support for both of these activities and the interaction between them
Verification that requirements at different abstraction levels are consistent with each other shall be supported	The upcoming standard ISO 26262 requires verification of requirements with respect to consistency with requirements at a higher (i.e. "earlier") abstraction level, so such verification needs to be supported by the RTP
Validation that the sets of requirements on all levels of abstraction are complete shall be supported	The documentation and to a certain extent also the formalisation of all expectations on the system is crucial input to the development process. Otherwise, important functionality may be omitted or has poor performance
Methods for analysis of causal relationships between (component-level) faults and (system-level) failures, such as FMEA, FTA, Markov models etc. shall be supported, at arbitrary abstraction levels from conceptual design to detailed implementation level	Such analysis methods are necessary for efficient categorisation, handling and prevention of hazardous events
Modelling according to the EAST-ADL framework shall be supported	EAST-ADL is an emerging de facto standard for modelling automotive systems at higher abstraction levels
Modelling according to the AUTOSAR framework shall be supported	AUTOSAR has gained wide acceptance of being a de facto standard for modelling automotive systems at the implementation abstraction level
Component-based development shall be supported	To have well specified components is a condition for reuse
Variation management shall be supported	In order to smoothly cope with product families where most parts are common, but there are differences in details, support is needed for this in the modelling language

As expected, some requirements were application dependent or company best-practices dependent, but using this approach commonalities within the requirements were enhanced. As example, in Tables 9.3–9.5 some high-level requirements are described, reporting a brief description, the rationale and the industrial domain.

9.4 Process of Evaluation of the CESAR Technical Assets

In every repetition of the innovation cycle for the project, the pilot applications were used with two different goals: at the beginning of each cycle the aim was identifying

Table 9.4 Examples of requirements in the aerospace domains

Description	Rationale
The RTP shall provide a meta-model which can support SysML features	
The RTP shall enable transfer of data and data extracts between the integrated tools and components	This is the core functionality of a backbone that provides the communication between different tools
The RTP shall be able to establish relations between data objects that are stored in the same or in different databases and managed by the same or different tools	Is required to connect data objects which are related. The linking of data objects enables the tracing of related objects Frequently data from different data sources are closely related. Thus association of application data enables clarification of dependencies and can facilitate change management. Required to establish a fully integrated data view
Requirements shall be well-formed	A "well-formed" requirement is defined as "a constraint applied to a property of an object [of the system] when a condition occurs (event) or is achieved (state)"
RTP shall provide a requirements capture language that expresses both informal (plain text) and formalized constraints (pre-/post conditions and invariants) of CBD component modules	Integrate the informal and formal specifications in a seamlessly understandable format by different engineering specialists: design, V&V, testers ...
The RTP shall support an architecture build up based upon CBD paradigm and formalism	Improve emergence of several alternative solutions to a set of requirements
The RTP shall provide configuration management capability (to manage in configuration all the artefacts:models, requirements, documents, problem reports, links ...) coupled with a change management capability to add automatically evolution information for each modified or created artefact	Capability to manage under configuration all the components of the development process and allow their reuse following CBD approach. When an item is automatically generated or updated, the RTP shall link evolution information to the item (in the configuration management or within the item code, file or model ...)

and describing the industrial partners needs through the requirements with respect to the RTP; at the end of each cycle the provided platform was evaluated, performing the practical industrial applications, already defined, to refine the needs and to improve the next version of the RTP.

In this section, we describe some practical aspects of the evaluation phase, starting from the instantiation of the generic RTP into a "domain" RTP (in Sect. 9.4.1), examining a tool-chain application example (in Sect. 9.4.2) and finally reporting the evaluation of the project technical assets (Sect. 9.4.3). We decided to illustrate these results at industrial domain level, instead of describing the evaluation results for each application example in details. In this way we would offer to the reader an overview of the major achievements of the project, from the point of view of the "end users".

Table 9.5 Examples of requirements in the automation and rail domains

Description	Rationale
The RTP shall support methods to analyze risks covering bottom-up as well as top-down approaches (e.g., HazOp, FMEA, FTA)	Risk and hazard analysis required by safety regulations
Traceability between requirements, use cases, design elements, test cases, test results, source code should be supported	This is a key requirement in the safety development process. Both forward and backward traceability should be supported
The RTP shall support a design process compliant with ISO 13849 (PL = e) – Safety of machinery (General principles for design) – to the extent applicable	Confermance to standard
The RTP shall support methods to analyze risks covering bottom-up as well as top-down approaches (e.g., HazOp, FMEA, FTA)	Risk and hazard analysis required by safety regulations
The RTP shall support the integration of new development and existing tools, by providing a documented and public interface	Tools that are not integrated in the RTP during the project shall be integrable into the RTP. All partners may not use all development tools available on the market. Companies that are using tools not used by any CESAR partner shall have not problem to integrate their tool of choice in the RTP

9.4.1 Process of Tailoring

As described in Chap. 6 (in particular in Sect. 6.6) and in Chap. 8, CESAR has defined an interoperability specification, which addresses all relevant aspects of tool integration. This specification defines how tools interact, by defining communication standards and protocols, basic infrastructure services, and data definition and handling. So the CESAR approach is vendor neutral, provides a comprehensive view on the tool integration dimensions and allows the flexible creation of development solutions which take the specifics of the stakeholder's development environment, organizational structures and existing tool landscape into account. The reader may refer to Chaps. 6–8 for details.

The adapting step regroups all the activities required for the instantiation of a dedicated tool-chain for a specific purpose. In this Section, we recall and briefly illustrate the practical activities required for the adapting of the generic assets (e.g., meta-model, tool adapters, data backbone). The reader may refer to Sect. 6.6 for details on RTP instantiation and tailoring.

- The first step of the adapting activities consists of the definition of the development process and the decision of the required development steps and respecting tools. This includes further the deployment of the data backbone (the so-called *RTP-Bus*, for example ModelBus – see Sect. 7.2.1) and its configuration according to the tools to be integrated, the design of a workflow as interaction pat-

tern of the platform services (*orchestration*) and model-transformation services required.
- The second step concerns the definition of the product meta-model and regroups two sub-activities. The first sub-activity is the identification of the data structure according to the included development steps, the respective viewpoints modelled and the tools involved. This data structure is a union of all the data structures of the tools involved. The second sub-activity is the identification of the required traceability links. These links (especially across the tools) define the relations between the development steps and therefore the required model transformations.
- The third step consists of the implementation of the specific tool adapters. Such tool adapters provide standard services for the integration to the RTP-Bus (e.g., check-in, check-out), as well as syntactic and semantic transformation in order to map the tool-specific data to the RTP instance.
- The last step concerns the evaluation of the integrated tool-chain according to typical product development activities.

The CESAR meta-model (CMM) (see Sect. 6.4) is intended to be used as a shared common model of data in an interoperability platform for tools with heterogeneous data sets. Thus, one crucial issue to enable interoperability among tools in a adapted RTP is the integration of foreign meta-models and to make them compatible with the common meta-meta of the RTP, the CMM. The most important step of such a meta-model integration in the adaptation process is to define a mapping between the foreign meta-model and the CMM. This enables common understanding of the foreign meta-model within the RTP using the CMM. The reader may refer to Sect. 7.4.2 for an example on mapping between EAST-ADL and CMM

Combining all entities to an *integrated development environment* takes several practical steps

- The first step is the setup of the generic RTP-Bus platform. It operates as data backbone connecting all internal services (like transformation services, orchestration, and security) and external services (mainly the tools connected by tool-adapters) necessary for the desired development process. These services and the RTP-Bus itself are usually deployed on a dedicated server machine.
- In the second step tool adapters are provided to the tools fulfilling the development process. Tool adapters can extend tools by acting as plug-ins or being standalone applications connecting the data management of the tool (usually running on the computers of the development engineers) with the RTP-Bus platform using a local computer network.
- The third step is the configuration of the platform services. Providing the necessary transformation rules to the transformation services is the most important step to enable the model-based data exchange. This also includes the configuration of the traceability management. To automate certain parts of the development process and platform background activities, a workflow is designed by a BPMN orchestration scheme as interaction pattern of the platform services.

Now the tool-chain setup is completed.

9.4.2 An Example: A PL Tool-Chain Applied to the SDSC System.

As example, we report here details on the application of the Product Line tool-chain, introduced and described before in previous chapters, to the Doors Management System, introduced in Sect. 9.2.2. In fact, the monitoring system takes into account the number of doors, the sensors and actuators included in each door, and any other internal component that is needed to keep the aircraft safety restrictions. We decided to report here only some practical details; for other specific aspects (related on how this kind of decision model may support the decision process), the reader may refer to Chaps. 2 and 3, in particular to Sect. 3.6.

Some of the identified variability points are:

1. The capacity of the aircraft could be medium or large. Selecting the large capacity implies a higher number of doors, and thus, a higher number of sensors to monitor doors.
2. An optional feature is the possibility to have Real time Communication to ground, but only in the case of a large aircraft.
3. For each system component (sensors, actuators, central units), different suppliers can be selected. Their selection implies changes in the component features (e.g. mass, costs, failure rate).

These variability points belong to the different input models. The first two variability points correspond to the input model of the *high level detail*, whereas the third one corresponds to the variability of the *detailed level.* An example of the impact of the decisions from one level to the other can be seen in Fig. 9.10. Two possible scenarios are represented: a medium and a large aircraft configuration. The selection of the capacity will define the number of doors of the aircraft (2 for the case of the medium 1, and 4 for the large configuration). Each door needs a specific number of sensors and HW/SW components. So, the selection of the aircraft capacity needs to be propagated to the detailed level.

This fact is reflected in the different number of HW/SW components of the detailed level (4 in the case of the medium configuration compared to 8 in the large version).

If the aircraft design includes Real Time Communication to ground system (solved in the high level design), new components need to be configured within the detailed level design. Moreover, there are variability assets that are solved only in the detailed level, such as the selection of the supplier.

In this example, the PL tool-chain shows how the different variability points can be managed at different levels and propagating them along the different phases. This tool-chain shows the feasibility of the solution in a real scenario.

High Level Variability Management

The first activity to be carried out is the definition of the features of the system. In this scenario that will be followed to show the tool chain functioning/process is based on two main features at vehicle level:

- Large or Medium capacity aircraft (of which one choice has to be made)
- Real-time to ground communication (which is optional but only available for the large capacity aircraft).

The possible configurations are:

- Medium Aircraft
  - By default (implicit variability and commonalities)
    - 250 seats
    - Small number of doors (i.e. 2)
    - Low-end controllers
    - Small number of SW & HW components
- Large Aircraft
  - The Real-time communication to ground system is optional
  - By default (implicit variability and commonalities)
    - 350 seats
    - Large number of door (i.e. 4)
    - High-end controller
    - High number of SW & HW components

The information flow and work products of the High Level Design may be summarized as follows:

1. Definition of the *decision model* depending on the constrains and variability specification of the domain. The main goal is to identify the variability within the model.
2. From the Decision Model, new instances (*application models*) can be obtained in order to define different plane configurations (application models). The variability is solved by the user, and the constraints/dependencies defined in the SDSCS system are automatically executed. As a result depending on the decisions made we obtain different kinds of products as depicted in the following Figure
3. Once the user has selected the plane configuration, is time to execute the transformation to GET the final results. This transformation manages all the variability solving within the product line architecture.

   As result of this phase, there are two main outputs:

- The “High level” architecture models with variability solved representing a concrete product. In the case of the SDSCS example this is a Papyrus + EAST-

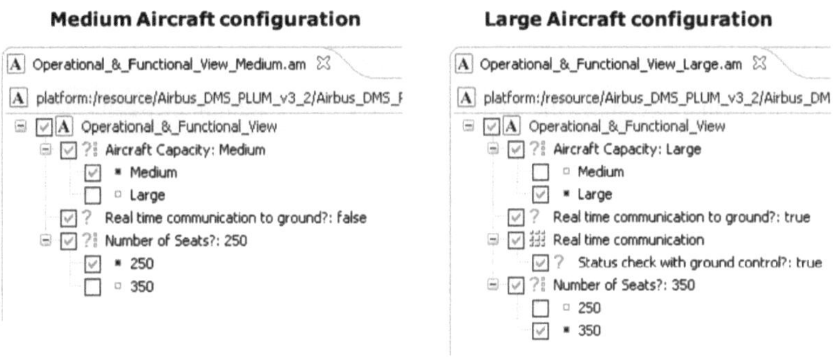

Fig. 9.10 Example of Application Models representing concrete products

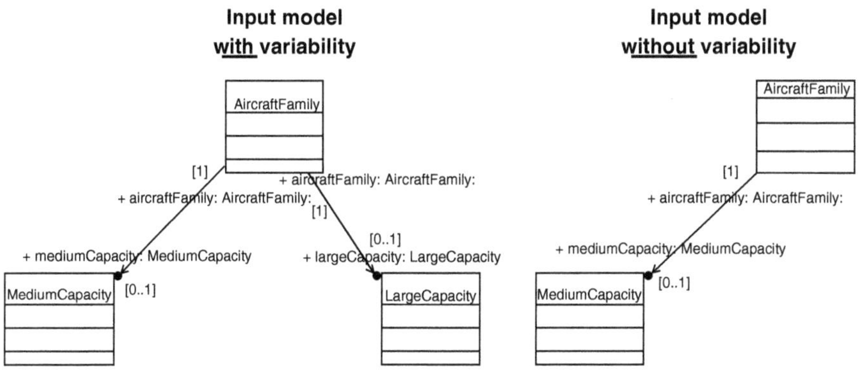

Fig. 9.11 Input and output model from the High Level design resolution phase

ADL model without variability vs the initial "High Level" model containing variability of the product line as shown in Fig. 9.11
- The propagation of "High Level" decisions into "Detailed Level" variability model. The values of the high level decisions made with the PLUM tool are propagated to the detailed level, concretely to the CVL tool variability Model.

The CVL tool is capable of actually executing all the previous defined bindings that have been defined at this level of the system. When talking about binding, we express that for each variability defined in the EAST-ADL UML model with CVL tool we can describe exactly how the design changes based on the decisions propagated from the High Level design. CVL based on this information can automatically resolve de low detail variability by executing the actions required based on the decision for each specific variability defined in the low level design. There are specific low level design variability decisions that may need to be done within CVL Tool prior to obtaining the final product, resulting in another more detail low level design.

9.4.3 *Results of Evaluation by the Industrial Domains*

As declared on page 325, the evaluation of CESAR technical assets is reported from the point of view of the industrial domains "end users". For each domain (automotive, aerospace, railway, automation) we report a summary of the evaluation results, obtained by using the RTP in the whole set of application examples. In this way we allow the reader to have a look on the evaluation of the whole set of the project technical assets. Moreover, by comparing the evaluation in different domains, the reader may notice the various points of view (depending on the industrial domain) with respect to the provided methods, tools and tool-chains. Some CESAR outcomes received, in fact, a very good evaluation in every industrial domain. So are, among others, the ontology-based requirements engineering methods and tools, the traceability, the product line capabilities.

9.4.3.1 Summary of Automotive Evaluation

The tools and the tools integration were addressed in the evaluation, but little additional evaluation of 'methods' has been performed except for the cases when a method is tightly related to a tool. The only evaluation item that could (at least loosely) be described as a method is the functional safety framework that has been developed in CESAR based on the ISO 26262 standard.

The use of the ModelBus as the backbone of the RTP appeared to be a good choice. It allows several tools to access and manipulate the same system model(s). However, actually the RTP[9] does not take full advantage of this since tool interoperability is typically not based on a common meta model but rather based on point-to-point tool interconnections, sometimes not using the Modelbus at all. Furthermore, some of the tools are stand-alone i.e. there is no interoperability provided by the RTP for these tools. The interoperability between the tools is therefore limited both in terms of scope and capability.

The tools in the RTP cover most parts of a classical V model for system development, from requirements tools like DOORS and DODT, via modelling and simulation tools such as Rhapsody, Papyrus, SCADE and Simulink, to verification tools like the BTC Tool Suite, Metrino and AVL InMotion. Furthermore, the RTP includes dependability-oriented tools such as OpenFTA, FaultTree+, BPA-FT9 and BPA-SD9. Finally, a number of tools for keeping track of the overall process and artefacts are included, like Link Repository, Permeter and RTP-Desktop.

Regarding the suitability of the individual tools (and one method) that have been evaluated in the automotive domain subproject of CESAR, the following are our main observations:

[9] With *actual release* of the RTP we mean the final release during the CESAR project.

- The DODT tool and the associated boilerplates methodology have been found to be highly useful and valuable. There is a learning threshold that could mean that the efforts for formulating requirements are increased compared to conventional techniques, but this seems to be counteracted by the higher quality of the requirements and the corresponding savings in subsequent stages of the development process including reduced possibilities for misinterpreting requirements. The possibility to define and use ontology, supported by the tool is expected to lead to improved quality of the requirements.
- The modelling of system architecture in Papyrus according to the EAST-ADL methodology leads to unambiguous models and simplifies the interaction between different teams in a development project. The formalization of the system architecture based on EAST-ADL enables automated analyses to be performed, for example checks that all components are mapped to requirements and test cases. Many of the weaknesses in Papyrus 1.12 have been solved in the newer version Papyrus MDT, which therefore represents a significant improvement over version 1.12.
- The Link Repository works quite well and may be used for managing traceability between artefacts created by different tools.
- The RTP-Desktop tool still seems quite immature and user-unfriendly (and difficult to install). However, if these deficiencies are rectified the tool is expected to provide significant benefits concerning the efficiency of the development process.
- The MBSA tool (Model Based Safety Analysis) can analyse Stateflow models with respect to whether some defined condition can be fulfilled or not. It can also generate fault trees and FMEAs from Stateflow models in which the intended function and various fault conditions can be represented. However, we have not yet been able to assess the actual benefits associated with the fault tree generation.
- The RTP supports Matlab/Simulink/Stateflow by providing tool-chains where information can be exchanged between Matlab and other tools. Since Matlab is extensively used in the automotive industry, this support is very valuable.
- The Permeter and its concept are very promising. For project aspects represented in the CMM format, the tool facilitates monitoring of project progress.
- The functional safety framework, based on the automotive standard for functional safety ISO 26262, provides a ‘blueprint’ for the functional safety work and supports in the functional safety planning and follow-up in development projects.
- The CESAR Meta Model (CMM) lacks support for most aspects highlighted in the ISO 26262 standard. This obviously reduces the value of the CMM for the automotive domain.

9.4.3.2 Summary of Aerospace Evaluation

The evaluations led by the aerospace industrial partners allowed to assess engineering methods, engineering tools and tool chains. It’s worth noting that one third of the technical assets was evaluated “ready for the industrialization” and that more than half of the technical assets have been ranked “interesting, but new research activities are necessary before starting the industrialization”.

Only 8 % of the high priority aerospace requirements assessed were not conclusive. It means all the industrial partners could conduct the assessments they have scheduled at the beginning of the project with minor adjustments.

The experiments conducted by the aerospace partners mainly demonstrated:

- The component based approach supported by the model based component tool chain is relevant but necessitates to be extended and matured before deployment in an industrial context
- The ontology-based requirements engineering is very promising if an appropriate domain ontology is defined
- The link repository provides a powerful generic traceability mechanism that can be used to link elements of different types, created in different tools throughout the whole engineering lifecycle
- The CESAR multi-views modelling approach based on the IEEE 1471 standard allows supporting model based system concurrent engineering. Nevertheless, some methodological adaptations have to be done to make the approach fully operational
- The safety capabilities provided in the CESAR project can support a classical safety process from a dysfunctional model generating FTA and FMEA. Nevertheless, more integration of the different tools used is necessary to become fully efficient and bought-in by the safety designers
- The product line capabilities delivered in the frame of CESAR based on PLUM/CVL are promising even if the maturity of these assets do not allow deploying the underlined technologies immediately
- The various modelling tools available on the market (COTS and Open Source) can support a multi-views design modelling approach. Nevertheless, developing specific DSL (Design Specific Language) is an industrial approach followed successfully by some partners
- Some capabilities of test case generation on C code showed their relevance. Indeed they permit to verify automatically requirement and to generate and simulate Test case
- The Model Bus interoperability features proposed are relevant. But, industrialization step is missing to be able to deploy the solutions promoted by CESAR.

9.4.3.3 Summary of Automation and Rail Evaluation

The major achievements of the RTP are related, in the opinion of automation and rail domains partners, mainly to the requirements engineering methods, practices, guidelines and tools and to the product line engineering methods and tools. In details, the experiments pointed out the relevance of

- Implementation and use of the domain ontology design method, which contribute to avoid misunderstandings that could rise between the involved stakeholders;

the DODT tool shows promising properties when it comes to writing clear and consistent requirements. Using ontology to define and describe major concepts improves understanding and reduces misconceptions among system designers, software engineers and testers. In addition the tool provides several mechanisms to help improve requirements construction like for example the Boilerplate methodology. Moreover, it's a common belief that the disciplinary effect of being forced to take an active standpoint to each and every concept being used, in long term will improve the requirements engineering process.

- Analysis of requirements (e.g. consistency, completeness verification) which contributes to reduce the number of errors usually present and thus to find and solve defects at early stages.
- Requirements management and related traceability between high level and low level requirements increase the awareness of the reported changes, and improves the reaction time to changes.
- The Investigation using the Papyrus/CVL tool has demonstrated the advantages of using a product line engineering methodology to improve the development process where a product have several variants depending on the context in which the product is going to be deployed.

These evaluated methods and tools, at this stage, represent a high potential for the usability within the companies and for the improvement of the usually implemented development processes.

As far as it concerns the process configuration, tailoring and related management activities, the RTP integrates tools and services that consent to deploy correctly the development process, by providing a knowledge base for educating teams on the methods applicable to the roles for which they are responsible, as well as process engineering capabilities in order to show how to apply these methods throughout the defined development lifecycle.

Although, the technology have to be matured, and the related method content has to be extended and adapted to specific domains, it represents a potential for management of centralized multiuser support database allowing a complete share of the process information, thus improving the integration and communication between the involved parts (e.g., stakeholder, requirement engineer, system engineer, test engineer, project manager, sales engineer etc.).

Chapter 10
CESAR Results

Eric Armengaud, Gerhard Griessnig, Ingrid Kundner, and Fabien Paganelli

CESAR is among the larger projects executed in the European Information Society Technologies program. Around 450 experts from 55 institutions have been working tightly together for over 3 years to provide a large number of outcomes in the areas of requirements engineering, component-based engineering and tool integration for the automotive, aerospace, rail and automation domains. These outcomes have been documented across 195 deliverables in the project. To understand the outcomes and their impact, in this chapter we (1) summarize the CESAR results and provide high-level metrics for characterizing the CESAR project, and (2) perform a mapping between CESAR outcomes and the strategic goals stated in the technical annex and in the ARTEMIS multi-annual strategic program (see Fig. 10.1). The high-level metrics discussed in this chapter provide a way of quantifying the following:

1. **Technical outcomes** such as new or improved tools or methods,
2. **Expertise** such as publications, training,
3. Outcomes having a **market or community impact** such as establishment of ecosystem for tool integration.

Figure 10.1 illustrates the high-level metrics and their organization. Metrics (left) are identified and described according to their definition, value, and impact. These metrics are then grouped, as stated above, with respect to technical outcomes, expertise or market and community impact (middle). In the sections that follow, we describe and discuss the different metrics shown in Fig. 10.1.

E. Armengaud (✉) · G. Griessnig · I. Kundner
AVL List GmbH, Graz, Austria
e-mail: eric.armengaud@avl.com

F. Paganelli
Sagem Défense Sécurité, Paris, France

A. Rajan and T. Wahl (eds.), *CESAR - Cost-efficient Methods and Processes for Safety-relevant Embedded Systems*, DOI 10.1007/978-3-7091-1387-5_10,

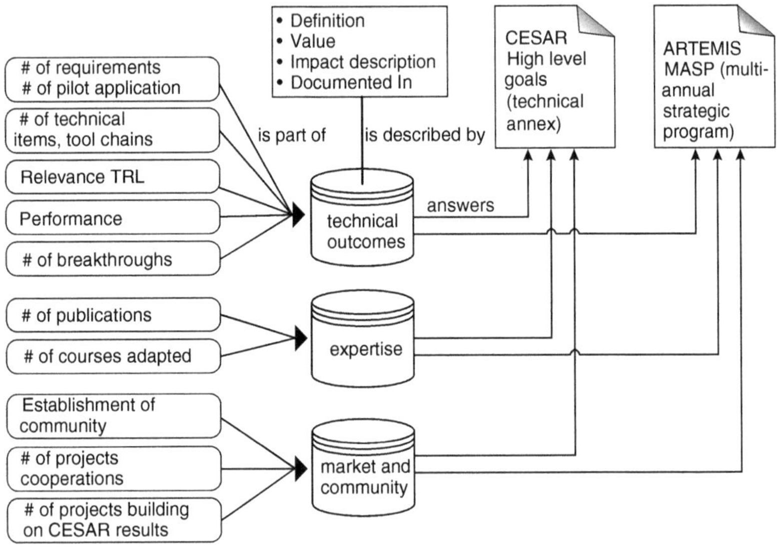

Fig. 10.1 Organisation of the CESAR metrics

10.1 Technical Outcomes

The first group of metrics quantifies the technical outcomes of CESAR. These metrics highlight the (1) technical expectations of the project, (2) total number of outcomes, (3) impact of the proposed solutions for the development of embedded systems, (4) correspondence between expectations and results, and finally, (5) convergence of the proposed technical solutions beyond domain boundaries.

10.1.1 M1: Number of Requirements on CESAR *Technology*

This metric quantifies requirements elicited by the various domain sub-projects (SP5/SP6/SP7) and respective task forces. It also quantifies the number of pilot applications that have been used to describe the needs of the end-users. The value of the metric is shown in Table 10.1.

It is clear from the value of M1 in Table 10.1 that the CESAR project identified a large number of needs from the application domains addressed (Automotive, Aerospace, Rail and Automation). It is also worth noting that 45 high-priority requirements were identified on account of to their relevance, feasibility, and commonality across domains.

Table 10.1 Value of metric M1

	Automotive	Aerospaace	Rail and automation	Total
Num. of pilot apps.	7	12	4	23
Num. of reqs.	138	376	130	644

10.1.2 M2: Requirements Coverage

Metric M2 quantifies the coverage achieved for all CESAR requirements and for high-priority requirements. The requirement can be classes as being addressed ("passed", "partially passed" or "inconclusive"), not addressed ("failed" or "not implemented") or not assessed. This metric describes the fulfilment of end-user expectations. Figure 10.2 illustrates the results for requirements coverage.

The results in Fig. 10.2 indicate that requirements coverage overall for CESAR requirements is low (37 %). This can be attributed to the large number of requirements as well as the broad areas of application they cover. The coverage of high-priority requirements, on the other hand, is high (>70 %), illustrating the importance placed on focusing on fewer, high relevance requirements in order to steer the project towards main user needs.

10.1.3 M3: Number of Technical Items (TI), Tools, Tool Chains

This metric represents the number of technical items. A technical item represents the smallest entity of technical outcomes. It may be a new method, new process, part of a tool enhancement, new tool, etc. The metric also quantifies the number of tool chains (tools are integrated into tool chains according to CESAR technology).

During the course of the CESAR project, a total of 137 technical items have been developed. Among these, 69 technical items have been selected for integration into tool chains. The tool chains represent the technologies after integration and are the main outcome of the CESAR project. The project developed 14 tool chains.

10.1.4 M4: Detailed Technical Outcomes

This metric is used to assess the RTP. Figure 10.3 illustrates evaluation of the tool chains. The Technical Readiness Level (TRL) is an indicator of the maturity of the respective TI or tool chain – how far they can be integrated within a company's internal development process for customer projects. TRL of 6 or 7 is usually required for use in customer projects. The relevance is an indicator of the industrial interest in this technology. Combined with the TRL metrics, this provides

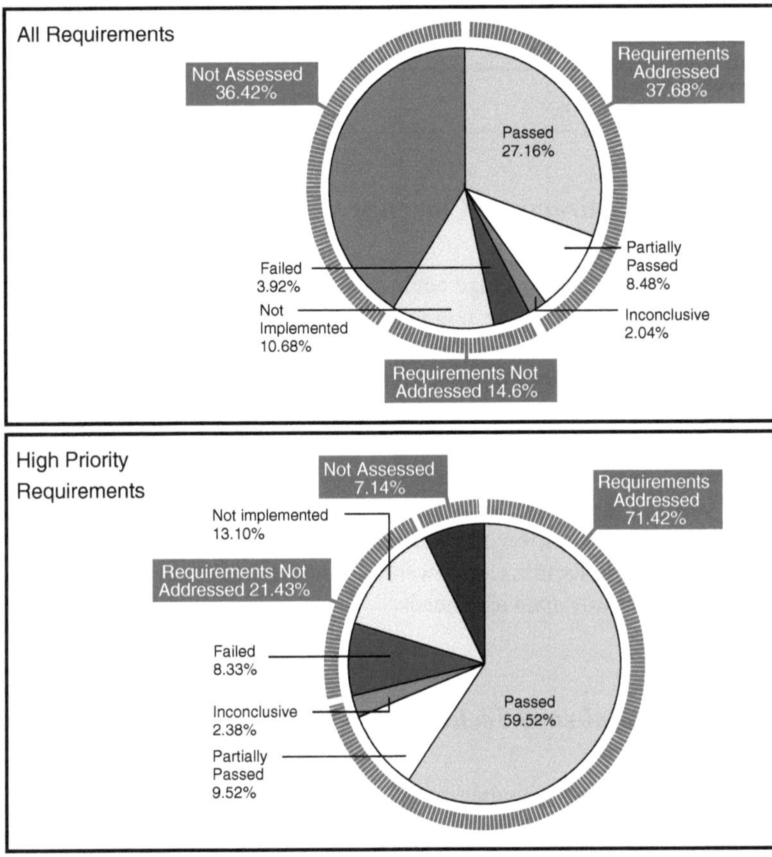

Fig. 10.2 M2 – requirements coverage

information about the industrial interest within the company in integrating this outcome and/or continuing research on this topic.

The cost/efficiency/quality performance describes the respective improvement provided by each technical item on the development of the embedded system.

The cross-domain acceptance describes the number of outcomes used by more than one application domain, thus illustrating the convergence across the domains by the methods proposed by CESAR.

10.1.5 M5: Cross Domain Acceptance

Cross domain acceptance represents the acceptance of the TIs by different domains and is defined as number of technical items evaluated by more than one domain.

Tool chain	TRL	Relevance	Cross domain acceptance	Cost performance	Efficiency performance	Quality performance	Number of assessor
Tool chain #1 Safety & Interoperability Specification	4.15	2.18	2	2.28	2.92	2.46	4
Tool chain #2 Product Lines Management	4.60	2.08	2	2.70	2.68	2.65	4
Tool chain #3 Automotive Test Management	5.46	2.16	2	2.76	2.72	2.84	5
Tool chain #4 SCADE Modelling	7.00	3.00	1	3.00	3.00	3.00	1
Tool chain #5 X-MAN Modelling and verifying	4.18	2.04	2	2.65	2.61	2.68	4
Tool chain #6 Detailed Design	6.13	2.63	2	2.88	3.00	2.88	3
Tool chain #7 Architecture exploration	3.00	2.00	1	3.00	3.00	3.00	1
Tool chain #8 Architecture optimization	3.00	2.00	1	3.00	3.00	3.00	1
Tool chain #9 Model checking of safety critical behavioural models	4.38	2.09	1	2.25	2.22	2.19	3
Tool chain #10 Automotive domain Airbag (VHDP, VFDP)	6.16	2.58	2	3.29	3.44	3.49	4
Tool chain #11 Model-based Safety Analysis	5.01	2.09	2	2.46	2.43	2.50	4
Tool chain #12 Requirements Engineering	5.49	2.23	3	2.80	2.98	3.16	6
Tool chain #13 Product Family Management	6.00	2.00	2	3.00	3.00	3.00	2
Tool chain #14 Test Case Management	6.00	2.50	1	3.00	2.00	3.50	2

Fig. 10.3 M4 – detailed technical outcomes

Figure 10.4 shows that 50 % of the CESAR outcomes have an impact on at least two application domains thus illustrating the convergence over the domains by the methods proposed by CESAR.

10.1.6 M6: Technical Readiness Level (TRL)

Technical Readiness level was previously defined in M4 detailed outcomes. The goal of this metric is to provide a technical item centric view (instead of tool chain centric) and illustrate the overall maturity of the technical outcomes. Two ranges of TRL have been identified to illustrate the technical readiness of the CESAR technology:

- Mature prototype: equivalent to TRL 4 to TRL 6, illustrating a technology that can be evaluated in an industrial context.
- Ready for industrialization: equivalent to TRL >7, illustrating possible integration in customer projects.

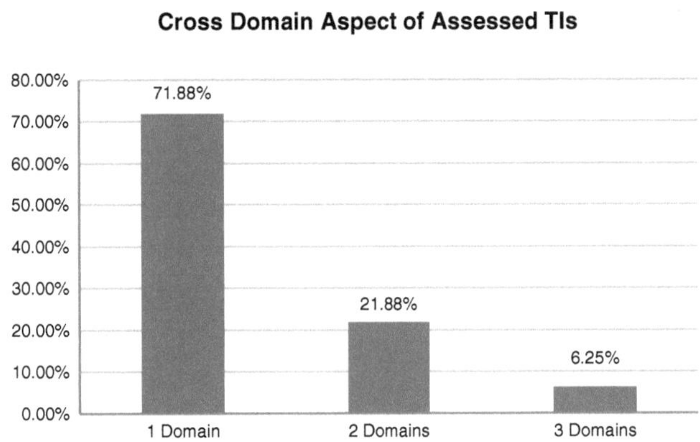

Fig. 10.4 M5 – cross domain acceptance

Figure 10.5 illustrates that the great majority (>70 %) of the CESAR outcomes have a TRL between 4 and 6, thus illustrating the maturation of the CESAR outcomes from early concept to mature prototypes. Additional efforts are required to industrialise these outcomes.

It is worth noting that 17 % of the TIs have a TRL of 7, indicating they are ready for industrialisation. These TIs have great potential for increasing market share and improving quality of support tool.

10.1.7 M7: Cost Performance

Cost performance is defined as the percentage of TIs leading to a decrease in development costs (either level 3 – positive but not so far as requested, level 4 – according to technical annex, or level 5 – overachieved). This metric provides a technical item centric view (instead of tool chain centric). Note that TIs with an average cost performance level of 2, 5 have been taken into account (it indicates that on average, at least half of the partners gave an evaluation result of 3 or 4).

Figure 10.6 shows that more than 68 % of TIs have an impact on the reduction of cost for the embedded systems development, illustrating the strong focus of the CESAR project on cost reduction. More than 15 % of the TIs support the cost target set at the beginning of the CESAR project.

10.1.8 M8: Efficiency Performance

Efficiency performance is defined as the percentage of TIs leading to an increase of the development efficiency (either level 3 – positive but not so far as requested,

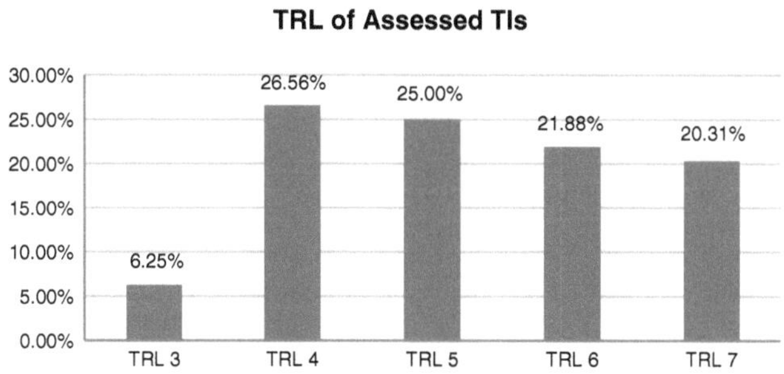

Fig. 10.5 M6 – Technical Readiness level (TRL)

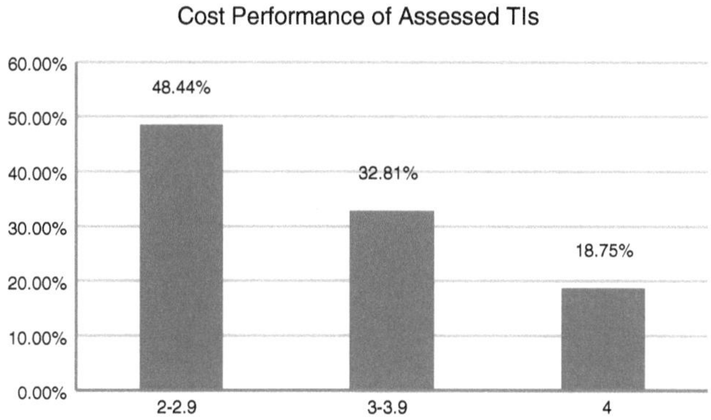

Fig. 10.6 M7 – cost performance

level 4 – according to technical annex, or level 5 – overachieved). This metric provides a technical item centric view (instead of tool chain centric).

Figure 10.7 shows that more than 84 % of TIs improve the efficiency of embedded systems development. This result showcases the strong focus of the CESAR project on efficiency improvement. More than 15 % of the TIs support the efficiency target set at the beginning of the CESAR project.

10.1.9 M9: Quality Performance

Quality performance is defined as the percentage of TIs leading to an increase of the development quality (either level 3 – positive but not so far as requested, level 4 – according to technical annex, or level 5 – overachieved) . This metric provides a technical item centric view (instead of tool chain centric).

As seen in Fig. 10.8, more than 72 % of TIs have an impact on the improvement of the development quality of embedded systems, illustrating the strong focus of

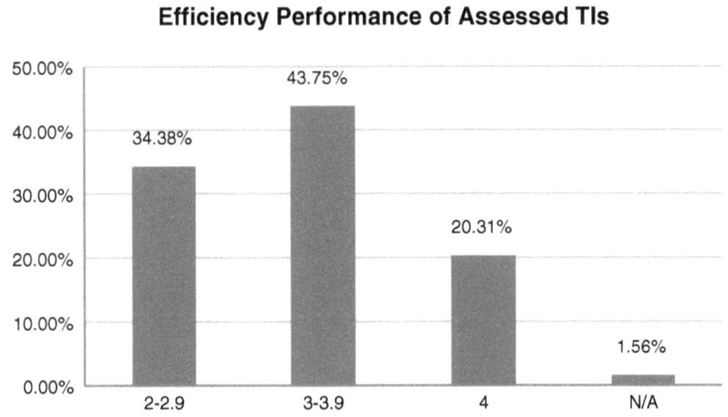

Fig. 10.7 M8 – efficiency performance

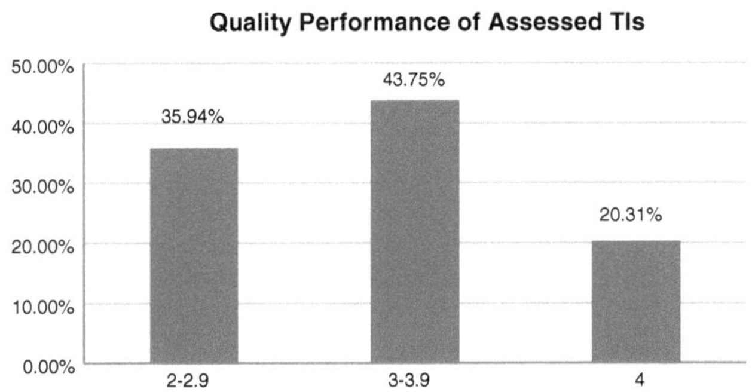

Fig. 10.8 M9 – quality performance

the CESAR project on quality improvement. More than 17 % of the TIs support the quality target set at the beginning of the CESAR project.

10.1.10 M10: Interoperability Specification Document

Creation of interoperability specification document is the first step towards the creation of an industry de-facto standard on interoperability. The Interoperability specification document summarizes the return of experience within CESAR for the interoperability topic and proposes a first step on which further research projects can build on.

10.1.11 M11: Number of Major Breakthroughs

This metric quantifies major breakthroughs showing a visible impact on the ES industry and on the processes for the development of safety-critical systems.

There were 37 major breakthroughs illustrating high innovation and relevance of the CESAR outcomes. It also shows potential for long term impact on market share for embedded systems industry, as well as on the number of technology providers/SMEs for these technologies.

10.2 Expertise

To quantify expertise, we use metrics that illustrate the impact of CESAR on growth and management of knowledge. The CESAR consortium provided a unique constellation of (1) renowned scientists, and (2) large scope of expertise. The combination led to numerous innovations and identification of new research directions.

The first metric we use is "**Number and Impact of Publications**". Number of publications in the form of Bachelors, Masters or PhD thesis is 44. Number of peer-reviewed publications is 65. The high number of publications illustrates the scientific impact of the CESAR project. Number of citations on average for a peer-reviewed publication was 5.8. These citations indicate the acceptance of CESAR outcomes in the scientific community. Finally, the distribution of the publications between academics/ industry partners/mixed sources is shown in Fig. 10.9. It can be seen that almost 50 % of the publications have been performed conjointly by academics/industrials illustrating the strong cooperation and know-how transfer between academics and industry partners.

The second metric we use is the "**Number of universities that have updated their courses in the context of CESAR**". There were seven such universities. Almost all the universities involved in CESAR updated their courses during the project lifetime. Adaptation of courses to train new engineers on the CESAR technology is important (a) for long-term sustainability of the main CESAR concepts, (b) for the further development of these methods and (c) for support during industrialization of main results.

10.3 Market and Community Impact

Given the relevance of the project consortium and the high budget, the expectations on CESAR are especially high with regards to impact on European embedded systems industry, specifically in the area of safety relevant applications, such as automotive, aerospace, rail and industry. Measuring the impact of the CESAR project on the European embedded systems market is a challenging task for the following reasons,

- The project primarily focused on improvement of methods and processes for existing product development. The direct relation to market impact is difficult to evaluate and measure. Additionally, this information is usually kept confidential.

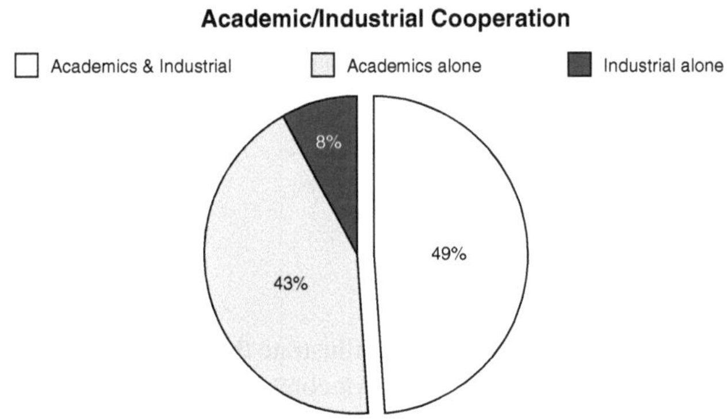

Fig. 10.9 M12 – number and impact of publications

- Industrialization of the results (either in term of improvement of company internal development or in terms of improvement of new development tool) is a long term process that will continue after CESAR end.
- The project has led to creation of new cooperation/strengthening of existing cooperation that will have an impact on the European market in the long term. Nevertheless, these results are not measurable now.

For these reasons, we have focused the CESAR metrics on the impact on interest groups, namely how sustainability of the achieved results has been guaranteed, and how the outcomes will be further enhanced to reach measurable impacts on the market in the long term.

The first metric in this context that was measured was '**Number of CESAR deliverables**'. This metric relates to the amount of information and knowledge generated during the CESAR project. The total number of CESAR deliverables was 195 with varying levels of availability. Twenty seven of these deliverables were publicly available (available at the CESAR website). Twenty one deliverables were publicly restricted (available via the process for exchange) for European research activities. The remaining 147 deliverables were only available within the CESAR consortium. This metric has a direct impact on the sustainability of the CESAR results.

The next notable metric is '**Established project cooperation**' which refers to the number of research projects that requested access to the CESAR deliverables. The total research budget for these projects was MEUR 76.9. This metric illustrates the impact of CESAR outcomes on the European research on critical embedded systems and the regrouping of projects into ecosystems.

The final metric we discuss is '**Projects building on CESAR results**'. The research projects building on CESAR results is shown in Fig. 10.2. There are ten such projects with a research budget of MEUR 181 from approved projects and a budget greater than MEUR 70 from projects in preparation. The large number of research projects building on the CESAR results as well as the size of these projects

Table 10.2 M18 – projects building on CESAR results

Acronym	Phase	Total investment	Coordinator
MBAT (ARTEMIS)	Running	MEUR 34.5	Daimler
pSAFECER (ARTEMIS)	Running	MEUR 10.5	Volvo
nSAFECER (ARTEMIS)	Running	MEUR 16.3	Volvo
VETESS (ARTEMIS)	Running	MEUR 19	Infineon UK
iFEST (ARTEMIS)	Running	MEUR 16	ABB Norway
CRYSTAL (ARTEMIS)	Proposal	>MEUR 70	AVL List GmbH
SPES 2020XT (National German)	Running	MEUR 25	TU Munich
ARAMiS (National German)	Running	MEUR 37	KIT
OpenCOSS (FP7)	Running	MEUR 11.7	Tecnalia
SAFE (ITEA 2)	Running	MEUR 14	Continental

(investment) and relevance of the respective consortiums illustrate (1) the strong impact of CESAR results in this domain and (2) the relevance of these topics for the industry (Table 10.2).

10.4 Lessons on Project Management

Thus far, we provided a technical synthesis of the CESAR outcomes. Nevertheless, we feel this work would not be complete without providing some concluding remarks on the management of CESAR. Managing this large research and technology project gave us several insights that we believe will be useful to similar initiatives in the future.

In order to meet the project goals, "convergence" was the key mission of the project management in its objective to lead a multi-domain project with different stakeholders. Convergence refers to the ability of a critical mass of partners to cooperate and synchronize as necessary to be able to share and meet the project objectives as defined at the beginning, while taking into account the evolution of the latter over the life of the project. Convergence had be achieved in a cooperative context for CESAR in the presence of the following aspects: (1) partners with potentially different interests running into difficulties when requesting others for contributions that they may not be keen on (non directive management), (2) evolution of objectives in the course of the project that result in issues to be addressed that were not identified at the beginning, (3) potential difficulties due to customer-supplier relationships outside the project, (4) industrial or cultural differences between project actors due to their business specifics, (5) project complexity owing to the large volume of different technical topics to be addressed, (6) continuous consideration of tangible measurements of the objectives.

In order to ensure the convergence of the distributed multi-domain project in such a context, some KSF (Key Success Factors) have been identified and some management tools have been set-up. Below we highlight ten KSFs,

1. The first KSF is the necessity of a ***tightly synchronized and experienced management team*** with well-defined roles and responsibilities. It is necessary to achieve shared and robust decisions in order to get the best chance of application anywhere in the project. Further, it ensures that people in charge of actions get the necessary management support. This leads to internal reliability in the decision process and trust to emerge and get reinforced between project actors. It requires a lot of effort, energy and passion from the management team. On this basis, management efforts must be planned adequately before project start.
2. The second KSF is ***transversal communication***: a lack of communication between partners in charge of the evaluation and partners in charge of providing innovative technologies might create gaps and issues in the expectations leading to a decreased ability of the project to meet its objectives; undermining the return of investment of all stake holders. In case the original project structure did not plan an adequate organization for explicitly making this transversal communication to occur, it is necessary to create one or more with different life times as contextually relevant.
3. The third KSF is a ***working collaboration platform at project start***. In particular, at the beginning of the project a shared view about the contact points of the different organizations and the names of the contributors to the different tasks is essential to overcome the communication barrier. It is not sufficient to provide a "data storage center" but it is necessary to establish a "project management center". Not to forget the effort of maintaining, updating and encouraging the culture of keeping track of documents, contact points, access rights etc.
4. The fourth KSF is to build a project requirement baseline with a ***"low" number of clear, shared and high priority requirements rather than a large volume of requirements***, e.g. through a requirements capturing, validation and prioritization light-weight process. Too many requirements with the same priority level lead to exponential analysis time that may disrupt the connection between technology providers and partners in charge of evaluation, the latter usually providing these requirements. The more partners, the more industrial domains represented, the more difficult it is for technology providers to handle too many requirements. Therefore it is important to have a phase of prioritization and a phase of explanation (hand-over) of the requirements performed conjointly in small groups composed of technology providers and evaluators. The technology providers may need some explanations in order to really understand the meaning and relevance of a requirement. They also need to explain their own constraints in order to make sure the requirement baseline will be able to be technically covered. This is a matter of project efficiency while consensus on this between project stakeholders will help to converge.
5. The fifth KSF is to find proper trade-off between organisation and innovation. Hence, ensuring quality by defining communication rules and ***light-weight well understood process*** is strongly required. It is usually difficult to create a common understanding in such a heterogeneous environment. On the other

side, the process shall not be too complex/heavy and restrain the innovation character of the R&T project – some flexibility is required to be able to address new research topics. Nevertheless, such flexibility calls for appropriate management and monitoring in order to ensure the process will deliver expected outputs. Such an approach has been applied successfully in CESAR on different kind of processes. An example is the high priority requirement selection process which was implemented and monitored by the Convergence working group. Another example was the deliverable review process applied to each project deliverable with appropriate reviewers and monitored by the project management team.

6. The sixth KSF is to apply a stepwise approach in the construction of the project outcomes. Building 3 evolutions of the same content, such as the CESAR RTP, is fruitful in terms of lessons learned (technical and organizational such as issues as met in the deployment of technologies on the field). It helps the project a lot to converge. Otherwise, one single big outcome is very likely not to be in line with expectations. Below is a picture adapted from what we have applied in CESAR. This is a simple closed loop process that shows each iteration of project outcomes, being analyzed and provided as a feedback and lessons learned for the next iteration, see Fig. 10.10. Again, this revealed to be a useful approach to help in the convergence of the project. Testing of solutions provided by technology providers must be organized in order to obtain at least a sufficient level of confidence in the workability of the technologies as provided before evaluation. In this spirit, some integration testing should be thought out and applied. Otherwise, integration could be costly, undermining evaluation efficiency. In the end, this could impact the ability of the project to meet its objectives. With management support, having a light transverse structure composed with field people can be extremely useful in quickly detecting forthcoming issues and ensuring some fast loop feedback between evaluation people and techno providers.
7. The seventh KSF is ***guidelines and training on methods and technologies*** by stakeholders and skilled people. For complex integrated technical outcomes such as the CESAR RTP, the involvement of the stakeholders and skilled people for deployment in evaluation environments is highly recommended, e.g through the participation of IT department representatives.
8. The eighth KSF is a ***systematic breakdown of innovation contributions to the objectives***. In order to better structure and monitor the technical progress it is useful in an early project phase to divide the very "high level" project objective into tangible sub objectives and map these with the intended contributions by the partners. A top-down resp. bottom-up traceability method should be taken into consideration.
9. The ninth KSF is taking into account all contributions from stakeholders, i.e. also reviewers of the project. They may reveal to have fruitful ideas and their critics can be very helpful for management to move the project in a relevant direction. Obviously, ideas and critics must be discussed to reach a relevant trade-off between expectations and what is practically affordable by the project.

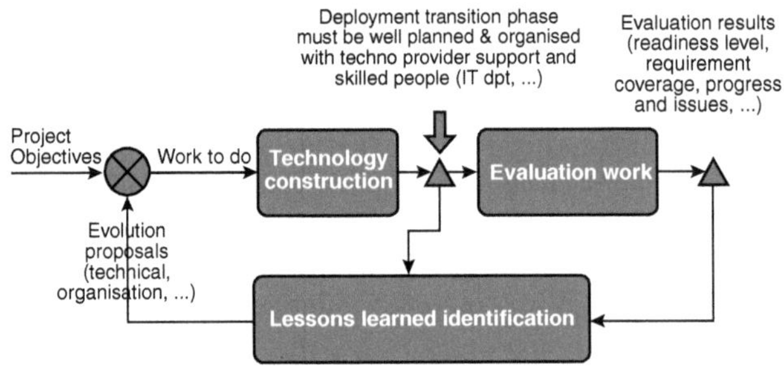

Fig. 10.10 CESAR iterative process

In this respect, it is suggested to build an "open issue" list that will be carefully followed-up all along the project. This is in some way the memory of the project and a good communication tool between the project and its reviewers in a constructive spirit.

10. The final KSF is to organize events during which stakeholders can get acquainted with one another and exchange ideas in a less formal environment. As everything relies first on people, it is important to ***create human links*** between the latter and maintain them over project life time. This is especially true at management level. This helps a lot to solve issues and misunderstandings that might arise during the project life and also generate useful ideas that can be discussed more formally in the project as relevant.

The above list of KSF, although not exhaustive, provides key insights gathered over three intense years of technical and management activity on the Artemis CESAR project.

List of Partners

Participant organisation name	Short name	Country
AVL list GmbH	AVL	AT
AIRBUS Operations GmbH	A-D	DE
AIRBUS Operations SAS	A-F	FR
ABB AS	ABB NO	NO
ABB AB	ABB SE	SE
AbsInt Angewandte Informatik GmbH	AbsInt	DE
ACCIONA Infraestructuras S.A.	ACCIONA	ES
Ansaldo STS S.p.A.	ASTS	IT
ASTRIUM SAS	ASTRIUM Satellites	FR
AIRBUS Operations Limited	A-UK	UK
Aristotle University of Thessaloniki	AUTH	GR
Commissariat à l'Énergie Atomique	CEA	FR
Centre National de la Recherche Scientifique	CNRS	FR
Centro Ricerche Fiat S.C.p.A.	CRF	IT
Critical Software S.A.	CSW	PO
Danieli Automation S.p.A.	DA	IT
Delphi France SAS	Delphi	FR
Deutsches Zentrum für Luft- und Raumfahrt e.V.	DLR	DE
Dassault systemes	DS	FR
European Aeronautic Defence and Space Company Deutschland GmbH - Defence Electronics	EADS-DE	DE
European Aeronautic Defence and Space Company Deutschland GmbH - Innovation Works	EADS-IW	DE
ELSAG DATAMAT[a]	ED	IT
Tecnalia Research and Innovation	TECNALIA	ES
ESTEREL Technologies SA	EST	FR
Fraunhofer Gesellschaft zur Förderung der Angewandten Forschung e.V.	Fraunhofer	DE

[a] Left project by June 2010
[b] Left project by December 2009

A. Rajan and T. Wahl (eds.), *CESAR - Cost-efficient Methods and Processes for Safety-relevant Embedded Systems*, DOI 10.1007/978-3-7091-1387-5,

Participant organisation name	Short name	Country
Auvation Software Limited	AUVATION	UK
Geensys[b]	Geensys	FR
Hellenic Aerospace Industry S.A.	HAI	GR
Infineon Technologies Austria AG	IFAT	AT
Infineon Technologies AG	IFX	DE
Institut National de Recherche en Informatique et en Automatique	INRIA	FR
ATHENA - Industrial Systems Institute	ISI	GR
Kungliga Tekniska Högskolan	KTH	SE
Norwegian University of Science and Technology	NTNU	NO
National Technical University of Athens	NTUA	GR
OFFIS e.V.	OFFIS	DE
Office National d'Etudes et de Recherches Aérospatiales	ONERA	FR
BTC - Embedded Systems AG	BTC-ES	DE
Oxford University	Oxford	UK
Sagem Défense Sécurité	SAGEM	FR
AleniaSIA Spa[c]	SIA	IT
Siemens AG	Siemens	DE
Stiftelsen SINTEF	SINTEF	NO
Quintec Associates (Thales Consulting and Engineering)	TC&E	UK
Thales Communications S.A.	TCF	FR
Thales Avionics S.A.	THAV	FR
Thales	TRT	FR
Alma Mater Studiorum - Università di Bologna	UNIBO	IT
The University of Manchester	UNIMAN	UK
Universita degli Studi di Trieste	UniTS	IT
The Virtual Vehicle Competence Center	ViF	AT
Volvo Technology Corporation	VOLVO	SE
Hispano-Suiza S.A.[d]	HS	FR
Messier-Bugatti S.A.	MB	FR
TURBOMECA	TM	FR
SNECMA SAS[e]	SN	FR
Geensoft[f]	Geensoft	FR
Selex Sistemi Integrati[g]	SSI	IT
AleniaAermacchi[h]	Alenia	IT

[c] Joined project by January 2012, being taken over by Alenia Aermacchi
[d] Left project by the end of the first period
[e] Joined project by the beginning of the second period
[f] Joined project by 1st of January 2010, taken over by Dassault Systemes by 1st of July 2011 including all Geensoft activities since 1st of January 2010.
[g] Joined project by 1st of July 2010, taking over 100 % of ED activities
[h] Joined project by 1st of December 2011, taking over 100 % of SIA activities

List of Authors

This book is based on the deliverables of the CESAR project, which have been written by the following authors.

Detlef Streitferdt	ABB AS
Frank Reichenbach	ABB AS
Tormod Wien	ABB AS
Christian Ferdinand	AbsInt Angewandte Informatik GmbH
Marc Langenbach	AbsInt Angewandte Informatik GmbH
Reinhold Heckmann	AbsInt Angewandte Informatik GmbH
José Luis Burón	ACCIONA Infraestructuras S.A.
Patricio Moreno	ACCIONA Infraestructuras S.A.
Andreas Lehmann	AIRBUS Operations GmbH
Gert Döhmen	AIRBUS Operations GmbH
Sylvain Metge	AIRBUS Operations SAS
Cecile Schindler	AIRBUS Operations SAS
François Pouzolz	AIRBUS Operations SAS
Guillaume Bouchez	AIRBUS Operations SAS
Marie-Line Valentin	AIRBUS Operations SAS
Odile Laurent	AIRBUS Operations SAS
Alain Rossignol	ASTRIUM SAS
Ana-Elena Rugina	ASTRIUM SAS
Jean-Francois Rolland	ASTRIUM SAS
Jean-Paul Blanquart	ASTRIUM SAS
Giuseppe Bonifacio	Ansaldo STS S.p.A.
Livia Biondi	Ansaldo STS S.p.A.
Luigi Velardi	Ansaldo STS S.p.A.
Salvatore de Martino	Ansaldo STS S.p.A.
Christopher Dean	AIRBUS Operations Limited

A. Rajan and T. Wahl (eds.), *CESAR - Cost-efficient Methods and Processes for Safety-relevant Embedded Systems*, DOI 10.1007/978-3-7091-1387-5,

Coralie Bernard	AIRBUS Operations Limited
Marcelin Fortes da Cruz	AIRBUS Operations Limited
Stephen Wright	AIRBUS Operations Limited
Eirini Karapistoli	Aristotle University of Thessaloniki
Georgios Chasapis	Aristotle University of Thessaloniki
Ioannis Papadimitriou	Aristotle University of Thessaloniki
Alan Crutcher	Auvation Software Limited
Ben Duncan	Auvation Software Limited
Barbara Wallner	AVL List GmbH
Christian Kreiner	AVL List GmbH
Gerald Sammer	AVL List GmbH
Gerhard Stempfer	AVL List GmbH
Heinz Laußegger-Rauch	AVL List GmbH
Josef Affenzeller	AVL List GmbH
Peter Ebner	AVL List GmbH
Richard Schneider	AVL List GmbH
Vincent Benda	AVL List GmbH
Alfred Wallner	AVL List GmbH
Christian Schyr	AVL List GmbH
Eric Armengaud	AVL List GmbH
Gerhard Griessnig	AVL List GmbH
Gerhard Hauser	AVL List GmbH
Horst Pflügl	AVL List GmbH
Ingrid Kundner	AVL List GmbH
Michael Maletz	AVL List GmbH
Quentin Bourrouilh	AVL List GmbH
Ralf Jurleit	AVL List GmbH
Roland Mader	AVL List GmbH
Udo Brockmeyer	BTC – Embedded Systems AG
Benjamin Blanc	Commissariat à l'Énergie Atomique
Jean-Yves Pierron	Commissariat à l'Énergie Atomique
François Terrier	Commissariat à l'Énergie Atomique
Hubert Dubois	Commissariat à l'Énergie Atomique
Jean-Pierre Gallois	Commissariat à l'Énergie Atomique
Nicolas Rapin	Commissariat à l'Énergie Atomique
Patricia Mouy	Commissariat à l'Énergie Atomique
Mamoun Filali	Centre National de la Recherche Scientifique
Manel Khlif	Centre National de la Recherche Scientifique
Marc Pantel	Centre National de la Recherche Scientifique
Marcus M. Shawky	Centre National de la Recherche Scientifique
Silvano Dal Zilio	Centre National de la Recherche Scientifique

Stéphane Bonnet	Centre National de la Recherche Scientifique
Fulvio Tagliabò	Centro Ricerche Fiat S.C.p.A.
Sandra Torchiaro	Centro Ricerche Fiat S.C.p.A.
Carlos Cacador	Critical Software S.A.
Andre Passos	Critical Software S.A.
Diogo Vicente	Critical Software S.A.
Joaquim Tojal	Critical Software S.A.
Fulvio Romano	Danieli Automation S.p.A.
Morayo Adedjouma	Delphi France SAS
Raphael Pezet	Delphi France SAS
Wojciech Machnik	Delphi France SAS
Álvaro Catalá-Prat	Deutsches Zentrum für Luft- und Raumfahrt e.V.
Frank Köster	Deutsches Zentrum für Luft- und Raumfahrt e.V.
Jan Gacnik	Deutsches Zentrum für Luft- und Raumfahrt e.V.
Maike Eichbaum	Deutsches Zentrum für Luft- und Raumfahrt e.V.
Matthias Grimm	Deutsches Zentrum für Luft- und Raumfahrt e.V.
Silke Köhler	Deutsches Zentrum für Luft- und Raumfahrt e.V.
Uday Biswas	Deutsches Zentrum für Luft- und Raumfahrt e.V.
Didier Simoneau	Dassault Systemes
Raphael Schoenig	Dassault Systemes
Thierry Gueguen	Dassault Systemes
Benoit Perrot	Dassault Systemes
Tatiana Prosvirnova	Dassault Systemes
Wladimir Schamai	EADS
Andreas Kröber	EADS Deutschland GmbH – Defence Electronics
Ralf Bogusch	EADS Deutschland GmbH – Defence Electronics
Roland Scherer	EADS Deutschland GmbH – Defence Electronics
Silke Gerlach	EADS Deutschland GmbH – Defence Electronics
Tobias Sorg	EADS Deutschland GmbH – Defence Electronics
Ulrich Vögele	EADS Deutschland GmbH – Defence Electronics
Uwe Kühne	EADS Deutschland GmbH – Defence Electronics
Alexandre Arnold	EADS Deutschland GmbH
Andreas Keis	EADS Deutschland GmbH
Andreas Mitschke	EADS Deutschland GmbH
Antoaneta Kondeva	EADS Deutschland GmbH
Christian Härdt	EADS Deutschland GmbH
Parham Vasaiely	EADS Deutschland GmbH
Ralf Mauersberger	EADS Deutschland GmbH
Torge Kummerow	EADS Deutschland GmbH
Wolfgang Meyerle	EADS Deutschland GmbH
Luca Spagnolo	ELSAG DATAMAT

Rodolfo Grimani	ELSAG DATAMAT
Riccardo Quacquarelli	ELSAG DATAMAT
Carmen Alonso Montes	TECNALIA
Cristina Lopez Saratxaga	TECNALIA
Jason Mansell	TECNALIA
Sergio Campos	TECNALIA
Adrian Noguero	TECNALIA
Francois-Xavier Dormoy	ESTEREL Technologies SA
Christian Hein	Fraunhofer Gesellschaft zur Förderung der Angewandten Forschung e.V.
Klaus-Dietrich Engel	Fraunhofer Gesellschaft zur Förderung der Angewandten Forschung e.V.
Tom Ritter	Fraunhofer Gesellschaft zur Förderung der Angewandten Forschung e.V.
Adeline De Sousa Silva	Fraunhofer Gesellschaft zur Förderung der Angewandten Forschung e.V.
Becker, Martin	Fraunhofer Gesellschaft zur Förderung der Angewandten Forschung e.V.
Christian Schäfer	Fraunhofer Gesellschaft zur Förderung der Angewandten Forschung e.V.
Ganesh Pai	Fraunhofer Gesellschaft zur Förderung der Angewandten Forschung e.V.
Karina Villela	Fraunhofer Gesellschaft zur Förderung der Angewandten Forschung e.V.
Michael Eisenbarth	Fraunhofer Gesellschaft zur Förderung der Angewandten Forschung e.V.
Michalis Anastasopoulos	Fraunhofer Gesellschaft zur Förderung der Angewandten Forschung e.V.
Pablo Oliveira Antonino de Assis	Fraunhofer Gesellschaft zur Förderung der Angewandten Forschung e.V.
Thiago Burgos	Fraunhofer Gesellschaft zur Förderung der Angewandten Forschung e.V.
Thomas Kuhn	Fraunhofer Gesellschaft zur Förderung der Angewandten Forschung e.V.
Thomas Patzke	Fraunhofer Gesellschaft zur Förderung der Angewandten Forschung e.V.
Thorsten Keuler	Fraunhofer Gesellschaft zur Förderung der Angewandten Forschung e.V.
Angelos Avrasoglou	Hellenic Aerospace Industry S.A.
Nikolaos Priggouris	Hellenic Aerospace Industry S.A.

Herbert Zojer	Infineon Technologies Austria AG
Michael Breunig	Infineon Technologies Austria AG
Nora Gerber	Infineon Technologies Austria AG
Stefan Farfeleder	Infineon Technologies Austria AG
Frank Badstuebner	Infineon Technologies AG
Georg Pelz	Infineon Technologies AG
Pierre-Olivier Nury	Infineon Technologies AG
Albert Benveniste	Institut National de Recherche en Informatique et en Automatique
Benoit Caillaud	Institut National de Recherche en Informatique et en Automatique
Cauê Avila Clasen	Institut National de Recherche en Informatique et en Automatique
Jean-Pierre Talpin	Institut National de Recherche en Informatique et en Automatique
Regis Gascon	Institut National de Recherche en Informatique et en Automatique
Robert de Simone	Institut National de Recherche en Informatique et en Automatique
Roopak Sinha	Institut National de Recherche en Informatique et en Automatique
Yves Sorel	Institut National de Recherche en Informatique et en Automatique
John Gialelis	ATHENA - Industrial Systems Institute
Carl Johan Sjöstedt	Kungliga Tekniska Högskolan
DeJiu Chen	Kungliga Tekniska Högskolan
Magnus Persson	Kungliga Tekniska Högskolan
Martin Törngren	Kungliga Tekniska Högskolan
Matthias Biehl	Kungliga Tekniska Högskolan
Sagar Behere	Kungliga Tekniska Högskolan
Tahir Naseer	Kungliga Tekniska Högskolan
Aurélie Soularue	Messier-Bugatti S.A.
Gabriel Scolan	Messier-Bugatti S.A.
Jean-Charles Breuvart	Messier-Bugatti S.A.
Marianne Felix	Messier-Bugatti S.A.
Quentin Bourgeteau	Messier-Bugatti S.A.
Sébastien Duboelle	Messier-Bugatti S.A.
Sébastien Iriarte	Messier-Bugatti S.A.
Thierno Barry	Messier-Bugatti S.A.
Hien Nam Le	Norwegian University of Science and Technology
Inah Omoronyia	Norwegian University of Science and Technology
Olawande Daramola	Norwegian University of Science and Technology

Rolv Braek	Norwegian University of Science and Technology
Tor Stålhane	Norwegian University of Science and Technology
Panos Marantos	National Technical University of Athens
Andreas Baumgart	OFFIS e.V.
Bernhard Josko	OFFIS e.V.
Christoph Etzien	OFFIS e.V.
Eckard Böde	OFFIS e.V.
Markus Oertel	OFFIS e.V.
Philip Rehkop	OFFIS e.V.
Stefan Häusler	OFFIS e.V.
Thomas Peikenkamp	OFFIS e.V.
Werner Damm	OFFIS e.V.
David Chemouil	Office National d'Etudes et de Recherches Aérospatiales
Julien Brunel	Office National d'Etudes et de Recherches Aérospatiales
Ajitha Rajan	Oxford University
Daniel Kroening	Oxford University
Maria Kourkouli	Oxford University
Michael Tautschnig	Oxford University
Nannan He	Oxford University
Pamela Farries	Oxford University
Philipp Ruemmer	Oxford University
Jean-Claude Derrien	Sagem Défense Sécurité
Julie-Anne Pruvost	Sagem Défense Sécurité
Fabien Paganelli	Sagem Défense Sécurité
Franck Giraud	Sagem Défense Sécurité
Guillaume Allain	Sagem Défense Sécurité
Jean-Louis Dufour	Sagem Défense Sécurité
Jean-Pierre Velu	Sagem Défense Sécurité
Marc Malot	Sagem Défense Sécurité
Nicolas Muller	Sagem Défense Sécurité
Philippe Baufreton	Sagem Défense Sécurité
Anna Todino	AleniaSIA Spa

Domenico Meola	AleniaSIA Spa
Francesco Lanteri	AleniaSIA Spa
Giulio Eula	AleniaSIA Spa
Ivo Viglietti	AleniaSIA Spa
Luciano Lanfranco	AleniaSIA Spa
Massimo Cifaldi	AleniaSIA Spa
Paolo Serena Guinzio	AleniaSIA Spa
Pier Luigi Ferraris	AleniaSIA Spa
Andreas Ulich	Siemens AG
Andreas Ulrich	Siemens AG
Erwin Reyzl	Siemens AG
Klaus Beetz	Siemens AG
Michael C. Jaeger	Siemens AG
Naoufel Boulila	Siemens AG
Peter Graubmann	Siemens AG
Rainer Ersch	Siemens AG
Reiner Schmid	Siemens AG
Sten Wery	Siemens AG
Tobias Gross	Siemens AG
Winfried Seidel	Siemens AG
Neil Loughran	Stiftelsen SINTEF
Øystein Haugen	Stiftelsen SINTEF
Rayner Vintervoll	Stiftelsen SINTEF
Thor Myklebust	Stiftelsen SINTEF
Ulrik Johansen	Stiftelsen SINTEF
Daniel Bognar	SNECMA SAS
Hugues Granier	SNECMA SAS
Jacques Auriol	SNECMA SAS
Valeria Fontana	Selex Sistemi Integrati
Graham Jolliffe	Quintec Associates (Thales Consulting and Engineering)
Emmanuel Leroy	Thales Communications S.A.
Jean-Philippe Jaquet	Thales Communications S.A.
Alain Lambert	Thales Avionics S.A.
Benoit Pierson	Thales Avionics S.A.
Hélène Misson	Thales Avionics S.A.
Jean-Luc Voirin	Thales Avionics S.A.
Loïc Petit	Thales Avionics S.A.
Marion Morel	Thales Avionics S.A.
Nicolas Meledo	Thales Avionics S.A.
Vincent Ibanez	Thales Avionics S.A.

Bernard Wendling	TURBOMECA
Philippe Vallart	TURBOMECA
Gérard Cristau	Thales
Joseph Machrouh	Thales
Nicolas Sordon	Thales
Andrea Paoli	Alma Mater Studiorum – Università di Bologna
Gabriella Perrozzi	Alma Mater Studiorum – Università di Bologna
Matteo Sartini	Alma Mater Studiorum – Università di Bologna
Cuong Minh Tran	The University of Manchester
Faris M. Taweel	The University of Manchester
Ioannis Ntalamagkas	The University of Manchester
Kung-Kiu Lau	The University of Manchester
Stephan Petr	The University of Manchester
Gianfranco Fenu	Universita degli Studi di Trieste
Markus Zoier	The Virtual Vehicle Competence Center
Martin Krammer	The Virtual Vehicle Competence Center
Nadja Marko	The Virtual Vehicle Competence Center
Cecilia Ekelin	Volvo Technology Corporation
Daniel Karlsson	Volvo Technology Corporation
Olof Bridal	Volvo Technology Corporation
Ramin Tavakoli Kolagari	Volvo Technology Corporation
Thomas Söderqvist	Volvo Technology Corporation
Jan Endresen	ABB NO
Erik Carlson	ABB NO
Michael Wagner	FORUS

Glossary

Abstract Interpretation Theory for formally constructing conservative representations (i.e. enforcing soundness) of the semantics of programming languages

Acceptable Risk That part of identified risk that is allowed to persist without taking further engineering or management action to eliminate or reduce the risk, based on knowledge and decision making.

Acceptance Acknowledgement by the certification authority that a submission of data, argument, or claim of equivalence satisfies applicable requirements.

Accident An unintended event, or sequence of events, that causes harm (i.e., death, physical injury or damage to the health of people, or damage to property or the environment).

Activity Set of cohesive tasks of a process.

Adaptability [Component adaptability] The extent to which a component can be modified for reuse, either by adding or subtracting capability.

Adaptation The activity of modifying an adaptable component.

Airworthiness The condition of an item (aircraft, aircraft system, or part) in which that item operates in a safe manner to accomplish its intended function.

Anomalous Behaviour Behaviour that does not fulfil the specified requirements

Applicant A person or organization seeking approval from the approval authority.

Application/Application Domain (CESAR term). A manufacturing industry sector from which companies are involved in the CESAR project.

Application Engineering The process of product line engineering in which the products of the product line are built by reusing assets and exploiting the product line variability.

Approval The agreement that an item is complete and suitable for its intended use.

Approval Authority Organization or person responsible within the state, country, or other relevant body concerned with the approval of a product in accordance with requirements.

Approved Accepted by the certification authority as suitable for a particular purpose.

A. Rajan and T. Wahl (eds.), *CESAR - Cost-efficient Methods and Processes for Safety-relevant Embedded Systems*, DOI 10.1007/978-3-7091-1387-5,

Approved Product An item that has been certified or otherwise recognized by an Approval Authority.

Architecture The fundamental organization of a system embodied in its components, their relationships to each other, and to the environment, and the principles guiding its design and evolution.

Architectural Description A collection of products to document an architecture.

Architectural Mitigation Use of hardware and/or software architectural features to eliminate, reduce, or control risk associated with an identified failure in a system.

Artefact A work product of an activity in a software development process.

Assessment Examination of a characteristic of an item or element.

Asset An item, such as architecture, design, specifications, source code, documentation, test suites, manual procedures, or any other useful result of a software production process.

Assumption An assertion about the system, its operating environment or modes of use, employed without proof, although justification may be required.

Assurance A statement, or process, intended to provide confidence on the condition or status of a system, process, activity, or material.

Authority The organization or person responsible within the state or country concerned with the certification of compliance with applicable requirements.

Auto-code Generator A coding tool that automatically produces source code or object code from low-level requirements.

Availability The ability of an item to be in a state to perform a required function under given conditions at a given instant of time or over a given time interval assuming that the required external resources are provided

Baseline Released version of one or more work products, hardware or software items or elements configurations which is under configuration management and used as a basis for further development through change management process

Branch In configuration management, a secondary series of versions of an item, derived from the primary trunk.

Byzantine Failure Failure model, where the failed component may exhibit arbitrary behaviour.

Cartography A mapping from pilot application scenarios to main activities, expressing which main activities are exercised by a certain scenario.

Cascading Failure Failure for which the probability of occurrence is substantially increased by the existence of a previous failure.

Causal Analysis Analysis of how and why a particular hazard may come into existence.

Certification Authority Organization or person responsible for granting approval on behalf of the nation of manufacture.

Change Control/Configuration Control The process of recording, evaluating, approving or disapproving, and coordinating changes to configuration items after formal establishment of their configuration identification or to baselines after their establishment.

Commercial Off-the-Shelf (COTS) (Software) Commercially available applications sold by vendors through public catalogue listings.
Commonality A property that is true for all members of a product line.
Common Cause Analysis (CCA) Encompasses Zonal Analysis, Particular Risks Analysis and Common Mode Analysis.
Common Cause Failure Failure which bypasses or invalidates redundancy or independence.
Common Mode Failure An event which affects a number of elements otherwise considered to be independent.
Compliance Successful performance of all mandatory activities; agreement between the expected or specified result and the actual result.
Component A part of a system, with defined interfaces to other components.
Configuration (1) The instantiation of predefined variation within an asset.
Configuration Identification (1) The process of designating the configuration items in a system and recording their characteristics.
Configuration Item (1) One or more hardware or software components treated as a unit for configuration management purposes.
Configuration Management A management process for establishing and maintaining consistency of a product's performance, functional and physical attributes with its requirements, design, and operational information throughout its life.
Confirmation Measure Confirmation review, audit or assessment concerning functional safety.
Conformance Established as correct with reference to a standard, specification or drawing.
Conformity Corresponds to the fulfilment of specified requirements by a product, process, service, system, person or body.
Conformity Assessment The process demonstrating whether specified requirements relating to a product, process, service, system, person or body have been fulfilled.
Connector The entity that realizes a connection between components.
Consequence Analysis Analysis of events which are likely to happen after an accident has occurred.
Conservative Representation A simplified or abstract representation in a formal notation of life-cycle data (such as requirements or code) such that statements that are true for this representation will always be true for the life-cycle data itself.
Contract In law, a contract is a (binding) agreement creating and defining the obligations between two or more parties. In Component Based Engineering, contracts provide a way to partition and characterise groups of properties that component instances must exhibit. A contract does not specify how a component has to be implemented, but merely what needs to be supported by realizing instances.
Coverage Analysis The process of determining the degree to which a proposed software verification process activity satisfies its objective.

Coverage Criteria Criteria defining how much functionality of the application code has been covered with existing test suites

Crash Failure Failure mode such as the service is halted (the external state becomes constant, i.e., system activity, if there is any, is no longer perceptible to the users)

Criticality Indication of the hazard level associated with a function, hardware, software, etc., considering the most severe possible consequences of an abnormal behaviour alone or in combination with external events.

Customer Organization or person that receives a product or service.

Data Collection Station No global CESAR definition. See Automotive domain specific definition below.

Data Dictionary The detailed description of data, parameters, variables, and constants used by the system.

Data Flow Analysis A static analysis technique for tracing how data is propagated in a program.

Deactivated Code Executable object code (or data) which is traceable to a requirement and by design is either (a) not intended to be executed (code) or used (data), or (b) is only executed (code) or used (data) in certain configurations of the target computer environment.

Dead Code Executable object code (or data) which exists as a result of a software development error (e.g., in a requirements definition, design, coding, or integration process) but cannot be executed (code) or used (data) in any operational configuration of the target computer environment.

Decision Model A product line artefact that captures relationships among variation points in a set of assets.

Demonstration Evidence of the properties of a system, or an element of a system, achieved by testing, trials or operational execution.

Dependability The ability of a system to deliver a service that can be justifiably trusted.

Deployment The set of activities that make a system available for use. By extension, a process used to articulate the steps and stakeholders of a given process for transforming a system from a packaged form to an operational state.

Derived Requirements Additional requirements resulting from design or implementation decisions during the development process

Design The process of defining the architecture, components, interfaces, and other characteristics of a system or component.

Design Authority The body responsible for the formulation of a design solution to fulfil the specified requirements and for overseeing the subsequent development and setting-to-work of a system in its intended environment

Detection Loop No global CESAR definition. See Automotive domain specific definition below.

Development Activity A selected and limited part of the pilot application with the purpose of highlighting an important aspect with respect to the development and evaluation of the RTP.

Development Assurance All of those planned and systematic actions used to substantiate, at an adequate level of confidence, that development errors have been identified and corrected such that the system satisfies the applicable certification basis.

Diagnosability Ability of a system to support the identification of information related to its real or potential faults.

Diagnostic Coverage Proportion of a hardware element failure rate that is covered by the safety mechanisms implemented.

Disapproval Conclusion by the appropriate authority that an item submitted for approval is either incomplete or not suitable for its intended use.

Domain Engineering The process of product line engineering in which the commonality and the variability of the product line are defined and realised.

Domain Specific Language (DSL) Programming language – or an executable specification language – that offers expressive power focused on a particular problem domain through dedicated notations and abstractions.

Dual Point Failure See Automotive domain specific definition below

Dual Point Fault See Automotive domain specific definition below

E/E/PE System System based on electrical (E) and/or electronic (E) and/or programmable electronic (PE) technology.

EUC Equipment under control.

Eclipse Modelling Framework (EMF) A modelling framework and code generation facility for building tools and other applications based on a structured data model.

eCore Meta-model definition language, supported by the Eclipse Modeling Framework project.

Electronic Data Plate Identification information of an item, provided under electronic format.

Element Part of a product that has been determined to be a basic unit or building block

Embedded Identifier Identification attributes of the software (e.g., creation date, part number, linker integrity verification checksum or cyclic redundancy check (CRC), or version identification) included in the target executable object code.

Emergency Condition when potentially catastrophic or critical hazardous events have occurred, and immediate and pre-planned safing action is possible and is mandatory in order to protect personnel.

Emulator A device, computer program, or system that accepts the same inputs and produces the same output as a given system using the same object code.

End-to-End Numerical Resolution Measure of the numerical precision resulting from computations through the integrated system.

Engineering The disciplined application of scientific knowledge to resolve conflicting constraints and requirements for problems of immediate practical significance.

Equipment A functional physical item.

Equivalence Class The partition of the input domain of a program such that a test of a representative value of the class is equivalent to a test of other values of the class.

Equivalent Safety See Aeronautics domain specific definition below.

Error Discrepancy between a computed, observed or measured value or condition and the true, specified, or theoretically correct value or condition.

Evaluation The process of determining whether an item or activity meets specified criteria.

Event Tree Graphical model of an accident scenario that yields multiple outcomes and outcome probabilities.

Evolution The process of modifying a software system or component after delivery to correct faults, improve performance or another attribute, or adapt to a changed environment.

Exposure Time State of being in an operational situation that can be hazardous if coincident with the failure mode under analysis.

External Measure Measure separate and distinct from the item that reduces or mitigates the risks resulting from the item.

Fail-Operational A characteristic design which permits continued operation in spite of the occurrence of a discrete malfunction.

Fail-Safe A characteristic of a system whereby any malfunction affecting the system safety will cause the system to revert to a state that is known to be within acceptable risk parameters.

Failure A situation or an event where the delivered service deviates from the correct service.

Failure Condition (FC) A condition with an effect on the system and its users or passengers, both direct and consequential, caused or contributed to by one or more failures, considering relevant adverse operational or environmental conditions.

Failure Effect (FE) A description of the operation of a system or an item as the result of a failure; i.e., the consequence(s) a failure mode has on the operation, function or status of a system or an item.

Failure Mode Manner in which an element or an item fails.

Failure Modes and Effects Analysis (FMEA) A bottom-up procedure for analysis of failure modes within a system for determination of the effect of failures on the system.

Failure Modes, Effects and Criticality Analysis (FMECA) A bottom-up procedure for analysis of failure modes within a system for determination of the effect of failures on the system, including the evaluation of its criticality (in terms of severity and likelihood).

Failure Rate Density of probability of failure divided by probability of survival for a hardware element

Fault The adjudged or hypothesized cause of an error.

Fault Detection Time A mechanism that limits or prevents the propagation of anomalous behaviour to other system/software components.

Fault Forecasting Means to estimate the present number, the future incidence, and the likely consequences of faults

Fault Prevention Means to prevent the occurrence or introduction of faults

Fault Removal Means to reduce the number and severity of faults

Fault Tolerance Means to avoid service failures in the presence of faults

Fault Tree Analysis (FTA) A top-down failure analysis in which an undesired state of a system (top level event) is analyzed using Boolean logic to combine a series of lower-level events (basic events).

Feature An end-user visible characteristic of a system.

Field Programmable Gate Array (FPGA) An integrated circuit designed to be configured by the customer or designer after manufacturing.

Formal Analysis Case Combination of the property to be analysed or proved, all assumptions for the formal analysis, and expected results, such as to verify correct functionality and to establish conditions that reveal potential errors; analogous to a test case.

Formal Methods Descriptive notations and analytical methods used to construct, develop and reason about mathematical models of system behaviour. A formal method is a formal analysis carried out on a formal model.

Formal Model Abstract representation of a given set of aspects of a system that is used for analysis, simulation, code generation, or any combination thereof

Formal Notation Description technique that has both its syntax and semantics completely defined.

Formal Specification A mathematical description of software or hardware that may be used to develop an implementation.

Formal Verification The use of mathematical reasoning to guarantee that properties are always satisfied by a formal model.

Function The action or actions that a product is designed to perform.

Functional Concept Specification of the intended functions and their interactions necessary to achieve the desired behaviour.

Functional Hazard Assessment (FHA) A systematic, comprehensive examination of functions to identify and classify failure conditions of those functions according to their severity.

Functional Safety Absence of unreasonable risk due to hazards caused by malfunctioning behaviour of E/E systems.

Functional Safety Concept Specification of the functional safety requirements, with associated information, their assignment to architectural elements, and their interaction necessary to achieve the safety goals.

Functional Safety Requirement Specification of implementation-independent safety behaviour, or implementation-independent safety measure, including its safety-related attributes.

Gap Analysis An engineering activity designed to identify the differences between a present state and the intended future state.

Glue Code New code needed to integrate COTS code into a larger system

Guidance Material that becomes a means of compliance to the regulations upon being recognized by the authorities

Guidelines Procedure that informally describes how a certain activity is to be performed.

Harm Death, physical injury or damage to the health of people, or damage to property or the environment.

Hazard Potential source of harm.

Hazard Analysis The process of describing in detail the hazards and accidents associated with a system, and defining accident sequences.

Hazard and Operability (HAZOP) Technique for identifying and analyzing the hazardous event of a system. It looks for hazards resulting from identified potential deviations in design operational intent.

Hazard Log The document in which all safety management activities, hazards identified, decisions made and solutions adopted, are recorded or referenced.

Hazardous Event Combination of a hazard and an operational situation

Host Computer The computer on which the software is developed.

Human-Robot Interaction (HRI) A field of study dedicated to understanding, designing, and evaluating robotic systems for use by or with humans.

Impact Analysis Activity of identifying what to modify to accomplish a change or of identifying the potential consequences of a change.

Independence (functional) Freedom from any mechanism which can affect the correct operation of more than one function as a result of either systematic or random failure

Independence (organisational) Freedom from involvement in the same intellectual, commercial and/or management entity.

Independence (physical) Freedom from any mechanism which can affect the correct operation of more than one system/sub-system/equipment as a result of random failures

Information Flow A description of the dependencies between an output of a software module and its inputs.

Infrastructure-to-Vehicle Communication System See Automotive domain specific definition below

Initial Automotive Safety Integrity Level (ASIL) See Automotive domain specific definition below

Initiating Event (IE) Failure or undesired event that initiates the start of an accident sequence

Inspection A static analysis technique that relies on visual examination of development products to detect errors and faults.

Instrumentation Code in an application to keep track of some attribute of an executing program.

Integral Process A process which assists the software development processes and other integral processes and, therefore, remains active throughout the software life cycle.

Integration Process A sub-process within the software development process that deals with the combination or merger of components to form a larger component with increased functionality and the integration of the software on the target hardware.

Integration Testing The verification that the merging of the components has been successfully performed. This verification activity includes software to software integration testing and software to hardware integration testing of the merged components.

Integrity The degree to which a system or component prevents unauthorized access to or modification of computer programs or data.

Intelligent Transportation System (ITS) See Automotive domain specific definition below

Interchangeability The ability to substitute one item for another within a system and have the system performs in accordance to its specification.

Interface A shared boundary across which information is passed.

In-Vehicle ITS See Automotive domain specific definition below

Item See specific definitions below

ITS Node See Automotive domain specific definition below

Latent Fault See Automotive domain specific definition below

Life Cycle Evolution of a system, product, service, project, or other human-made entity from conception through retirement.

Low-Level Requirements Software requirements developed from high-level requirements, derived requirements, and design constraints from which source code can be directly implemented without further information.

Maintainability The probability that a given active maintenance action, for an item under given conditions of use can be carried out within a stated time interval when the maintenance is performed under stated conditions and using stated.

Maintenance The combination of all technical and administrative actions, including supervision actions, intended to retain a physical item in, or restore it to, a state in which it can perform its required function.

Malfunction

Malfunctioning Behaviour Failure or unintended behaviour of the item with respect to the design intent for this item.

Mass Customization The large-scale production of goods tailored to individual customers' needs.

Mean Time Between Failures (MTBF) Mathematical expectation of the time interval between two consecutive failures of a system or element.

Means of Compliance The intended method(s) to be used by the applicant to satisfy the requirements stated in the certification basis for an aircraft or engine.

Meta-Model The analysis, construction and development of the frames, rules, constraints, models and theories applicable and useful for modelling a predefined class of problems.

MICROSAR Run-time Environment See Automotive domain specific definition below

Minimal Cut-Set (MCS) The smallest combination of basic events (e.g. component, equipment failures, dependant system faults, external events etc.) which, if they all occur, will cause the top event to happen.

Mishap Unplanned event or series of events resulting in death, injury, occupational illness, damage to or loss of equipment or property, or damage to the environment.

Mitigation Action taken to reduce the risk presented by a hazard, by modifying the hazard in order to decrease the mishap probability and/or the mishap severity.

Model A semantically closed abstraction of a software or hardware system, i.e. a simplification of reality that gives a complete description a system from a particular perspective.

Model Based Development (MBD) Formalized application of modelling to support system requirements, design, analysis, verification and validation activities beginning in the conceptual design phase and continuing throughout development and later life cycle phases.

Model-Checking Model checking explores all possible behaviours of a formal model to determine whether a specified property is satisfied.

Monitoring (1) [Safety] Functionality within a system which is designed to detect anomalous behaviour of that system.

Multiple Point Fault Individual fault that, in combination with other independent faults, leads to a multiple point failure.

Multiple Version Dissimilar Software Two or more software components that satisfy the same functional requirements, but are intentionally different from one another.

Negation Enforcement of a safe state following detection of a hazardous fault.

Negation Time Time span which begins when the existence of a fault is detected and ends when a safe state is enforced

New Development Process of creating an item having previously unspecified functionality, a novel implementation of an existing functionality, or both.

Node In configuration management, a particular version in the series of versions of an item, from which two different versions (and subsequent series) are derived.

Non-conformance Non-fulfilment of a specified requirement.

Non-functional Requirement Requirement that defines a system property or a condition under which the system is required to operate or exist.

Object Code A low-level representation of the computer program not usually in a form directly usable by the target computer but in a form which includes relocation information in addition to the processor instruction information.

Observer An observer is a node taking as inputs all the flows relevant to the safety property to be specified , and computing a single Boolean flow which is true as long as the observed flows satisfy the property.

Open Source Software The term "Open source software" was defined by OSI (Open Source Initiative) in 1998. In the definition presented by OSI, open source software must comply with the following criteria: Free redistribution, source code is available, derived works, integrity of source code, no discrimination against persons, groups or fields of endeavour. The License must not be specific to a product, not restrict other software and have to be technology-neutral.

Operating Mode Perceivable functional state of the system of element.

Operating Point For a dynamic system, it defines the system's overall state at a given time.

Operational Situation See Automotive domain specific definition below

Parametric Requirement Generic requirement within a domain. It has defined parameters that must be set when the requirement is instantiated for a specific project.

Particular Risk Risk associated with those events or influences that are outside the systems and items concerned, but which may violate failure independence claims.

Partitioning Separation of functions or elements.

Patch Modification to executable object code in which one or more of the planned steps of re-compiling, re-assembling, or re-linking is bypassed. This does not include embedded identifiers.

Pattern Consistent description of a solution to a recurring problem in a certain context.

Perceived Fault See Automotive domain specific definition below

Performance A quantitative measure characterizing a physical or functional attribute relating to the execution of an operation or function. Performance attributes include quantity, quality, coverage, timeliness, and readiness.

Pilot Application (PA) Refers to the development process of an example application, which has the intention to serve as a source of inspiration in terms of requirement elicitation and evaluation of the Reference Technology Platform.

Platform A set of sub-systems and interfaces that form a common structure from which a set of derivative products can be efficiently developed and produced.

Post-condition A constraint that must be true at the completion of an operation.

Practice A software engineering or management activity that contributes to the creation of the output (work products) of a process or enhances

Pre-condition A constraint that must be true when an operation is invoked.

Pre-existing Software Software developed prior to the application currently in question.

Preliminary System Safety Assessment (PSSA) A systematic evaluation of a proposed system architecture and its implementation, based on the functional hazard assessment and failure condition classification, to determine safety requirements for all items in the architecture.

Procedure Set of tasks that must be performed to accomplish an operation.

Process Defines how development is – or should be – performed, i.e. the specific activities that need to be conducted. In order to perform product line engineering, a single system development process needs to be adequately adapted.

Product A deployed physical system or software.

Product Assurance Discipline devoted to the study, planning and implementation of activities intended to assure that the design, controls, methods and techniques in a project result in a satisfactory level of quality in a product.

Product Liability The onus on a producer or others to make restitution related to personal injury, property damage or other harm caused by a product.

Product Line A set of software-intensive systems sharing a common, managed set of features that satisfy the specific needs of a particular market segment or mission and that are developed from a common set of assets in a prescribed way.

Product Line Architecture Description of the structural properties for building a group of related systems, typically the components and their interrelationships.

Product Line Engineering A paradigm to develop product lines and to produce products in a cost-effective way using platforms and large-scale production, tailored to individual customers' needs.

Product Line Infrastructure A set of assets and the related decision model that cover all life-cycle stages considered during application engineering.

Product Process Recognition The acceptance of the evidence that a recognized development process was applied to a PDS product.

Product Service History A contiguous period of time during which the software is operated within a known environment, and during which successive failures are recorded.

Programmable Logic Controller A microcomputer-based controller that uses stored instructions in programmable memory to implement logic, sequencing, timing, counting and arithmetic functions through digital or analog input/output modules, for controlling machines and processes.

Property The representation of a requirement (functional or non-functional) in a formal notation.

Protection The use of a mechanism to ensure that a tool function cannot adversely impact another tool function

Proven in Use See Automotive domain specific definition below

Qualification Process Process to demonstrate the ability to fulfil specified requirements.

Qualitative Analysis Analysis or evaluation based on qualitative values.

Quality (1) The degree to which a system, component, or process meets specified requirements.

Quality Control Part of quality management focused on fulfilling quality requirements.

Quality of Service In the field of computer networking, Quality of Service (QoS) is a set of quality requirements on the collective behaviour of one or more objects. This is generally obtained by providing a mechanism to assign different priority to different applications, users, or data flows, or to guarantee a certain level of performance to a data flow.

Quantitative Analysis Analysis or evaluation based on numerical values and/or mathematical calculations

Random Failure Integrity The degree to which a system is free from hazardous random faults.

Random Fault Unpredictable occurrence of a fault.

Random Hardware Failure Failure that may occur unpredictably during the lifetime of a hardware element and that follows a probability distribution

Reasonably Foreseeable Misuse Use of a product, process or service under conditions or for purposes not intended by the supplier, but which can happen, induced by the product, process or service in combination with, or as a result of, common human behaviour.

Redundancy Existence of means, in addition to the means which would be sufficient for a functional unit to perform a required function or for data to represent information.

Reference Technology Platform (RTP) A cross-domain standardised platform that provides meta-models, methods, and tools for safety-relevant hard real-time system development.

Refinement Operation that adds detail to an abstraction while preserving the semantics of the abstraction.

Regression Testing Any type of software testing that seeks to uncover software errors by partially retesting a modified program.

Release Particular version of a configuration item that is made available for a specific purpose (for example, test release).

Reliability The probability that an item will perform a required function under specified conditions, without failure, for a specified period of time.

ReqTool Any requirement management tool that could be instantiated within the RTP thanks to the RTP interoperability features.

Requirement (1) A condition or capability needed by a user to solve a problem or achieve an objective.

Residual Fault See Automotive domain specific definition below

Residual Risk Risk remaining after the deployment of safety measures.

Reusability (a) The degree to which an asset can be used in more than one software system, or in building other assets.

Reuse The use of an asset in the solution of different problems.

Re-verification Verification of corrective actions to ensure that previously reported errors have been eliminated and to detect and report new errors that may have been introduced by the corrective actions.

Reverse Engineering The process of developing higher level software data from existing software data.

Review Examination of a work product, where the level of detail is up to the reviewer.

Risk Combination of the probability of occurrence of harm and the severity of that harm.

Risk Time The period of time during which an item must fail in order to cause the failure effect in question. This is usually associated with the final fault in a fault sequence leading to a specific failure condition.

Road Side Intelligent Transportation System See Automotive domain specific definition

Robustness (software) The extent to which software can continue to operate correctly despite abnormal inputs and conditions.

Safe Fault See Automotive domain specific definition below

Safe State Operating mode of an item without an unreasonable level of risk.

Safety Absence of unreasonable risk

Safety Acceptance The safety status given to a product by the final user.

Safety Application Rules (SAR) The definition of conditions, which must be fulfilled while using a system.

Safety Approval The safety status given to a product by the requisite authority when the product has fulfilled a set of pre-determined conditions.

Safety Assessment Report The document containing the conclusions of the assessment performed by an assessment body on the system under assessment.

Safety Authority The body responsible for delivering the authorization for the operation of the safety related system.

Safety Case Argument that the safety goals for an item are complete and satisfied by evidence compiled from work products of the safety activities during development.

Safety Critical Term applied to any condition, event, operation, process, or item whose proper recognition, control, performance, or tolerance is essential to safe system operation and support.

Safety Goal See Automotive domain specific definition below

Safety Integrity The ability of a safety-related system to achieve its required safety functions under all the stated conditions within a stated operational environment and within a stated period of time.

Safety Integrity Level (SIL) Discrete level (one out of a possible four), corresponding to a range of safety integrity values, where safety integrity level 4 has the highest level of safety integrity and safety integrity level 1 has the lowest

Safety Life Cycle The additional series of activities carried out in conjunction with the system life-cycle for safety-related systems

Safety Management The application of organizational and management principles in order to achieve safety with high confidence.

Safety Management Plan The organizational structure, processes, procedures and methodologies that enable the direction and control of the activities necessary to meet safety requirements and safety policy objectives

Safety Measure Activity or technical solution to avoid or control systematic failures and to detect random hardware failures or control random hardware failures, or mitigate their harmful effects.

Safety Mechanism Measure implemented by a E/E functions or element, or in other technologies, to detect or control failures in order to achieve a safe state of the item, or maintain a safe state of the item, or both. Safety mechanisms are implemented within the item to prevent faults from leading to single point failures or residual failures and to prevent faults from being latent.

Safety Monitoring Means of protecting against specific failure conditions by directly monitoring a function for failures that would result in a failure condition.

Safety Objectives Requirements on likelihood of occurrence of failures. They are generally stated, in qualitative or quantitative terms, for specific feared events or per category of severity.

Safety Operating Limits Any functional limit imposed on the system to meet the safety requirements.

Safety Plan Plan to control and guide the safety activities of a project including dates, milestones, tasks, deliverables, responsibilities and resources.

Safety Process The series of procedures that are followed to enable all safety requirements of a product or product line to be identified and met.

Safety Task A maintenance action and/or crew checks intended to detect latent failures which would, in combination with one or more specific failures or events, lead to hazardous or catastrophic failure conditions.

Safety Validation Assurance, based on examination and tests, that the safety goals are sufficient and has been achieved.

Safety Viewpoint Organised set of information on a system related to its safety.

Safety-Related Element Element which has the potential to contribute to the violation of a safety goal.

Scenario A selected and limited part of the pilot application with the purpose of highlighting an important aspect with respect to the development and evaluation of the RTP.

Scoping The process of determining the boundaries of the product line engineering activity. This can be performed on three levels: product portfolio, domain and assets.

Service Experience Intervals of time during which the software is operated within a known relevant and controlled environment, during which successive failures are recorded.

Service Experience Data Data obtained on successful and failed operation during the execution of the software over the service experience intervals.

Severity Measure of the extent of harm to an individual in a specific situation.

Similarity See specific definitions below

Simulator A device, computer program or system used during software verification, that accepts the same inputs and produces the same output as a given system, using object code which is derived from the original object code.

Single Event Upset (SEU) Random bit flip in data that can occur in hardware.

Single Point Fault See Automotive domain specific definition below

Software Architecture Structure or structures of the system, which consists of software components, the externally visible properties of those components, and the relationships among them.

Software Assurance The planned and systematic actions necessary to provide confidence and evidence that a software product or process satisfies given requirements.

Software Change A modification in source code, object code, executable object code, or its related documentation from its baseline.

Software Conformity Review A review, typically conducted at the end of a software development project, for the purpose of assuring that the software life cycle processes are complete, software life cycle data is complete, and the Executable Object Code is controlled and can be regenerated.

Software Development A set of activities that result in software products.

Software Development Standard A standard that provides uniform requirements to be applied during the software development.

Software Engineering The application of formal methods in the cost-effective development (analysis, design, construction, and evolution) of cost-effective software systems. (Note that this term explicitly redefines the [IEEE610] definition, especially because that neglected cost-effectiveness)

Software Integration The process of combining code components.

Software Integrity Assurance Standard A standard that provides criteria to evaluate and measure a software product and/or a process to provide assurance that the product and/or process satisfies given requirements and can be relied upon to work correctly in its intended environment.

Software Life-Cycle Environment The software life cycle process context in which the tool operates. It includes interfaces to adjacent tools and manual process.

Software Partitioning The process of separating, usually with the express purpose of isolating one or more attributes of the software, to prevent specific interactions and cross-coupling interference.

Software Platform A set of software sub-systems and interfaces that form a common structure from which a set of derivative products can be efficiently developed and produced.

Software Requirement A description of what is to be produced by the software given the inputs and constraints.

Software Tool A computer program used to help develop, test, analyze, produce or modify another program or its documentation.

Software On-line Support Tool Software tool that can directly influence the safety-related system during its run time.

Software Off-line Support Tool Software tool that supports a phase of the software development lifecycle and that cannot directly influence the safety-related system during its run time.

Soundness A characteristic of a logical argument or method that holds if and only if it is impossible to reach a false conclusion from true statements.

Specification A document that explicitly states essential technical attributes and requirements for a product and procedures to determine that the product's performance meets its requirements and attributes.

Stakeholder Individual or organization having a right, share, claim or interest in a system or in its possession of characteristics that meet their needs and expectations.

Standard A rule or basis of comparison used to provide both guidance in and assessment of the performance of a given activity or the content of a specified data item.

Standard Operating Procedure (SOP) Describes a best-practice approach to executing tasks related to the production and maintenance of hardware and software, as well as to incident and change management.

State Space The allowed values for a set of attributes.

Static Analysis A software technique that helps to reveal certain properties of a program without executing it.

Stress Profile The degree and number of external influences which a product can withstand whilst performing its required functionality.

Structure A specified arrangement or interrelation of parts to form a whole.

Success Criterion A conditional state of development progress of a pilot application that, if reached, indicates successful evaluation of the RTP with respect to a certain scenario.

Syntax Structure of expressions or the rules for putting symbols together to form an expression.

System (a) An integrated composite that consists of one or more of the processes, hardware, software, facilities, and people, that provides a capability to satisfy a stated need or objective.

System Architecture The structure of the hardware and the software selected to implement the system requirements.

System Failure Mode and Effect Analysis Inductive analysis that focuses on potential failure modes associated with the functionality of a system caused by design. It regards possible malfunctions of product systems as possible failures.

System Life Cycle The series of activities occurring during a period of time that starts when a system is conceived and ends at decommissioning when the system is no longer available for use

System Safety Assessment (SSA) A systematic, comprehensive evaluation of the implemented system to show that the relevant safety requirements are met.

System Safety Assessment Process See Aeronautics domain specific definition below

Systematic Failure Failure, related in a deterministic way to a certain cause, which can only be eliminated by a modification of the design or of the manufacturing process, operational procedures, documentation or other relevant factors

Systematic Fault An inherent fault in the specification, design, construction, installation, operation or maintenance of a system, sub-system or equipment.

Technical Safety Concept Specification of the technical safety requirements to implement and to allocate them to the system design.

Technical Safety Report Documented technical evidence for the safety of the design of a system/sub-system/equipment.

Technical Safety Requirements Requirements derived from the associated functional safety requirements to provide their technical implementation.

Test A procedure to prove performance using stated objective criteria with pass or fail results.

Test Case A set of test inputs, execution conditions, and expected results developed for a particular objective, such as to test a particular program path or to verify compliance with a specific requirement.

Test Case Sequencing A set of several test cases for a component or system under test, with associated functionality where the post condition of one test is often used as the precondition for the next one.

Test Coverage Analysis The process of determining the degree to which a proposed verification process activity satisfies its objective.

Test Item Software or system item that is an object of testing.

Test Management Description of the scope, approach, resources and schedule of intended test activities.

Test Oracle Mechanism, different from the program itself that can be used to check the correctness of the output of the program for the test cases.

Test Procedure Detailed instructions for the set-up and execution of a given set of test cases, and instructions for the evaluation of results of executing the test cases.

Test Scenario A set of test cases that ensure that the business process flows are tested from end to end.

Test Suite A collection of test scenarios and/or test cases that are related or that cooperate with each other.

Test Vector A set of inputs for testing new designs.

Testing Process of planning, preparing and executing or exercising a system or system component to verify that it satisfies specified requirements, to detect errors, and to create confidence in the system behaviour.

Theorem Proving Deductive methods involving mathematical arguments, such as mathematical proofs, for establishing a specified property of a formal model.

Third Party Vendor A person or organization that supplies component systems to a reuse-driven engineering business, either pre-existing or developed under contract.

Timed Automata Extension of finite state automata with clock variables that works with timed sequences of events.

Time Limited Dispatch See Aeronautics domain specific definition

Tool A computer program or a functional part thereof, used to help develop, transform, test, analyze, produce, or modify another program, data, or its documentation.

Tool Development Environment The environment where the tool is developed.

Tool Operational Environment The environment where the tool is used. It includes workstation, operating system, and external dependencies.

Tool Operational Requirements The Tool Operational Requirements can be viewed as the "system requirements" for a tool. Tool Operational Requirements are from the perspective of the user and may not provide all requirements necessary to develop a tool.

Tool Qualification See Aeronautics domain specific definition

Tool Requirements The Tool Requirements are the requirements used to develop and verify a tool. They include all functional, interface, and safety-related requirements that will be implemented in the tool. Tool Requirements may be presented in multiple levels of requirements.

Tool Verification Environment The environment where the tool in its executable form is verified.

Traceability Ability to trace the history, application or location of what is under consideration.

Trace Data Data providing evidence of traceability of development and verification processes

Traffic Control Centre See Automotive domain specific definition

Transition Criteria The minimum conditions, as defined by the software planning process, to be satisfied to enter a process.

Trunk In configuration management, primary series of versions of an item, from which may be derived secondary series called branches.

Type A set of entities which share a common set of properties.

Type Consistency When a value of the subtype of a given type can be substituted for a value of the given type without changing the behaviour of the system, i.e., the subtype fulfils all the requirements of the given type.

Type Design (1) The drawings and specifications, and a listing of those drawings and specifications, necessary to define the configuration and the design features of the product shown to comply with the requirements for that product;

Type Safety A type system to prevent certain undesirable program behavior. A property of some programming languages.

Unbounded Recursive Algorithm An algorithm that directly invokes itself (self-recursion) or indirectly invokes itself (mutual recursion) and does not have a mechanism to limit the number of times it can do this before completing.

Unintended Function A function that is visible at the system level and was neither intended nor a predicted fault condition in the PSSA.

Universal Remote Station See Automotive domain specific definition

Unreasonable Risk Risk judged to be unacceptable in a certain context according to valid societal moral concepts.

Untraceable Code Code (or data) which exists as a result of a software development error (e.g., in a requirements definition, design, coding, or integration process). It is not traceable to any system or software requirement and may or may not be executable.

Validation The process of evaluating a system or component during or at the end of the development process to determine whether it satisfies specified requirements.

Variability A property that is different among the members of a product line. Alternatively: A property which is not yet fixed during the design, e.g. for purposes of architecture exploration.

Variability Management The activities required for managing differences between products of a product line.

Variability Mechanism A construct that leads to variation in assets.

Variable Message Sign (VMS) See Automotive domain specific definition

Variant A selectable property which makes products distinguishable.

Variation Point The representation of a certain variation determined by several variants. A variation point represents those places in assets where variation occurs.

Verification Determination of completeness and correct specification or implementation of requirements from a previous phase.

Verification and Validation (V&V) The process of determining whether the requirements for a system or component are complete and correct, the products of each development phase fulfil the requirements or conditions imposed by the previous phase, and the final system or component complies with specified requirements.

Verification Review Verification activity to ensure that the result of a development activity fulfils the project requirements, or technical requirements, or both.

Version An initial release or re-release of a configuration item. An instance of the system that differs in some way from other instances.

View A representation of a whole system from the perspective of a related set of concerns

Viewpoint A specification of the conventions for constructing and using a view. A pattern or template from which to develop individual views by establishing the purposes and audience for a view and the techniques for its creation and analysis. (Also known as a meta-view)

Virtualisation General technique of abstracting low level resources at some higher level with a software module.

Weather Station See Automotive domain specific definition below

Well-Formed Expression An expression that only contains elements from a defined set of primitive operands and operators and meets all of the rules for valid combinations of these primitives.

Wireless Sensor Network (WSN) A set of autonomous nodes, each of which typically comprising one or more sensors, a radio transceiver, a microcontroller and an energy source, that cooperatively monitor one or more physical magnitudes.

Work Product Reference to the complete information concerning the results of associated requirements.

Worst Case Execution Time (WCET) The maximum length of time a computational task could take to execute on a specific hardware platform.

Wrapper Code Software used to encapsulate some functionality of a system for the purpose of isolating it from the remaining functionality of the system.

Zonal Safety The safety of a system with respect to installation, interference between systems, and potential maintenance errors.

References

1. Airbus Report, Typical pilot application description for aerospace domain. Cesar Deliverable I_SP6_R1.5_M1
2. A. Albinet, S. Begoc, J.-L. Boulanger, O. Casse, I. Dal, H. Dubois, F. Lakhal, D. Louar, M.-A. Peraldi-Frati, Y. Sorel, Q.-D. Van, The MeMVaTEx methodology: from requirements to models in automotive application design, in *ERTS'08*, Toulouse, France, 2008
3. ANSI/IEEE Std 500-1984 P & V, *IEEE Standard Reliability Data for Pumps and Drivers, Valve Actuators, and Valves* (IEEE, New York, 1984)
4. Aquintos, PREEvision. www.aquintos.com
5. Architecture analysis & design language, http://www.aadl.info/aadl/currentsite
6. ARINC Aeronautical Radio Inc., ARINC specification 429 mark 33 digital information transfer system (DITS), Part 1: functional description, electrical interface, label assignments and word formats (2004), http://www.arinc.com/
7. ARINC Aeronautical Radio Inc., ARINC specification 429 mark 33 digital information transfer system (DITS), Part 2: discrete data standards (2004), http://www.arinc.com/
8. ARINC Aeronautical Radio Inc. ARINC Specification 429 Mark 33 Digital Information Transfer System (DITS), Part 3: file data transfer techniques (2009), http://www.arinc.com/
9. A. Arnold, G. Point, A. Griffault, A. Rauzy, The altarica formalism for describing concurrent systems. Fundam. Inform. **40**(2–3), 109–124 (1999)
10. ARTEMIS-IA, *Artemis Strategic Research Agenda 2011*, 1st edn. (2011)
11. AUTOSAR AUTOmotive Open System ARchitecture, http://www.autosar.org/. Retrieved on 18 Dec 2011
12. AUTOSAR GbR, *Virtual Function Bus*, V2.1.0, R4.0 Rev 2 (2010), http://www.autosar.org/download/R4.0/AUTOSAR_EXP_VFB.pdf
13. A.T. Bahill, J. Daniels, Correction to "quantitative methods for tradeoff analyses". Syst. Eng. **8**(1), 93–93 (2005)
14. P.G. Bassett, *Framing Software Reuse: Lessons from the Real World* (Prentice-Hall, Upper Saddle River, 1997)
15. A. Baumgart, A common meta-model for the interoperation of tools with heterogeneous data models, in *Proceedings of the Third Workshop on Model Driven Tool and Process Integration (MDTPI)*, Paris, France, 16 June 2010, ed. by C. Hein, M. Wagner, R. Mader, A. Keis, E. Armengaud (Fraunhofer Verlag, Stuttgart, 2010), pp. 31–40
16. A. Baumgart, C. Ellen, M. Oertel, P. Rehkop, S. Farfeleder, S. Schulz, A reference technology platform with common interfaces for distributed heterogeneous data, in *Proceedings of the Embedded World 2012 Exhibition and Conference*, Nuremberg, March 2012
17. A. Baumgart, C. Ellen, M. Oertel, P. Rehkop, S. Farfeleder, S. Schulz, A reference technology platform with common interfaces for distributed heterogeneous data, in *Proceedings of the Embedded World 2012 Exhibition and Conference*, Nuremberg, March 2012

A. Rajan and T. Wahl (eds.), *CESAR - Cost-efficient Methods and Processes for Safety-relevant Embedded Systems*, DOI 10.1007/978-3-7091-1387-5,

18. D. Benavides, S. Trujillo, P. Trinidad, On the modularization of feature models, in *First European Workshop on Model Transformation*, Rennes, France, 2005
19. T. Berners-Lee, Linked Data Design Issues, http://www.w3.org/DesignIssues/LinkedData.html
20. M. Biehl, C.-J. Sjöstedt, M. Törngren, A modular tool integration approach: experiences from two case studies, in *3rd Workshop on Model-Driven Tool and Process Integration (MDTPI2010)*, Paris, France, 2010. QC 20101209
21. E. Böde, W. Damm, J. Hoyem, B. Josko, J. Niehaus, M. Segelken, Adding value to automotive models, in *Automotive Software – Connected Services in Mobile Networks*. volume 4147 Lecture Notes in Computer Science. ARTIST and NSF Workshop on Automotive Software Development (Springer, Berlin, 2006), pp. 86–102
22. G. Bonifacio, P. Marmo, A. Orazzo, I. Petrone, L. Velardi, A. Venticinque, Improvement of processes and methods in testing activities for safety-critical embedded systems, in *Computer Safety, Reliability, and Security*, ed. by F. Flammini, S. Bologna, V. Vittorini. Lecture Notes in Computer Science, vol. 6894 (Springer, Berlin/Heidelberg, 2011), pp. 369–382
23. M. Broy, A. Deimel, J. Henn, K. Koskimies, F. Plasil, G. Pomberger, W. Pree, M. Stal, C. Szyperski, What characterizes a software component? Softw. Concepts Tools **19**(1), 49–56 (1998)
24. M. Broy, M. Feilkas, M. Herrmannsdoerfer, S. Merenda, D. Ratiu, Seamless model-based development: from isolated tools to integrated model engineering environments. Proc. IEEE **98**, 526–545 (2010)
25. Bureau d'Enquêtes et d'Analyses pour la Sécurité de l'Aviation Civile (FR), Accident to Turkish Airlines DC-10 TC-JAV in the Ermenonville Forest on 3 March 1974 – Final report, http://www.aaib.gov.uk/cms_resources.cfm?file=/8-1976%20TC-JAV.pdf, February 1976. Translation made by the French State Secretariat. Posted by the Air Accidents Investigation Branch of the Department of Trade (UK). Retrieved on 18 Dec 2011
26. J. Burge, J. Carroll, R. McCall, *Rationale-Based Software Engineering* (Springer, Berlin/London, 2008)
27. L. Capretz, Y: A new component-based software lifecycle model. J. Comput. Sci. **1**(1), 76–82 (2005)
28. D. Carrizo, O. Dieste, N. Juristo, Study of elicitation techniques adequacy, in *11th Workshop on Requirement Engineering*, España, 2008, pp. 104–114
29. B. Christiansson, L. Jakobsson, I. Crnkovic, CBD process, in *Building Reliable Component-Based Software Systems*, ed. by I. Crnkovic, M. Larsson (Artech House, Boston, 2002), pp. 89–113
30. H. Chuping, L. Peiqiong, Y. Yping, The application of failure modes and effect analysis for software in digital fly control systems, in *16th Digital Avionics Systems Conference (DASC) 1997*, AIAA/IEEE, vol. 1 (IEEE, 1997)
31. M. Cifaldi, N. Priggouris, O. Laurent, V. Ibanez et al. Architecture modelling methodology for RTP V3. Technical report D_SP3_R2.3_M3, The CESAR consortium, 2011
32. P. Cimiano, J. Völker, Text2Onto – a framework for ontology learning and data-driven change discovery, in *Proceedings of the 10th International Conference on Applications of Natural Language to Information Systems, NLDB 2005*, Alicante, 15–17 June 2005
33. E. Clarke, D. Kroening, F. Lerda, A tool for checking ANSI-C programs, in *Tools and Algorithms for the Construction and Analysis of Systems (TACAS 2004)*, ed. by K. Jensen, A. Podelski. Lecture Notes in Computer Science, vol. 2988 (Springer, Berlin/New York, 2004), pp. 168–176
34. Compositional variability management, http://www.cvm-framework.org/
35. J.O. Coplien, *Multi-paradigm Design for C++* (Addison-Wesley Longman, Boston, 1999)
36. I. Crnkovic, M. Chaudron, S. Larsson, Component-based development process and component lifecycle. J. Comput. Inf. Technol. **13**(4), 321–327 (2005)
37. I. Crnkovic, M. Chaudron, S. Larsson, Component-based development process and component lifecycle, in *Proceedings of International Conference on Software Engineering Advances*, Tahiti, 2006, pp. 44–53

38. W. Damm, A. Votintseva, A. Metzner, B. Josko, T. Peikenkamp, E. Böde, Boosting re-use of embedded automotive applications through rich components, in *Proceedings, FIT 2005 – Foundations of Interface Technologies*, San Francisco, 2005
39. W. Damm, H. Hungar, S. Henkler, I. Stierand, B. Josko, P. Reinkemeier, A. Baumgart, M. Bueker, T. Gezgin, G. Ehmen, R.Weber, Spes2020 architecture modeling, 2010. SPES2020 Deliverable D3.5
40. W. Damm, H. Hungar, B. Josko, T. Peikenkamp, I. Stierand, Using contract-based component specifications for virtual integration testing and architecture design, in *Design, Automation & Test in Europe Conference & Exhibition (DATE)* (IEEE, 2011), pp. 1–6
41. J. Daniels, P.W. Werner, A.T. Bahill, Quantitative methods for tradeoff analyses: regular papers. Syst. Eng. **4**(3), 190–212 (2001)
42. V. Debruyne, F. Simonot-Lion, Y. Trinquet, EAST-ADL: an architecture description language, in *Architecture Description Languages*. IFIP The International Federation for Information Processing (Springer, Boston, 2005), pp. 181–195
43. R.M. Dijkman, Consistency in multi-viewpoint architectural design. Ph.D. thesis, University of Twente, 2006
44. J. Doerr, D. Kerkow, T. Koenig, T. Olsson, T. Suzuki, Non-functional requirements in industry-three case studies adopting an experience-based NFR method, in *Proceedings of 13th IEEE International Conference on Requirements Engineering, 2005* (IEEE, Los Alamitos, 2005), pp. 373–382
45. dSPACE, SystemDesk, http://www.dspace.de/en/ltd/home/products/sw/system_architecture_software/systemdesk.cfm
46. EAST-ADL Association, http://www.east-adl.info/. Retrieved on 18 Dec 2011
47. Eclipse process library, http://www.eclipse.org/epf/
48. EN 50126:1999, *Railway Applications – The Specification and Demonstration of Reliability, Availability, Maintainability and Safety (RAMS)* (1999)
49. EN 50128:2001, *Railway Applications – Software for Railway Control and Protections Systems* (2001)
50. EN 50129:2003, *Railway Applications – Communication, Signalling and Processing Systems – Safety Related Electronic Systems for Signalling* (2003)
51. EN 50159:2010, *Railway Applications – Communication, Signalling and Processing Systems – Safety-Related Communication in Transmission Systems* (2010)
52. A. Engel, M. Winokur, G. Doehmen, M. Enzmann, Assumptions/promises – shifting the paradigm in systems-engineering, in *Proceedings of the 18th Annual International Symposium of the International Council on Systems Engineering*, Utrecht, 2008
53. Engineering software architectures, processes and platforms for system-families (1999), http://www.esi.es/esaps/
54. Epf practice library, http://www.eclipse.org/epf/downloads/praclib/praclib_downloads.php
55. ERTMS European Rail Traffic Management System pool, http://www.ertms.com. Retrieved on 18 Dec 2011
56. ESTEC QS Division (Reliability and Maintainability Section), Guidelines for considering a software intensive system with FMECA studies. ESTEC, January 1992. QS/91/247/082/RA
57. G. Harbour, G. Garcia, P. Gutierrez, D. Moyano, MAST: Modeling and analysis suite for real time applications, in *13th Euromicro Conference on Real-Time Systems* (IEEE Computer Society, Washington, DC, 2001), pp. 125–134
58. European Aviation Safety Agency, Certification specifications and acceptable means of compliance for large aeroplanes CS-25, http://easa.europa.eu/agency-measures/docs/certification-specifications/CS-25/CS-25%20Amendment%2011.pdf, July 2011. Retrieved on 18 Dec 2011
59. Extensible Markup Language (XML), http://www.w3.org/XML/
60. Fact-based maturity through institutionalisation lessons-learned and involved exploration of system-family engineering, http://www.esi.es/Families/
61. S. Farfeleder, T. Moser, A. Krall, T. Stulhane, H. Zojer, C. Panis, DODT: increasing requirements formalism using domain ontologies for improved embedded systems development, in

Design and Diagnostics of Electronic Circuits & Systems (DDECS) (IEEE, Piscataway, 2011), pp. 271–274
62. Firebird – The RDBMS that's going where you're going, http://www.firebirdsql.org/.
63. Formal notation suitable to express safety properties. ESACS report, Sept 2001
64. M. Fortes da Cruz, P. Raistrick, AMBERS: improving requirements specification through assertive models and SCADE/DOORS integration, in *The Safety of Systems, Proceedings of the 15th Safety-Critical Systems Symposium*, Bristol, UK, ed. by F. Redmill, T. Anderson (Springer, London, 2007), pp. 217–241
65. From concepts to application in system-family engineering, http://www.esi.es/Cafe/
66. D.D. Gajski, S. Abdi, A. Gerstlauer, G. Schirner, *Embedded System Design: Modeling, Synthesis, Verification* (Springer, Dordrecht/New York, 2009). ISBN 978-1-4419-0503-1
67. A. Gaufillet, B. Gabel, Avionic software development with TOPCASED SAM, in *Proceedings of the Embedded Real Time Software and Systems 2010*, Toulouse, 2010
68. M. Glinz, R. Wieringa, Guest editor's introduction: stakeholders in requirements engineering. IEEE Softw. **24**, 18–20 (2007)
69. GME: Generic Modeling Environment, http://www.isis.vanderbilt.edu/Projects/gme/
70. P.L. Goddard, Software FMEA techniques, in *Proceedings of the Annual Reliability and Maintainability Symposium*, Los Angeles, 2000
71. C. Gonzalez-Perez, B. Henderson-Sellers, Templates and resources in software development methodologies. J. Object Technol. **4**(4), 173–190 (2005)
72. G. Griessnig, R. Mader, T. Peikenkamp, B. Josko, M. Törngren, E. Armengaud, CESAR: cost-efficient methods and processes for safety relevant embedded systems, in *Embedded World 2010- ARTEMIS Session*, Nürnberg, 2010
73. G. Griessnig, I. Kundner, E. Armengaud, S. Torchiaro, D. Karlsson, Improving automotive embedded systems engineering at european level. Elektrotech. Informationstechnik **128**(6), 209–214 (2011)
74. N. He, D. Kroening, T. Wahl, K.-K. Lau, F. Taweel, T. Cuong, P. Ruemmer, S. Sharma, Component-based design and verification in x-man, in *Embedded Real Time Software and Systems (ERTS^2)*, Toulouse, 2012
75. T. Heath, C. Bizer, Linked data: evolving the web into a global data space, http://linkeddatabook.com/editions/1.0/
76. H. Hecht, A. Xuego, M. Hecht, *Computer-Aided Software FMEA* (SoHaR Incorporated, Culver City/Los Angeles, 2003)
77. G. Heineman, W. Councill (eds.), *Component-Based Software Engineering: Putting the Pieces Together* (Addison-Wesley, Boston, 2001)
78. A. Herrmann, D. Kerkow, J. Doerr, Exploring the characteristics of NFR methods: a dialogue about two approaches, in *Proceedings of the 13th International Working Conference on Requirements Engineering: Foundation for Software Quality* (Springer, Berlin/Heidelberg, 2007), pp. 320–334
79. IAEA, Vienna, *Component Reliability Data for Use in Probalistic Safety Assesment* (IAEA, Vienna, 1988)
80. IBM Software, ILOG, www-01.ibm.com/software/websphere/ilog/
81. IBM, Rational DOORS, http://www.ibm.com/software/awdtools/doors/
82. IBM, Rational rhapsody, http://www.ibm.com/software/awdtools/rhapsody/
83. IEC, *Standard IEC 61131 "Programmable controllers"*
84. IEC, IEC 61800 adjustable speed electrical power drive systems – part 5-2: safety requirements – functional, July 2007
85. IEC, IEC 61508 Edition 2.0, *Functional Safety of Electrical/Electronic/Programmable Electronic Safety-Related Systems* (International Electrotechnical Commission, Geneva, 2010)
86. IEEE, Software architecture: introducing ieee standard 1471, http://ieeexplore.ieee.org
87. Information technology for european advancement, http://www.itea2.org/
88. ISO International Organization for Standardization, ISO 9001:2008, Quality management systems – requirements, 2009

89. ISO International Organization for Standardization, ISO/TS 16949:2009, Quality management systems – particular requirements for the application of ISO 9001:2008 for automotive production and relevant service part organizations, 2009
90. ISO/IEC 9126-1, Software engineering – product quality – part 1: quality model. Technical report, ISO/IEC, Geneva, 2001
91. ISO/IEC 10303, Technical report, ISO/IEC, 2002
92. ISO/IEC 12207:2008, Systems and software engineering - software life cycle processes, Technical report, ISO/IEC, 2008
93. ISO, ISO/IEC 15288:2008, Systems and software engineering – system life cycle processes. Technical report, ISO/IEC, 2008
94. ISO, ISO/IEC/IEEE 29148:2011, Systems and software engineering – life cycle processes – requirements engineering. Technical report, ISO/IEC/IEEE, 2011
95. P. Johannesen, F. Tørner, J.Torin, Actual based hazard analysis for satfety critical systems, in *Proceedings of the 23th International Conference on Computer Safety, Reliability and Security*, Potsdam, Germany, Sept 2004
96. B. Josko, Q. Ma, A. Metzner, Designing embedded systems using heterogeneous rich components, in *Proceedings of the INCOSE International Symposium*, Utrecht, 2008, vol. 2008
97. H. Kaiya, M. Saeki, Ontology based requirements analysis: lightweight semantic processing approach, in *5th International Conference on Quality Software*, Melbourne, 2005, pp. 223–230
98. K. Kaur, H. Singh, Candidate process models for component based software development. J. Softw. Eng. **4**(1), 16–29 (2010)
99. M. Khlif, O. Tahan, M. Shawky, CO-SImulation trace analysis (COSITA) tool for vehicle electronic architecture diagnosability analysis, in *Intelligent Vehicles Symposium (IV)* (IEEE, Piscataway, 2010), pp. 572–578
100. G. Kotonya, I. Sommerville, S. Hall, Towards a classification model for component-based software engineering research, in *Proceedings of the 29th EUROMICRO Conference* (IEEE Computer Society, Los Alamitos, 2003), pp. 43–52
101. P. Kruchten, The 4+1 view model of architecture. IEEE Softw. **12**, 42–50 (1995)
102. S. Kuenzli, Efficient design space exploration for embedded systems. Ph.D. thesis, ETH Zurich, 2006
103. A. Lamsweerde, *Requirements Engineering: From System Goals to UML Models to Software Specifications* (Wiley, Chichester/Hoboken, 2009)
104. T.G. Lane, Studying software architecture through design spaces and rules. Technical report CMU/SEI-90-TR-18, System Engineering Institute, Carnegie Mellon University, Pittsburg, PA, USA, 1990
105. K.-K. Lau, Z. Wang, Software component models. IEEE Trans. Softw. Eng. **33**(10), 709–724 (2007)
106. K.-K. Lau, F. Taweel, Domain-specific software component models, in *Proceedings of the 12th International Symposium on Component-Based Software Engineering*, ed. by G. Lewis, I. Poernomo, C. Hofmeister. Lecture Notes in Computer Science, vol. 5582 (Springer, Berlin, 2009), pp. 19–35
107. K.-K. Lau, P. Velasco Elizondo, Z. Wang, Exogenous connectors for software components, in *Proceedings of the 8th International Symposium on Component-Based Software Engineering*, ed. by G. Heineman et al. Lecture Notes in Computer Science, vol. 3489 (Springer, Berlin/New York, 2005), pp. 90–106
108. K.-K. Lau, M. Ornaghi, Z. Wang, A software component model and its preliminary formalisation, in *Proceedings of the 4th International Symposium on Formal Methods for Components and Objects*, ed. by F. de Boer et al. Lecture Notes in Computer Science, vol. 4111 (Springer, Berlin/New York, 2006), pp. 1–21
109. K.-K. Lau, F. Taweel, C. Tran, The W Model for component-based software development, in *Proceedings of the 37th EUROMICRO Conference on Software Engineering and Advanced Applications* (IEEE, Piscataway, 2011), pp. 47–50

110. J.D. Lawrence, Software safety hazard analysis. *NUREG/CR-6430, UCRL-ID-1225 14*, Lawrence Livermore National Laboratory, California, Feb 1996
111. B. Li, M. Li, K. Chen, C. Smidts, Integrating software into PRA: a software related failure mode taxonomy. Risk Anal. **26**, 997–1012 (2006)
112. J.-H. Li, Q. Li, J. Li, The W-Model for testing software product lines, in *International Symposium on Computer Science and Computational Technology*, Shanghai, 2008, pp. 690–693
113. Isograph Inc., Faulttree+ in reliability workbench, www.isograph-software.com/ftpoverdgc.htm
114. MathWorks, Simulink – simulation and model-based design, http://www.mathworks.com/products/simulink/
115. Meta object facility, http://www.omg.org/mof/
116. B. Meyer, Applying design by contract. IEEE Comput. **25**(10), 40–51 (1992)
117. B. Meyer, *Object-oriented Software Construction*, 2nd edn. (Prentice-Hall, Upper Saddle River, 1997)
118. C. Croarkin, P. Tobias, NIST/SEMATECH e-handbook of statistical methods, *NIST/SEMATECH*, http://www. itl. nist. gov/div898/handbook (July 2006)
119. N. Noy, D. McGuinness, *Ontology Development 101: A Guide to Creating Your First Ontology* (2001)
120. D. Oliver, T. Kelliher, J. Keegan, *Engineering Complex Systems with Models and Objects* (McGraw-Hill, New York, 1997)
121. OMG, Systems modeling language, http://www.sysml.org/
122. OMG, Software and systems process engineering metamodel specification (SPEM) Apr 2008, http://www.omg.org/spec/SPEM/2.0/
123. OSLC Core Specification, http://open-services.net/bin/view/Main/OslcCoreSpecification
124. Payrus UML, www.papyrusuml.org
125. Papyrus unified modeller project, http://www.eclipse.org/modeling/mdt/papyrus/
126. K. Pohl, G. Böckle, F. van der Linden, *Software Product Line Engineering: Foundations, Principles, and Techniques* (Springer, New York, 2005)
127. O. Preiss, A. Wegmann, Stakeholder discovery and classification based on systems science principles, in *Proceedings of the 2nd Asia-Pacific Conference on Quality Software*, Hong Kong, 2001 (IEEE, 2001), pp. 194–198
128. R.S. Pressman, *Software Engineering: A Practitioner's Approach* (McGraw Hill, Boston/London, 2001)
129. Project SPEEDS: WP.2.1 Partners, *SPEEDS L-1 Meta-Model*, May 2009. SPEEDS deliverable D.2.1.5, revision 1.0.1, http://speeds.eu.com/downloads/SPEEDS_Meta-Model.pdf
130. Protégé tool, http://protege.stanford.edu/
131. *Railway Applications – The Specification and Demonstration of Reliability, Availability, Maintainability and Safety (RAMS).* Part 2: Guide to the application of EN 50126-1 for safety (2007)
132. B. Ramesh, M. Jarke, Toward reference models for requirements traceability. IEEE Trans. Softw. Eng. **27**, 58–93 (2001)
133. Resource Description Framework (RDF), http://www.w3.org/RDF/
134. M. Sampath, R. Sengupta, S. Lafortune, K. Sinnamohideen, D. Teneketzis, Failure diagnosis using discrete-event models. Control Syst. Technol. IEEE Trans. **4**(2), 105–124 (1996)
135. C. Seguin, Formal notation suitable to express safety properties. ESACS technical report, 17 Sept 2001
136. H. Sharp, A. Finkelstein, G. Galal, Stakeholder identification in the requirements engineering process, in *Proceedings Tenth International Workshop on Database and Expert Systems Applications*, Florence, 1999, pp. 387–391
137. M.M. Shawky, M. Langenbach, O. Laurent, J.-P. Blanquart, S. Bonnet, M. Khlif, R. Pezet, T. Prosvirnova, M. Biehl, D. Chen, M. Persson, M. Törngren, Architecture exploration and validation methodology for RTP V2 – checking of safety, diagnosability, reliability properties. Technical report D_SP3_R2.5_M2, The CESAR consortium, 2010

138. G. Sindre, A. Opdahl, Eliciting security requirements by misuse cases, in *Proceedings 37th International Conference on Technology of Object-Oriented Languages and Systems, TOOLS-Pacific 2000* (IEEE, Los Alamitos, 2000), pp. 120–131
139. SINTEF, *OREDA Offshore Reliability Data*, 5th edn. (OREDA Participants/Der Norske Veritas, Høvik, 2009)
140. SINTEF, Generic language and tool for variability modeling (2010), http://modelingwizards.isti.cnr.it/wp-content/uploads/2010/10/SINTEF-A13505-Report.pdf
141. Society of Automotive Engineers (SAE), *Certification Considerations for Highly Integrated or Complex Aircraft Systems*, 1 Nov 1996. Product Code: ARP4754
142. I. Sommerville, *Software Engineering*, 7th edn. (Addison Wesley, Boston, 2004)
143. I. Sommerville, *Software Engineering*, 8th edn. (Addison-Wesley, Harlow/New York, 2007)
144. SPEEDS Methodologie – a white paper, Technical report, SPEEDS, 2006
145. J. Spencer, T. Blevins, *Interoperable Enterprise Business Scenario* (2002), www.opengroup.org/cio/iop/index.htm
146. A. Spillner, The W-MODEL – strengthening the bond between development and test, in *International Conference on Software Testing, Analysis and Review*, 2002, http://www.sqe.com/stareast
147. SQLAPI++ - C++ library for accessing SQL databases, http://www.sqlapi.com/
148. T. Stalhane, I. Omoronyia, F. Reichenbach, Ontology-guided requirements and safety analysis, in *6th International Conference on Safety of Industrial Automated Systems*, Tampere, 2010
149. N. Storey, *Safety Critical Computer Systems* (Addison Wesley, Harlow/Reading, 1996)
150. System Reliability Centre, Part failure mode distributions
151. Systems engineering measurement primer, Technical report, INCOSE Measurement Working Group, 1998
152. Technical Committee ISO/TC 22 Subcommittee SC 3 (ed.), *ISO/WD26262: Road Vehicles – Functional Safety*. (Automotive Standards Committee of the German Institute for Standardization, 2009)
153. Tecnalia, Product line unified modeller (plum), http://www.esi.es/plum/index.php
154. The ATESST Consortium, *EAST-ADL Domain Model Specification*, June 2010, ATESST Deliverable D4.1.1, Version 2.1 RC3, http://www.atesst.org/home/liblocal/docs/ATESST2_D4.1.1_EAST-ADL2-Specification_2010-06-02.pdf
155. The CESAR Global Glossary (2012), http://www.cesarproject.eu/index.php?id=33&L=0. (to appear)
156. The Institute of Electrical and Electronic Engineers, *IEEE Std. 610* (IEEE Standard Computer Dictionary, 1991)
157. The Open Group, *SOA Source Book*, http://www.opengroup.org/projects/soa-book/
158. The SoftIntegration Ch SDK, http://www.softintegration.com/products/sdk/chsdk/
159. The V-model, Development standard for IT-systems of the Federal Republic of Germany, IABG, http://www.v-modell.iabg.de
160. The World Wide Web Consortium (W3C), http://www.w3.org/standards/semanticweb/data
161. S.R. Trammel, B.J. Davis, Using a modified HazOp/FMEA methology for assesssing system risk, in *Proceedings of the Second International EMAT Workshop* (IEEE, Los Alamitos, 2001)
162. UK Ministry of Defence, MOD architecture framework (MODAF), www.modaf.org.uk/
163. UN ECE United Nations Economics Commission for Europe, UN ECE regulation 13: uniform provisions concerning the approval of vehicles of categories M, N and O with regard to braking (2011), http://www.unece.org/fileadmin/DAM/trans/main/wp29/wp29regs/r013r7e.pdf. Retrieved on 18 Dec 2011
164. Unified method architecture concept, http://epf.eclipse.org/wikis/openupsp/base_concepts/guidances/concepts/introduction_to_uma,_94_eoO8LEdmKSqa_gSYthg.html
165. Unified method framework concept, http://epf.eclipse.org/wikis/
166. UNISIG UNion Industry of SIGnalling consortium, http://www.unife.org/. Retrieved on 18 Dec 2011

167. R.J. Urbanic, W.H. ElMaraghy, Using a modified failure modes and effects analysis within the structured design recovery framework, in *The Future of Product Development* (2007), pp. 409–419
168. US Department of Defense, DoD architecture framework, http://dodcio.defense.gov/dodaf20.aspx
169. P. Velasco Elizondo, K.-K. Lau, A catalogue of component connectors to support development with reuse. J. Syst. Softw. **83**, 1165–1178 (2010)
170. J. Vlissides, R. Helm, R. Johnson, E. Gamma, *Design Patterns: Elements of Reusable Object-Oriented Software* (Addison-Wesley, Reading, 1995)
171. W3C, Linked data basic profile 1.0, http://www.w3.org/Submission/2012/SUBM-ldbp-20120326/
172. J. Wood, D. Silver, *Joint Application Development* (Wiley, New York, 1995)
173. Wordnet – a lexical database for English, http://wordnet.princeton.edu/
174. S. Zverlov, Comparison of architecture exploration tools at the design level. Master's thesis, Technical University of Munich, 2011

Index

A. Rajan and T. Wahl (eds.), *CESAR - Cost-efficient Methods and Processes for Safety-relevant Embedded Systems*, DOI 10.1007/978-3-7091-1387-5,

MIX
Papier aus verantwortungsvollen Quellen
Paper from responsible sources
FSC® C105338

If you have any concerns about our products,
you can contact us on
ProductSafety@springernature.com

In case Publisher is established outside the EU,
the EU authorized representative is:
Springer Nature Customer Service Center GmbH
Europaplatz 3, 69115 Heidelberg, Germany

Printed by Libri Plureos GmbH
in Hamburg, Germany